SAND, SNOW, AND STARDUST

SAND, SNOW, ★ AND ★ STARDUST

HOW US MILITARY ENGINEERS CONQUERED EXTREME ENVIRONMENTS

GRETCHEN HEEFNER

THE UNIVERSITY OF CHICAGO PRESS

Chicago

The University of Chicago Press, Chicago 60637
© 2025 by Gretchen Heefner
All rights reserved. No part of this book may be used or reproduced in any manner whatsoever without written permission, except in the case of brief quotations in critical articles and reviews. For more information, contact the University of Chicago Press, 1427 E. 60th St., Chicago, IL 60637.
Published 2025
Printed in the United States of America

34 33 32 31 30 29 28 27 26 25   1 2 3 4 5

ISBN-13: 978-0-226-83159-6 (cloth)
ISBN-13: 978-0-226-83160-2 (e-book)
DOI: https://doi.org/10.7208/chicago/9780226831602.001.0001

Library of Congress Cataloging-in-Publication Data

Names: Heefner, Gretchen, author.
Title: Sand, snow, and stardust : how US military engineers conquered extreme environments / Gretchen Heefner.
Description: Chicago : The University of Chicago Press, 2025. | Includes bibliographical references and index.
Identifiers: LCCN 2024041065 | ISBN 9780226831596 (cloth) | ISBN 9780226831602 (ebook)
Subjects: LCSH: Military engineering—United States. | Extreme environments. | Military meteorology. | Military geophysics.
Classification: LCC UG465.3 .H44 2025 | DDC 358/.220973—dc23 /eng/20240912
LC record available at https://lccn.loc.gov/2024041065

♾ This paper meets the requirements of ANSI/NISO Z39.48-1992 (Permanence of Paper).

CONTENTS

List of Figures and Maps vii
Cast of Characters ix

Introduction: Into the Extremes 1

PART I: SAND

Prelude 19
1. The Smallest of Details 25
2. Mud, Rain, Frost, and Wind 40
3. The Cactus Cult 55
4. The Desert Room 72
5. Extreme Wish Lists 87
6. Engineering Deserts 104
7. Analog Deserts 123

PART II: SNOW

Prelude 139
8. Top of the World 145
9. Arctica 163
10. Pituffik 177
11. Fieldwork 192
12. Limits 209
13. Secrets of the Ice 228

PART III: STARDUST

Prelude 245
14. Deserts in Space 249
15. Moondust 264
16. Analog Earth 280

Conclusion: Everyday Extremes 299

Acknowledgments 307
List of Abbreviations 313
Notes 315
Selected Bibliography 371
Index 385

FIGURES AND MAPS

FIGURE I.1   "Logistics Bridge to Mars" (1962) 2
FIGURE P1.1   Frontispiece to *A Pocket Guide to North Africa* (1942) 20
FIGURE P1.2   *Saturday Evening Post* map of the Sahara (1942) 22
FIGURE 1.1   President Roosevelt in the Oval Office, February 23, 1942 26
FIGURE 1.2   The Desert Training Center 30
FIGURE 1.3   GIs relaxing in the Desert Training Center 37
FIGURE 2.1   *Stars and Stripes* cartoon (1943) 41
FIGURE 3.1   The Joshua tree 57
FIGURE 3.2   Botanist Forrest Shreve in the field (1940) 59
FIGURE 3.3   Conservationist Minerva Hamilton Hoyt 66
FIGURE 4.1   Harvard Fatigue Lab test chamber 73
FIGURE 4.2   The copper man 85
FIGURE 6.1   F-86 Sabres roar off a runway 106
FIGURE 6.2   Military facilities across Libya 107
FIGURE 6.3   Ralph Bagnold, the "Sandman" 113
FIGURE 6.4   Strategic deserts of the world 121
FIGURE 7.1   Global environments 124
FIGURE 7.2   Sir Hubert Wilkins in the desert with color swatches 127
FIGURE 7.3   Analog desert maps 133
FIGURE P2.1   Explorer on the Greenland ice cap 140
FIGURE 8.1   Strategic map of the Arctic 148
FIGURE 8.2   Map of the Thule area, Greenland 150

VIII ★ LIST OF FIGURES AND MAPS

FIGURE 8.3   Logistical routes to the Arctic 157
FIGURE 9.1   Wilkins with map of the North Pole 164
FIGURE 9.2   Engineer reliability map 175
FIGURE 10.1 Greenlander with dog sledge in front of Air Force radar 178
FIGURE 11.1 Landing on the ice 193
FIGURE 11.2 Science and strategy meet on the ice 197
FIGURE 11.3 Snow classification 198
FIGURE 11.4 Equipment for the desert and Arctic extremes 203
FIGURE 11.5 US defense-related installations in northern Greenland 205
FIGURE 11.6 Aerial photo of Camp Tuto 206
FIGURE 12.1 Entering Camp Century 210
FIGURE 12.2 Life in buildings made of ice and snow 216
FIGURE 12.3 Blueprint for Camp Century 219
FIGURE 12.4 Construction of Camp Century 221
FIGURE 12.5 Around the clock excavation (1961) 226
FIGURE 13.1 Examining ice cores 230
FIGURE 13.2 Benson's Greenland transverse routes 236
FIGURE 13.3 Drilling room at Camp Century 237
FIGURE P3.1 Chesley Bonestell's *A Lunar Landscape* (1957) 246
FIGURE 14.1 The cover of the Corps of Engineers' *Lunar Construction* model and plan 250
FIGURE 14.2 Article on Percival Lowell's canal thesis 252
FIGURE 14.3 Midcentury Mars 254
FIGURE 15.1 US Army's proposed Moon base, late 1950s 266
FIGURE 15.2 Lunar environment simulator, the Extraterrestrial Engineering Center 268
FIGURE 15.3 The surface of the Moon 277
FIGURE 16.1 Apollo 17 astronaut on the lunar surface 281
FIGURE 16.2 Cinder Lake Crater Field 281
FIGURE 16.3 Eugene Shoemaker, astrogeologist 284
FIGURE 16.4 Astronaut geology school 290
FIGURE 16.5 NASA Mars conference in the Mojave 294

CAST OF CHARACTERS

SAND

MARY HUNTER AUSTIN, author, walker, and narrator of the Mojave
RALPH BAGNOLD, a.k.a. the "Sandman," British desert enthusiast and explorer in the style of his interlocutor D. H. Lawrence
ARTHUR CASAGRANDE, engineer with a love of classification
BRIG. GEN. DONALD A. DAVISON, the first go-to desert airfield engineer for the Army Air Forces
DAVID BRUCE DILL, self-experimenting physiologist who knew how to link his interests to military needs
GEN. DWIGHT D. EISENHOWER, lauded US military leader during World War II who, in 1952, became president of the United States
WELDON HEALD, alpine adventurer who consulted on desert rations and hydration
MINERVA HAMILTON HOYT, the John Muir of deserts
MAJ. GEN. GEORGE S. PATTON, larger-than-life World War II general who, for a time, was relegated to the Mojave Desert
ERNIE PYLE, war reporter who covered the average-GI
FRANKLIN DELANO ROOSEVELT, US president who used radio to tell Americans about a world at war
FORREST SHREVE, botanist whose plant maps were ignored, until they weren't
PAUL SIPLE, decorated Eagle Scout who spent his life promoting research in the extremes

VILHJÁLMUR STEFÁNSSON, polar explorer with an interest in Inuit clothing

LT. PAUL D. TROXLER, engineer who saw sand instead of political borders

SIR HUBERT WILKINS, problem solver with a penchant for color swatches and camouflage

FREDERICK WULSIN, desert explorer and inventor of an unusable metallic suit

SNOW

HENRI BADER, a.k.a. "The Iceman," Swiss glaciologist determined to discover a hidden message in snowflakes

BERNT BALCHEN, polar aviator and later military consultant who pops up in surprising places

CARL BENSON, young researcher who dug snow pits across Greenland's ice cap

SAMUEL W. BOGGS, geographer who worried that maps tell false tales

WALTER CRONKITE, genial TV persona

CHESTER LANGWAY JR., ice-core driller who hit bedrock in Greenland

JEAN MALAURIE, French anthropologist whose yearlong research with the Inughuit of northwestern Greenland coincided with the arrival of the Americans

ODAAQ, renowned Inughuit hunter and guide from northwestern Greenland

ROBERT PEARY, American explorer who nearly died trying to cross Greenland

HAROLD "OAKIE" PRIEBE, construction worker who returned to Greenland year after year to build military facilities

KNUD RASMUSSEN, Danish Greenlander rumored to have named the Thule area

COL. MORTON SOLOMON, engineer sent to the top of the world

PAUL-ÉMILE VICTOR, leader of French polar expeditions and collaborator with US defense programs

ALFRED WEGENER, German scientist-explorer who wanted to chart winter weather on top of the world

STARDUST

WILLY LEY, popularizer of all things space aged
CHESLEY BONESTELL, father of midcentury space art
ARTHUR C. CLARKE, sci-fi writer and head of the British Interplanetary Society
BRUCE M. HALL, builder of models for lunar construction
PERCIVAL LOWELL, believer that, to find Mars, one must go to the desert
EUGENE SHOEMAKER, the first astrogeologist whose ashes were placed on the Moon
FRED WHIPPLE, Harvard astronomer and early space enthusiast
WERNHER VON BRAUN, Nazi rocketeer who combined technological savvy with space-aged dreams

INTRODUCTION

INTO THE EXTREMES

If the engineers had their way, we would already have a bridge to Mars. Or perhaps at least to the Moon. Not an actual bridge, like the one pictured in the *Military Engineer* (fig. I.1), but a logistic bridge, one represented by the little blocks: "supplies," "plans," "training," "personnel," "communications." These were the building blocks to extraterrestrial success. They were how people would be housed and cared for once they had been hurled into the cosmos to colonize other planets. And the men of the US Army Corps of Engineers—who built this bridge in 1962, at least in their mind's eye—believed it could be done.[1]

They were confident it could be done because they had done it before. For two decades, the engineers had been setting down similar building blocks, not to space, but to some of Earth's most extreme environments. In 1960, for example, military engineers put the finishing touches on a base *underneath* the Greenland ice cap. It was powered by a portable nuclear reactor, allowing two hundred men to live year-round in a place assumed to have been uninhabitable. According to newsreel footage, the men at Camp Century had record players, reading lamps, and a pet dog named Mukluk—an oddly banal existence considering that they lived in subglacial caverns nestled just eight hundred miles from the North Pole.[2]

At the same time, four thousand miles to the south—where it was more than one hundred degrees warmer—thousands of airmen and

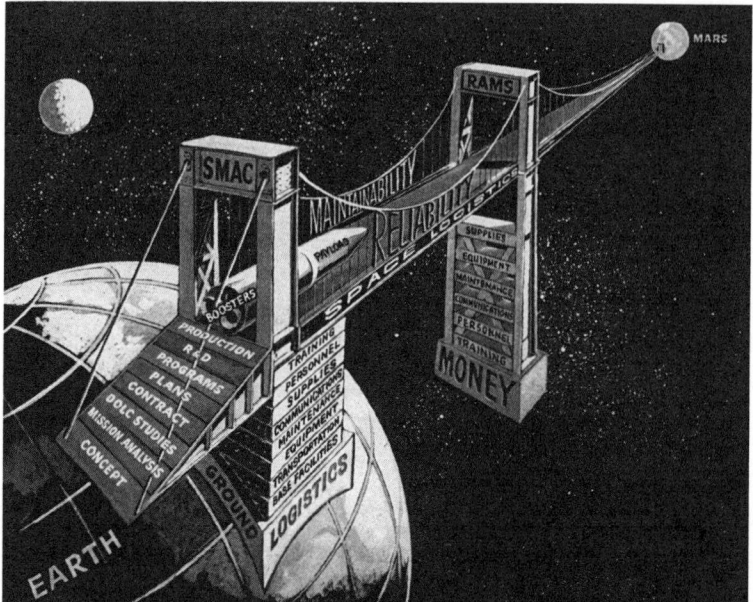

FIGURE I.1 "Logistics Bridge to Mars" (1962), appeared in the *Military Engineer*. The drawing highlights the logistical support and expertise that engineers were poised to provide as the US government prepared to send humans into outer space. Image originally appeared in the *Military Engineer* 54, no. 360 (July–August 1962). Reprinted with permission of the Society of American Military Engineers.

their families were living on a base on the edge of the Sahara. Wheelus Air Force Base, constructed in the early 1950s, had two outdoor swimming pools, a golf course, and a seemingly endless supply of bottled Coca-Cola, Tootsie Rolls, Ovaltine, and packets of Kodachrome film. The US ambassador to Libya called the base a "little America on the gleaming shores of the Mediterranean."[3] It wasn't Mars, obviously, but in 1962, these were successful construction programs in environments so extreme that they had long been written off as no-man's-lands.

Thus, it might have seemed possible to build something similar in outer space. The same magazine that published the bridge to Mars was explicit about the connections. In 1961, the *Military Engineer* described operating on the Moon as a mash-up of working in every extreme on Earth. Maneuvering there would be like "crossing the

Himalayas, descending to the bottom of the Mindanao Deep, establishing a post at the South Pole, crossing under the Polar ice cap, and traversing the Sahara—combined!" As the midcentury engineer and popularizer of all things space-aged Willy Ley wrote, engineers could plan for and envision almost anything. They had less control over whether the money and political will existed to see such programs through.[4]

The bridge to Mars was never built. But that it was presented as a possibility in 1962 was a sign of a hard-earned environmental confidence that would have been totally lacking among the same men just twenty years before. Back then, no one was thinking about space or fantastic subglacial bases; they were thinking about war. And in 1942, as they looked out on the world, they saw potential battlefields in places they had rarely bothered to imagine, let alone plan for. From the Aleutian Islands up in the Bering Sea to the sand seas of the Sahara, World War II was truly global, even as US forces were alarmingly parochial. "Our knowledge of the nontemperate areas of the world," military leaders admitted, was "either woefully weak or completely lacking."[5] In the 1940s, they had not yet envisioned the sort of logistic networks and supply lines—pathways and bridges, metaphorically—that could take them from the United States to any point in the world, and then beyond. Nor had they seriously considered the types of knowledge and expertise building those pathways would require. As a result, the engineers stumbled. In planning for battle in North Africa, for instance, military engineers got the environment wrong. A costly mistake they vowed would not be repeated.

Sand, Snow, and Stardust traces how the United States' defense establishment, in a very short amount of time, came to know and tried to master extreme environments. Places such as the desert, the Arctic, and outer space, places that exist out there, somewhere, at the edges of our maps. These are places that have long and generally been written off—wastelands, useless, remote, lifeless. But from the 1940s through the 1960s, the defense establishment set down bridges not just to and from these places, but between them as well.[6] In doing so, it transformed the way the world worked and what was possible in it. What we know about these environments today is directly related to the US military's priorities stemming from World War II, the Cold War, and beyond.

This is also the untold story of how the acquisition of environmental knowledge turned the United States into a planetary power. While the United States emerged in the second half of the twentieth century as the world's most powerful country, this book digs down into the particulars, examining in fine detail the places and materials—the sand, the snow, and the stardust—that had to be known before men and machines could be sent nearly anywhere.[7] To pave a modern runway on the Greenland permafrost, engineers had to first understand frozen ground; to keep planes operational in the Sahara, they had to devise systems to stymie blowing sands. The more sophisticated weapons and technologies became, the more environmental information they had to gather. The more intelligence was amassed, the more US military facilities and projects could spread out, snaking around all of Earth and reaching up into the skies. Research programs and scientific activities spread out from military facilities because increasingly more information was required.

Although guns and bombs are typical markers of military might, intimate environmental knowledge, the kind that allowed you to stay in place, was a prerequisite for true hegemony. The measure of planetary power became not the number of runways paved or barracks built, but the number of things put down in places long considered wastelands. Not just one such place; anybody might become the master of a single domain. Real authority came from lording over all of them and at the same time. After all, the Soviets were already considered to have mastered the Arctic given their experiences and facilities in Siberia. Plenty of European powers had operated in cold climates or in their desert colonial possessions. Building bases in the desert in the early 1950s, deploying men under the Greenland ice sheet by the 1960s, creating blueprints for lunar bases that would be populated within a decade (with Mars always just a few years off): the US defense establishment planned to do all those things. It saw them as markers of an unrivaled environmental expertise that—as the following chapters reveal—created the foundations for a new kind of planetary power.[8]

These practices were made possible because the US defense establishment created—and has subsequently maintained—a jaw-dropping worldwide infrastructure of military installations. From the first, the modern base structure was novel, massive, and global. Before World War II, the United States operated just fourteen overseas military bases; in 1960, over one thousand. Scholars have called this network of US military bases a "leasehold empire," a sign of "liberal imperialism," an "empire of bases," and, most recently, a "pointillist empire." As the qualifiers suggest, the postwar projection of US power and influence is seen as somehow distinct from what came before; this is not merely European-style imperialism with Uncle Sam at the helm. Of course, to many of the people living with the realities of US power abroad, the distinction hardly mattered.[9]

But there are ways in which the distinction has merit. Unlike the British, French, and Japanese empires, the United States in the 1940s and 1950s did not seek to acquire territory or incorporate new countries into its political union. Instead, it sought a network of military sites, small plots of extraterritorial space around the world where US military personnel and machines could operate outside the purview of foreign governments, citizens, and justice systems. These bases, and the threat of force they suggested, would be all that was needed to ensure the free flow of goods and material around the world, thus projecting and protecting US interests. As one State Department official warned in 1956, the United States was acquiring not entire colonies, but small "slices of sovereignty."[10] And this global system of bases demanded that the United States amass the sort of environmental knowledge it had long ignored.

Most of these military facilities were constructed in temperate areas, but some were set in the arid extremes. In 1951, for example, the Corps of Engineers began building a handful of air bases across North Africa, from Morocco to Libya. Each base required ten-thousand-foot paved runways, barracks, mess halls, roads, and more. Small cities in the middle of desert environments. At the same time, the Corps initiated work on air bases in Greenland, Iceland, and Alaska. For the northern locations, nearly all construction materials had to be ordered, boxed, and shipped from the East Coast. Engineers had to learn to pave over permafrost and insulate buildings from extraordinary heat and cold. They learned the hard way that a shipping container of eggs left out overnight within the Arctic Circle could not be used, and that concrete set strangely in intense heat.[11]

These out-of-the-way places were important for two reasons. First, they were ideal for hiding the burgeoning US military presence. Policymakers were aware of the political pitfalls of basing troops in foreign countries, where they were often unpopular. Defense officials and diplomats alike hoped that shunting bases off into remote areas would make them less obvious, perhaps even forgettable. That was, however, rarely possible. Everywhere the US military has gone—no matter how remote the location—it has rearranged land and environments, displaced people, shifted economies and cultures, muddied politics. The history of US overseas bases is full of violence, contestation, and resistance.[12] Second, from bases in these far-flung places, US forces could be deployed anywhere at any time. This was particularly important through the late 1940s and 1950s, when long-distance air travel (of both military and civilian varieties) remained time-consuming and laborious. To be effective as a deterrent, strategic bombers had to be stationed close to potential targets so that crews could take off, deliver their ordnance, and return safely to base without running out of fuel. New weapons of the 1960s changed the geographic calculus: not only could planes fly longer and higher; intercontinental ballistic missiles, many buried in silos in the United States, could reach targets halfway around the world in thirty minutes. But by then the structure of far-flung bases had been established and was slow to contract.

It was through the creation and maintenance of the base network that military leaders were able to amass expertise and knowledge about extreme environments. The chapters that follow track military engineers as they moved out to rearrange these wild and unexpected places, from wartime deserts, where survival was imaginable (part 1); to the High Arctic, where life seemed nearly impossible (part 2); and then to outer space, where man could exist only with herculean logistic support (part 3). In each environment, military engineers gained experience and skills that allowed them to ratchet up to the next, more extreme place. The insights that the defense establishment gleaned along the way range from the granular, how grains of sand and flakes of snow operate, to the planetary, or how world systems work. The materials of extreme environments came under the literal microscope as the landscapes were used increasingly to host simulations for global, and even interplanetary, exploration. Still today, plans for human space exploration and extraplanetary colonization are based on what is known about extremes on Earth.

This book actually began with a grain of sand, or lots of them. I was researching the construction of military bases in the Sahara in the 1950s, and sand kept filtering into the story. It found its way into every corner of every room, no matter how tightly the doors and windows had been shut. In 1952, the US ambassador to Libya wrote that each morning when his family woke up, "they could write [their] names on the bedside tables."[13] Sand was everywhere. The military engineers sent to build the bases complained of drifts that heaved across their runways, of winds that picked the sand up and hurled it against machines, scraping and gouging. Sand got into dinner. Sand ruined things.

I tried to ignore the sand. But when a grain of it is lodged in your shoe, sand becomes hard to ignore. The more you think about it, the more intriguing it becomes. Imagine a day at the beach. Wet sand can be packed into three-dimensional shapes like a solid; dry sand smoothly filters through fingers like a liquid. Add the right amount of water, and you can make a slurry perfect for drip castles that will harden into little pellets before they collapse into the rising tide. Sand is strange. It is not quite liquid, not quite solid. Physicists explain that granular materials like it are the "odd balls of matter." Some even believe that granular material should be its own category of matter.[14]

Sand was not the only material that filtered into the archive. Soon grains of sand were joined by flakes of snow and sheets of ice. As military engineers traveled, slide rules at the ready, they had to replicate the same tasks regardless of environmental peculiarities. Standard blueprints meant that a runway built in Arctic tundra was to look and behave precisely like a runway smoothed over desert sand. From the perspective of a military engineer, a portable nuclear reactor buried under the lunar surface needed to function just like one buried under the ice cap. Such facilities had to be implanted even as airplanes and machines became more technologically sophisticated and sensitive and as lists grew longer of what was needed to build and maintain bases. Engineers thus had to study ever more closely the materials that made up these extreme locations.[15]

The dogged efforts of US military engineers to squinch the world's varied environments into cookie-cutter shapes has left a rich documentary and environmental record. Across the files of the defense apparatus, sand turned to snow, which turned to stardust—sometimes called space dust, or moondust, or regolith, or even fairy dust—and then back to ice and sand. There was no literal transformation, of course; no engineering alchemy took place. But throughout the 1940s, 1950s, and 1960s, US military engineers swapped stories and memos about the strange, novel conditions they had found across the world's extremes, linking those traditionally isolated environments through their material oddities.[16]

The granular particulars mattered most of all in places little understood or rarely visited. How could military engineers design lunar structures—bases, really—without ever having seen or touched the Moon's surface? This was hardly an abstract question; throughout the late 1950s and into the 1960s, military engineers were repeatedly asked to weigh in on plans for a Moon base, first by the Air Force, then the Army, and finally by NASA. The makeup of the Moon was more central to the space program than popular accounts allow. Take the iconic first moonwalk on July 21, 1969. Most people are familiar with what Neil Armstrong said when he stepped onto the surface: "One small step for man, one giant leap for mankind." Less well known is what he said before that. "I'm at the foot of the ladder," he explained to the audience on Earth (more than six hundred million people) as he climbed out. He described how the craft had settled: "The [lunar module's] footpads are only depressed in the surface about 1 or 2 inches." As he stepped down another rung, he added, "The surface appears to be very, very fine grained . . . it's almost like a powder." Then and only then—assured that the Moon would hold his weight—did he drop off the ladder, imprinting his boot into the lunar dust and his sound bite into history.[17]

Such apparently mundane details were precisely the morsels of data that the engineers and scientists back at mission control wanted and what the Apollo astronauts had been trained to provide. Even the photos of astronaut footprints provided evidence of the makeup of the lunar surface. Not only did Armstrong confirm the widely held assumption that the Moon was solid; he also put to rest a host of other wild theories. Some experts had theorized that lunar dust would climb up the lander's legs, tethering the astronauts' spacecraft. Others thought the Moon was made of fluff that would suck the lander down into it. Neither of those things happened, obviously, allowing Armstrong and Buzz Aldrin to move about the surface, collecting eighty pounds of rocks and dust that they packaged in carefully designed bags to bring down to Earth.

This reconnaissance, investigation, and collecting on the lunar surface mirrored what engineering teams had been doing for years on Earth. When military engineers were sent to far-flung and little-known desert and Arctic construction zones, they conducted detailed reconnaissance—sometimes in person, but also increasingly from the air—and scooped up samples. The engineers ran tests and compared the materials they collected from around the world. They built engineering models for paving runways on permafrost and setting foundations into desert dunes. They ran more tests, took additional samples, and revised their models. Was desert sand like Arctic ice? Which terrestrial materials would best approximate the Moon? As they gathered environmental intelligence, they began to create planetary knowledge, linking seemingly disparate environments through observation and analysis. The men who built the bases of the US military empire—the engineers at the center of this story—understood that trying to master the tiny granular materials was vital to projecting not just international power but interplanetary control as well.[18] And that is how a grain of sand became a book about power on a planetary scale.

I came to study the extremes because of sand grains scattered in an archive, but I came to appreciate the ways that extreme environments can reflect larger processes and themes. As the environmental historian Steve Pyne succinctly explained: "In reducing experience, environment, and history to their minima," studying the extremes "sharpens our senses of how history elsewhere works." The nature writer Barry Lopez understood the possibilities of contemplating existence at the edges of the habitable world. In his bestselling *Arctic Dreams* (1986), he explained that the disorientation of the extremes leads to novel understandings of things closer to home. Sometimes it is the very strangeness of a place that makes it a good mirror for what we think we know. "Out there," the writer Mary Hunter Austin agreed, "where the borders of conscience break down[,] . . . where the boundary between soul and sense is as faint as a trail of a sandstorm, I have seen things happen that I did not believe myself." Austin was writing in 1909 about the deserts of the US Southwest, but the sense of estrangement and wonder was the same.[19]

In the desert, light twists and bends; mirages have been known to send people away from where they meant to go. Sand can shift without warning. Without perspective granted by trees or buildings, the naked eye is impotent in deciphering distances. Lack of perspective on the Moon was considered a danger for astronauts, who might have become disoriented and failed to find their way back to their spacecraft. In all the extremes, it seems, one can walk for hours without getting any closer to a speck on the horizon. Examining the extremes, then, may just change the way we see the world.

It certainly changed the way the US defense establishment viewed the world and what might be possible in it, although not in the ways that nature writers might have anticipated. The US defense apparatus and the people in its employ demonstrated blithe disregard for the places they manipulated and evaluated. Extreme environments were seen as expendable. They were written off as wastelands to be used and manipulated for strategic ends. Tank tracks made on the Mojave Desert floor during World War II training, for example, are still visible today. Plant populations disturbed then have not revived.

Even more obvious are the materials that the US military simply left behind: machinery and weapons; rearranged terrain; vast pits and redirected waterways; human, chemical, and sometimes even radioactive waste. In Greenland, for example, nearly everything the United States brought was on a one-way ticket. In 2016, authorities in Greenland complained of thirty abandoned sites in need of decontamination, including the location where a thermonuclear bomb went missing in 1968.[20] It has yet to be found.

Replicated thousands of times around the world, actions like these add up. As scholars have shown, the US military has been a major contributor to the greenhouse gases that are responsible for global warming. In 2017, the US military emitted more greenhouse gases than industrialized countries such as Sweden or Denmark. At the same time, the military's role is often to secure the flow of fossil fuels that underwrite rapid and often unsustainable economic growth. Nuclear weapons, too, have transformed the global environment by spewing radiation far from test sites.[21] Today, these forces are impossible to disentangle from the climate crisis, which—as evidence has made clear—has a disproportionate impact on the extremes. The Arctic is warming four times as fast as the rest of the globe. Snow-free periods there are expanding. In 2021, for the first time in recorded history, it rained on the Greenland ice sheet. Deserts, too, are getting hotter and drier. In the Mojave, ravens are filling branches, and skies have been left empty by collapsing populations of other birds, species that cannot adjust fast enough to the changes humanity hath wrought.

These contradictory realities—that the US military has done great harm to the Earth and has also gathered unprecedented and important information about the world's environments—need to be held at the same time, narrated in tandem. Both are crucial for understanding the world today and how we got here. At the heart of this contradiction—and thus central to the stories that follow—is an important but overlooked reality: strategic thinking defined the global environment in very specific and powerful ways. Within the US military, "environment" became a category used for planning and training, for parceling out the world in ways that would enhance the deployment of personnel and weapons. "Environment" was a characteristic to examine and manage, a set of circumstances that might intersect with deployments in unusual ways—but ways that could be codified for better management. More strategically, "environment" was a category that was both local and global, making it all the more powerful in projecting power around the world. In fact, the local became abstracted to the regional and the regional to the global. "Environment" came to encompass not the particulars of places but a set of characteristics that could be applied across different geographies that shared environmental features.

The military's work in the extremes highlighted the fact that to control the Earth, it had to be understood as the *whole* Earth. That is, not as discrete places or continents, not in terms of territorial borders or political alliances, but as a planet made entirely reachable, if not always welcoming. For strategic planners, the planetary perspective meant zooming out to see the blue marble floating in a dark sky and then imagining it webbed with military infrastructure and military knowledge. And that knowledge became power. Operating across the extremes transformed how the Earth was understood and altered the parameters of what could be known. As engineers set down new facilities and gathered data, they began to make links across the very environments they had long assumed were isolated. The connections started with the materials, to be sure, but were amplified as space technologies provided never-before-seen panoramic vistas. Cameras, and later humans in orbit, could take it all in, capture images of entire deserts, whole ice sheets, not just the slivers or sections that had been painstakingly traversed and charted. Still later, sensors would provide more data still, beaming down through ice and water, sand and snow.

This is not the way most of us are accustomed to thinking of "environment." But given the preponderant power and access that the US defense apparatus had to extreme environments, strategic understanding was important to defining *environment*. Throughout the 1940s and 1950s, military engineers and planners selected information about the extremes that they considered most strategic. That means that they chose some ideas or forms of expertise while overlooking others. Place-based and indigenous forms of expertise, not surprisingly, tended to be dismissed as too parochial or nuanced to fit the needs of a modern military. Instead, engineers organized the world around the types of operational and practical information that would allow them to move around that world seamlessly, regardless of the costs or the consequences they left behind.

It is difficult for us to grasp just how little was understood about the world's most out-of-the-way places in the mid-twentieth century. Today, of course, we can find endless amounts of information about any point on the globe in just a few seconds: the average daily April temperatures in Thule, Greenland (high of nine degrees Fahrenheit; low of negative six degrees Fahrenheit); expected summer precipitation in Death Valley, California (0.1 inches in July); when sea ice will begin to melt in Baffin Bay (May). We can even find out about conditions on Mars. NASA's Mars Reconnaissance Orbiter records snowfall on the Red Planet's surface, where, because the snow is carbon dioxide, it falls not as recognizable six-sided crystals, but as cubes. Even more dramatically, the Webb Space Telescope provides fantastic, Technicolor visions of dying stars and close-ups of formations like a Cosmic Seahorse, the Pillars of Creation in the Eagle Nebula, or the Phantom Galaxy.[22]

Yet at midcentury, military engineers not only had no access to this sort of information; it was simply not known or measured in ways that were useful to them. Any bits of data that had been accumulated were usually stored in inaccessible places. In the 1940s, there had been no concerted effort to collect and organize data on the world's extremes. So, although an engineer might know how to build military barracks to the right specifications, he might not be able to adapt those plans to a place without trees and wood, or one where it was so hot that concrete set before being poured.

Wartime needs put in motion a spectacularly broad effort to collect information about places that were considered unknowable. Extreme environments were question generators, and the curious rushed in. From the military services, such as the US Army, Navy, and Army Air Force, to nonmilitary agencies like the US Geological

Survey and the Weather Bureau, war pulled men and materials into the military's ambit with surprising speed. Academics who had traipsed across the world's deserts to look for new species and adventurers who had spent months riding sledges in the Arctic all found wartime homes in the research centers that were created to account for the unfamiliar. The US Army Air Force created the Arctic, Desert, and Tropic Information Center to help commanders plan for survival and training in such places. The Army's Quartermaster Corps, responsible for equipping troops, opened branches based on the extremes: the Desert Warfare Board and the Arctic Warfare Board. There, new boots were tested, uniforms designed, and food rations run through batteries of tests for taste and longevity. Operational questions dominated wartime thinking: Had anyone collected average monthly climate data? No, a continual problem in places with few or no residents. How much water did a man need for desert operations? Early one-quart-a-day experiments were quickly deemed dangerous. Would US soldiers wear shorts? The answer was no.

If global war led to new questions about operating in extreme environments, the Cold War multiplied the uncertainties. US policymakers looked at the world in the late 1940s and early 1950s and saw Soviet influence everywhere, which—to them—meant that everywhere was a place that the United States had to be prepared to defend. Defense budgets grew accordingly. Of interest here is the fact that, as the historian Ronald Doel shows, the Cold War amplified strategic interest and funding for fields devoted to extreme environments, so much so that by the late 1960s, "the earth sciences were at a zenith of federal concern." Previously marginal areas of expertise—oceanography, terrestrial magnetism, glaciology, and many more—emerged as the darlings of strategic investment. As the Pentagon explained, the government had a "vital interest in the environmental sciences since the military services must have an understanding of, and an ability to predict and even to control the environment in which it is required to operate."[23]

This dramatic increase in spending had implications for military construction and operations. Around the world, military engineers put down new nodes in a network of knowledge and power that pressed out and up, looping up into the air and down into the ground, circling the Earth and branching up to the Moon and beyond. While wartime knowledge was temporary and expedient, Cold War strategic plans required more permanent ways of knowing the world, of settling in. For example, air bases would be designed to last not for a few months, as they were in World War II, but for twenty years. This required an intimate understanding of local environments so that engineers could access any indigenous resources and military activities—radar facilities, testing ranges, training grounds—could sprawl into the surrounding areas. A little-known reality of US militarism is that a base, once established, tends to spread out. For engineers, this meant novel studies of fantastical-sounding concrete-like phenomena like "perma-crete," "snowcrete," and "mooncrete." Environmental intelligence required peering up into the sky and devising ways to go down under the ground, all while moving across terrain. At the same time, addressing those practicalities required a basic understanding of how grains of sand moved or how snow compacted into ice. As one engineer's report noted simply, "Usually military research does not have to reach so far down toward the foundations of science."[24]

That was an invitation. Indeed, across the strategic infrastructure, research programs were pulled along in the wake of the unparalleled military logistic and operational juggernaut. Scientific work on global environments has since moved through these outposts, forced to travel along the paths set by the US military. Civilian scientists simply could not get to Greenland to study ice and snow, or to the middle of the Sahara for weeks on end, without an organization to drop food and supplies or to airlift them out when it was time to go home. Military patronage shaped the character and content of emerging civilian scientific fields as they became tied to the national security bureaucracy. How could it not? As Naomi Oreskes has shown in her work on military funding for oceanography, there is no "free ride when it comes to science."[25] US strategic priorities drove the sorts of questions that were asked and programs initiated. That the US military has repeatedly treated extreme environments as expendable is intimately tied to the types of knowledge it produced there, and the military origins of that knowledge matter for what we know, how we know it, when we knew it (or were kept from knowing it), and what we can do with it going forward.

There was plenty of hubris that went into many military projects in the extremes, plenty of assumptions about how easily and quickly US power could be imprinted across wild terrain, about how easily environmental peculiarities could be overcome. There was a history of doing this, of course. Americans had long prided themselves on ingeniously unlocking latent natural potential, plowing what had been deemed infertile soil, building railways, damming rivers, digging canals. Nature could presumably be made to do man's work. Such confidence made the engineers reach out farther and farther into the extremes, sure that work in one place might lead to success in another.[26]

For a time, they were not wrong, and so they kept going. Yet it was not always possible to make nature yield to machines and men.[27] Extreme environments pushed and pulled against the levers of US power, changing the ability of the military to function as it wanted to. Even simple tools like the compass needed refiguring. So, too, did tires and gears. In Greenland, keyholes had to be eliminated because wispy ice crystals perpetually managed to blow in. Standard blueprints had to be amended, notes scribbled in the margins, text hastily crossed out.

Construction practices generated new ways of knowing the world in part because environments could resist even the best-laid plans. Design and construction were always works in progress. Engineers in the extremes did not act despite the environmental conditions they found themselves in but because of them.[28] As we will see, working in the extremes not only stymied US strategic plans but often channeled them in unexpected ways.

By the time Neil Armstrong landed on the Moon, the outlines of the environmental intelligence that would best support US power had been sketched out and practiced. Engineering teams had worked out ways of moving sand dunes, digging ice tunnels, and landing on unknown surfaces. But even though these things could be done, that did not mean that they should have been done. Operating in the extremes was expensive and time-consuming, terribly and constantly so. The logistic bridges had to be unendingly maintained and updated. But to what end? Even before Apollo 11 took flight, the ice tunnels in Greenland had been abandoned. It turned out that the ice deformed faster than anyone had anticipated, and the effort to keep tunneling and supporting crews living there—just keeping the structures viable—would have taken every bit of everyone's time. A Sisyphean task, to be sure.

Prestige and strategic thinking have short attention spans. Far-flung research programs and expanding military bases became harder to justify in the mid-1960s. War in Vietnam both commandeered defense dollars and changed the focus of environmental interest. Who needed Arctic uniforms or studies of desert water rations when many of the military's men were in the jungle? Changes in US Cold War strategy shifted the geography of military bases, too. So did local opposition, which forced the closure of US bases in Northern Africa. Intercontinental missiles and long-range bombers meant that the far-flung base network could contract. Thinking about outer space followed the same trajectory. Landing on the Moon came to seem like enough. December 13, 1972, during the Apollo 17 mission, was the last time a human stepped foot there. The quest for planetary power stalled out on limited strategic vision and lack of funds.

Yet the US defense establishment had put into place the logistic blocks and pathways that made aspirations of planetary discovery and control possible, if not always realizable. How that happened with such surprising speed, and how limited those visions ultimately were, is the subject of what follows. The military's efforts to master the material foundations of these places—sand, snow, stardust—determined how other forms of power and knowledge would flow. And it all began with the smallest of details. You had to understand the merest of particles, a grain of sand even, if you wanted to send a man to the Moon.

I. SAND

PRELUDE

No one had prepared them for the mud. Or the rain. Or the wet chill that settled deep into their bones. They had expected to be pouring sand, not muddy water, out of their boots. But from the moment the American GIs arrived in Tunisia, the war reporter Ernie Pyle explained, they had encountered "day after day" of rain. Not a refreshing tropical rain, but "a piercing, chill, England-like downpour." It was the kind of weather that made them think they might never be warm again.[1]

Worse, however, was the fact that the men had been led to expect the opposite. "Wear a sun helmet," "avoid sunstroke"—these were the bits of wisdom handed down from the War Department. Not until they were halfway across the Atlantic did the men learn where they were going: North Africa. Avoid the sun at midday, the soldiers were advised, when it was best to keep a "good roof" overhead.[2]

The men who headed to North Africa as part of Operation Torch were handed a small booklet, *A Pocket Guide to North Africa*, while their boats rocked from side to side, and they worried about German submarines. Thirty-five thousand men in all, shipped across the Atlantic in October 1942. Half as many more Americans shipped directly from England. The United States had entered World War II the previous December, and although troops were already fighting in the Pacific theater, Torch was the first time US forces would meet the German and Italian enemies. Combined British and American forces

FIGURE P1.1 Frontispiece to *A Pocket Guide to North Africa* (1942). The pocket guide was handed out to American GIs as they crossed the Atlantic before taking part in Operation Torch, the Allied invasion of North Africa. Across popular publications and military documents, like this one, the desert was assumed to be a sandy expanse. Published by the Special Service Division, Army Service Forces, US Army. Government Information Resources, Southern Methodist University.

were to kick the Axis powers out of Africa, chasing them across the Mediterranean into Europe. Expectations were high.

A Pocket Guide to North Africa opened with a cartoon image of an endless, flat expanse of desert, an oasis far off in the distance, and trucks of soldiers driving in the dust (fig. P1.1)—precisely what a child might draw if asked to draw a desert. If, as the cartoon's caption suggested, this was North Africa the way people "always imagined it," then the drawing signaled a collective failure of imagination that the GIs would not soon forget.[3]

It was hard to square what the GIs had been told with the reality they found on the ground. By December, Tunisian nights were bitterly cold, and the Army-issued windbreakers provided little protection against the punishing hail. Thigh deep in mud, frigid GIs angled their helmets down over their faces to avoid the angry skies. Pyle reported that soldiers increasingly "resented" the fact that their families back home wrote letters to them assuming that, because they were in Africa, they must be "suffering from heat prostrations." Brigadier General Theodore Roosevelt III (son of President Theodore Roosevelt) openly mocked pamphlets that had promised sunshine and warmth. With "customary dumbness," he declared, the military had considered all of Africa "a tropical country." What he wouldn't do, he told his men, for a little warm sun.[4]

All through the winter of 1942, US commanders in Algeria and Tunisia dashed off requests for heavier winter clothing and woolen coats, equipment no one had thought to include in their requisition lists just a few months earlier. Skeptical supply officers back in the United States "could not understand the need for clothing against extreme cold in Africa."[5] And in any case, there were no winter coats available because they had not been requisitioned.

We began "to look upon mud rather than the Axis forces as our enemy," admitted Brigadier General Donald A. Davison, who had larger concerns than his own clothing allowance. As one of the few aviation engineers for the US Army Air Forces, it was Davison's job to get the planes off the ground. But the same mud that anchored men up to their knees also fixed their machines to the earth. The only way to take off from their muddy field, a pilot recalled, was to "load a bunch of crew chiefs on the horizontal stabilizer to lift the nose until we picked up speed." As the plane took off, the men jumped.[6]

That winter in North Africa, as hopes for a quick victory faded into muddy skirmishes and watery runways, Davison was low on options. With the right equipment, his crews would have been able to navigate and manage the terrain, improve existing airfields, and build new ones. But just as troops lacked winter coats, Davison and his crews lacked the necessary equipment. Under the impression that they would be working with sand and dry dirt, the engineers had packed

FIGURE P1.2 *Saturday Evening Post* map of the Sahara (1942). The map of North Africa appeared in the *Post* on May 23, 1942, along with the article "Sahara War." The image and story echoed popular misconceptions about what Americans would find when they landed in North Africa later that year. Note the dramatic sun, camel, and palm tree oasis; there is no indication of the actual topography of the area. Illustration provided by Curtis Licensing.

"midget" bulldozers, not the heavy machinery they would need to operate in a muddy morass. No one had drawn pictures of mud when they thought of North Africa.

And that was the problem. Before the war, no one in the US defense establishment had given much thought to North Africa or to the world's deserts. They envisioned the entire continent something like the drawing in figure P1.2, a land of sun and palm trees. In midsummer of 1942, when President Franklin D. Roosevelt decided that the US Army would go to Algeria and Tunisia that fall, military commanders and engineers had just a few months to figure out what to requisition and how to prepare. In the absence of good intelligence, they relied on incomplete notions about what the desert was like.

In January 1943, General Dwight D. Eisenhower, supreme commander of the Allied forces in the Mediterranean, admitted that, during Operation Torch, the weather and conditions had slowed the Allies in their quest to take Tunis.[7] What he did not admit was the extent to which the Americans had planned for the wrong environment. The desert they seem to have imagined, the one they packed for, was over a thousand miles away. There, the British and Germans were locked in seesawing tank battles across the great sand seas and dune fields of the Sahara. On most maps, to be fair, everything looked like the same desert. But on the ground, it was a desert as different as different could be.

★ 1 ★
THE SMALLEST OF DETAILS

At 10 p.m. eastern time on February 23, 1942, the crinkling of newsprint joined the crackle of the radio as maps were unfolded in homes across the United States. The American people were tuning in for a geography lesson. President Franklin Delano Roosevelt was to speak to the nation "on the progress of the war." Sixty-one million Americans listened, nearly 80 percent of the adult population. For days leading up to the address, the White House asked people to prepare by digging out old maps or acquiring new ones so they could follow along with the president's tour of the world at war. That morning, newspapers from San Francisco to Boston printed their own charts, some a quarter page, others half. Residents of Chicago and Los Angeles were greeted with full-page world maps, printed above and below the fold.[1]

"I'm going to speak about strange places that many of [my listeners] have never heard of," Roosevelt told his aides as they worked on the speech; he was largely right (fig. 1.1). Over the previous few months, newspapers had described clashes in places like Tobruk, Sumatra, Sidi Barrani, and Bataan. The president's task that February evening was to somehow make the strange more familiar. To bring the stakes of the very foreign war into the homes of every American listening. But he would not do this by naming every potential battlefield; Roosevelt conspicuously avoided the particulars. His immediate aim was to subtly convince his audience that the whole

FIGURE 1.1 President Roosevelt in the Oval Office, February 23, 1942. While delivering his fireside chat "On the Progress of War," FDR gestured to a world map behind him to explain the first "truly world war." Nearly 80 percent of the country's adult population tuned in. Library of Congress Prints and Photographs Division; LC-USZ62-127693. Bettmann, Getty Images.

world mattered. "This is a new kind of war," he explained. "Different from all other wars of the past, not only in its methods and weapons but also"—he punctuated these last three words with particular vehemence—"in its geography." There was a lot for the American people, and its military leaders, to learn.[2]

Two minutes into his remarks, Roosevelt invited those listening "to take out and spread before you a map of the whole earth." There was no need for listeners to remember every detail of what they would cover, but he did want them to trace routes and lines around the world, across the vast oceans—once barriers, now battlefields—to distant shores. This was a war of connection, dependent on the maintenance of "vital lines" of communication and supply between places. The country's most important role was to keep the lines open. This was not like the last war, or any before that, Roosevelt explained.

This war was everywhere. As the *New York Times* informed its readers, "No war was ever so literally a world war as this." It was not long before the newspapers, and the president, began using the word *global* to describe this world of interconnected places.[3]

Even as the president prepared the country for the new global way of thinking, he knew that the country's knowledge about the world—and particularly its varied environments—was dangerously parochial. He could describe the invisible web of air routes and shipping lanes that the country was creating from shore to shore, point to point, but it was up to policymakers and military commanders to grapple with the reality that many of the places he gestured toward remained uncharted, literally not on their maps. Since the boys had come home from the previous war, there had been little institutional effort to accumulate and catalog intelligence about environments at the edges of the habitable world. Military planning had been largely defensive, focused on potential conflict in the continental United States or "similar climatic areas." No one had considered large-scale maneuvers or construction at Earth's extremes.[4] Yet those were precisely the places they now had to plan for.

Earlier that same month, Major General George Patton had declared to his troops, "The war in Europe is over for us." Never shy about speaking his mind, the commander then went on to predict that "our first chance to get at the enemy will be North Africa," which was an environment that the US military knew almost nothing about. The *New York Times* was more specific about the alarmingly "strange" places that the war had pried open. American GIs would soon be fighting in "Arctic snows, in desert wastes, and in equatorial jungles."[5]

Much has been written about how the United States lurched toward war, preparing incrementally for an undefined intervention. Until the Japanese attack on Pearl Harbor, the country lacked both the political will and the military capacity to fight. The country also lacked basic information about the world's environments that could have made the variables of fighting and the stakes of battle clearer. Although it is perhaps inaccurate to describe the interwar years as "isolationist," much of the country was remarkably incurious about the rest of the world.[6]

There were two interconnected reasons for the United States' ill preparedness. One had to do with lack of knowledge about the world in general. The other had to do with lack of knowledge about certain environments. The president did not mention specific environments in his fireside chat that evening of February 23, but his comments were intended to force the American people and military planners to begin to prepare themselves for all they did not know.

The war compelled American military planners and strategists to consider the world's environments as they never had before, not simply as exotic places to explore and visit, but as sites of conflict and construction. Throughout 1942, the defense apparatus staggered toward acquiring the information it needed about the strangest of these "strange places." The initial and urgent task was to collect and compile data. But that was just the beginning. In preparing to dig down and in, to anchor the infrastructure of US military power across the whole world, military leaders had to grapple with a stunning array of environmental information. Would men be able to acclimate to extreme heat or cold? How would equipment function? Would oils freeze or separate? Could engineers create runways on all terrain types, including sand, ice, and snow? What was a desert like? Were all deserts the same? What happened when military technologies moved from one place to the next? Where could military leaders begin to accumulate the types of environmental intelligence that would help them win a war?

Major General Patton needed information. Just two weeks after the president's chat, he was ordered to Southern California to take command of the new Desert Training Center (DTC), where the US military would attempt to make up for what it did not know about one extreme: the desert. Patton's first tasks were to locate the best area for the DTC and then to oversee a rigorous training regime. The seasoned commander knew what was at stake. He had been in the Army a long time. In 1916, he was part of the US Army's efforts to capture the Mexican revolutionary Francisco "Pancho" Villa; during World War I, he led tanks into battle; since then, he had been instrumental in thinking about how to incorporate tanks into US military training and doctrine. By 1942, then, he had a keen eye on US preparedness, and his assessment was grim. US troops were poorly trained, he thought. He was not wrong. Just two years earlier, in fact, the US Army ranked seventeenth in the world in terms of size and combat readiness; it was "a third-rate military power," with forces only slightly larger than the national army of Bulgaria.[7]

It took Patton just four days of surveying the Southwest to select the "perfect" area for training: a ten-thousand-square-mile patch of land that spanned the Mojave and Sonoran Deserts (the size of the training range doubled in 1943; fig. 1.2). The area was not totally unfamiliar to Patton, as he had lived for a time in California. But still, Patton marveled at the wide-open spaces. "The only restrictions as to movement," he proclaimed, "are those imposed by nature." Finally, after months of training men in cramped centers back east, hemmed in by towns, roads, and rivers, the trainees would be able to "open fire with live ammunition or drop bombs at any time and in any direction without endangering anyone." In the desert, Patton realized, there was "room to burn."[8]

While Patton was confident about his troops, he was less certain about the desert. "While I have played polo and navigated ships across the Pacific," Patton admitted in letter to Roy Chapman Andrews, the head of the American Museum of Natural History in New York, "I have a limited amount of knowledge about the desert."[9]

Patton needed information primarily to plan for the new training center, but also because of the ongoing fighting in North Africa. In fact, Patton spoke frequently about the possibility of joining the British there. Since June 1940, British and German armored divisions had been seesawing across what was called the Libyan, or Western, Desert, the part of the Sahara that stretches from the Nile in Egypt to Tripoli in Libya. US newspapers and magazines showed a Western Desert campaign marked by "clouds of dust," "rocky, dune-ribbed" terrain, and extreme heat. Articles delighted in setting the scene. The Sahara was a land that knew no moderation: immense expanses of sand, intense heat, endless sand dunes. "Nature's No Man's Land," the *Saturday Evening Post* declared.[10]

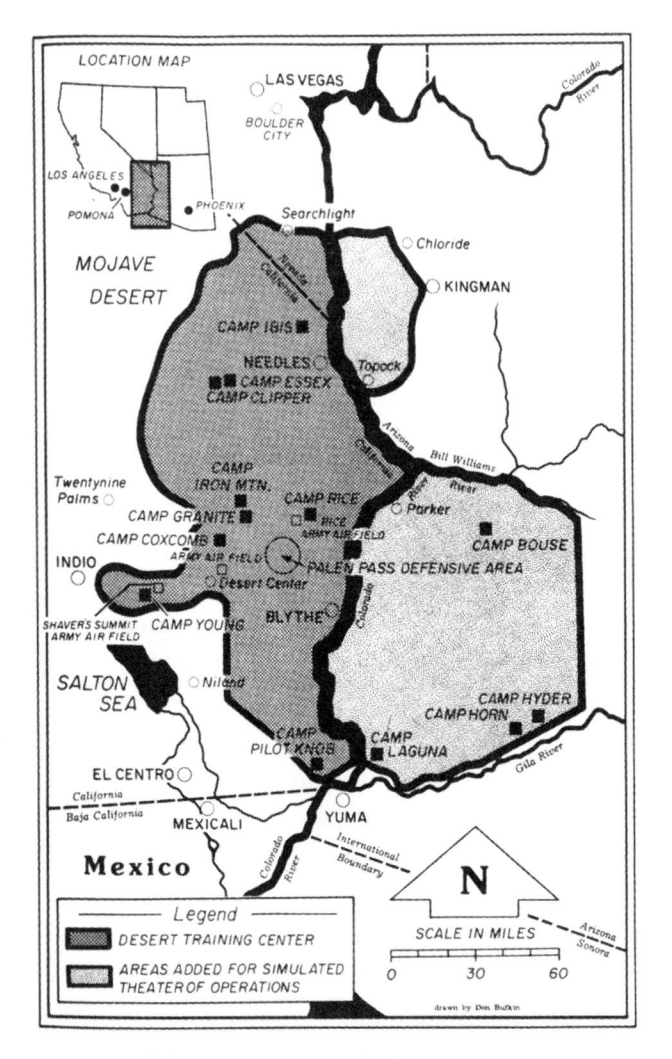

FIGURE 1.2 The Desert Training Center, 1942–1944. Eventually the DTC spread across twenty thousand square miles. The site was chosen for the area's relative emptiness and accessibility. Many of the surrounding communities were linked through rail and road networks; the Colorado River aqueduct ran right through the DTC, allowing access to water. Joshua Tree National Monument (now a national park) is where the bite-size chunk is taken out of the western edge of the DTC. Courtesy the Bureau of Land Management.

Back in the United States, Army commanders watched the tank battles with increasing resolve. Allied losses in the Mediterranean would threaten the very sea lanes and air routes that President Roosevelt had spent so much time outlining, not to mention access to oil. A firm decision to send US troops to North Africa as part of Operation Torch was not made until July 1942—a decision that remained top secret until November when the troops arrived—but the possibility of desert war animated discussions about preparation and training throughout the year. To be sure, all extreme environments were of considerable interest, but for a moment, the desert was supreme.

At the very moment of the president's fireside chat, plans were being drawn up to send equipment to Egypt. Three hundred Sherman tanks, self-propelled artillery pieces, and maintenance crews were scheduled to disembark in early summer. As a result of shipping and supply problems, they did not actually arrive until September.[11] The president may have been talking about seamless supply lines, but in reality, they were full of kinks and gaps. US commanders, too, traveled to Egypt to consult with their British counterparts and to study desert tactics and strategies. It was a start, to be sure, but that was as far as the Americans had gone in thinking about deserts. Which is why Patton turned to writing personal letters to acquaintances who might be able to share some insights into how to fight in sandy, dusty places.

Andrews, of the Museum of Natural History, offered as good a person to start as any other. The dapper midwesterner had taken his post at the American Museum of Natural History in 1934 after a series of well-publicized journeys and archaeological finds in the Gobi Desert in Asia. In 1923, he was on the cover of *Time* magazine. For the rest of the decade, stories of Andrews's exploits—many of dubious veracity—were splashed across national newspapers and magazines. Andrews was rumored to have outrun bandits, fought off poisonous snakes, amputated a man's hand, and escaped a pack of fierce Mongolian dogs. If he sounds a bit like Indiana Jones, some have suggested he was the model for the fictional archaeologist. Andrews himself wrote about keeping calm in sandstorms, the best way to drive across the desert, and the importance of a firm constitution for any and all expeditions into the unknown.[12] Like Patton, Andrews was frequently photographed with a gun on his hip. Where Patton had a signature ivory-handled six-shooter tucked into his holster, Andrews favored a rifle for hunting or a simple pistol for protection.

It is easy to see why Patton would have been attracted to Andrews's persona. These were the qualities Patton wanted instilled in his own men. "We must train millions of men to be soldiers," Patton understood. "We must make them tough in mind and body, and they must be trained to kill." But it is equally as likely that Patton simply had no idea who else to turn to for information. In 1942, there were not many desert experts at home or abroad, at least not in the US military. Nodding to his own ignorance, Patton concluded his letter to Andrews: "Do not hesitate to give me the most trivial details which from your experience, you might consider superfluous."[13]

A similar set of conversations was taking place throughout the military establishment as US commanders looked out on a world at war. The year earlier, Colonel Simon B. Buckner Jr. wondered about extreme cold. As newly appointed commanding officer of the Alaska Defense Command, he needed data about cold-weather combat and gear. Buckner queried anyone he could think of: "trappers, dog-sled drivers, old settlers from the gold rush days, guides, prospectors, itinerant doctors, outfitters and member of the Interior Department's Bureau of Native Arts and Crafts." His goal was to fill out the "limited knowledge and experience in our Army regarding this subject."[14] Although his list of potential sources was long, it was decidedly narrow in its reliance on the professions he knew and the populations he could most easily access. Buckner had no intention of visiting the villages of Alaskan Natives even though there he likely would have found the most pronounced Arctic expertise.

On the eve of war, when US military commanders like Patton and Buckner sought expertise and intelligence about these far-flung battlefields, there was no institutional repository of environmental knowledge, no Army library of the global extremes. Until that moment of need, in fact, the US government—its military and civilian agencies alike—had shown little interest in accumulating such information about the world. Even previous military engagements, such as those in Cuba, the Philippines, and Siberia, provided few transferable lessons for the 1940s. The sort of retrospective and practical guides that emerged in such abundance after World War II were conspicuously lacking in the lead-up to it. In the rare cases when such documents were available, files sat unopened. In 1944, the office of the quartermaster general admitted that, in planning for cold-climate combat during World War II, it had failed to consult materials from 1918, when fifteen thousand Americans were deployed to Siberia during the Russian Civil War.[15]

Given all that was not known and all that was poorly understood, Patton's first letter to Andrews—and additional letters he sent that spring—are less odd than they first might seem. Individual commanders were using personal connections to find out whatever they could about where US soldiers might need to fight. As Patton penned letters and Buckner queried local organizations, the defense establishment swung toward mobilization. The slow and idiosyncratic search for information would gradually turn more intentional and institutional. Planning for war set into motion a giant transfer and reordering of knowledge about the Earth's varied environments. Information was pulled from everywhere and anywhere; even picture postcards became visual source material.[16] New centers were established to account for new environments and the need to develop the right equipment and gear for them.

In 1942, the US Army's Quartermaster Corps, long charged with equipping troops, divided its wartime Special Services branch into environmentally specific boards: for desert warfare, Arctic warfare, and jungle warfare. Each board was given the responsibility to study and develop the best equipment and gear for troops to be deployed into the respective climate of focus. The creation of the Desert Training Center was part of this process. That same year, training areas were also established in Alaska and Colorado for cold-weather and high-altitude work.[17]

The US Army Air Force formed its own Arctic Research and Information Center, eventually renamed the Arctic, Desert and Tropic Information Center (ADTIC). The goal of the center was to examine the functioning of both man and air machines in extreme conditions. "Global war confronts us with a dual fight," *Air Force Magazine* explained in 1943. The ADTIC was created to provide the information required to "win wherever we fight—on the ground and in the air of the frigid Arctic, the arid desert and the steamy tropics."[18] For reasons not immediately obvious, work on extreme environments—cold, wet, hot, dry, high—was often lumped together in the military and across such publications. Even the various environmentally specific boards and centers shared information and plans. Sometimes, the extremes were also referred to as "exotic." This is in part a reflection of the small, rarefied world this sort of work encompassed. But it is also a recognition that the extremes mess with "normal" functions and equipment, if not in similar ways, then at least in ways that are equally as dramatic. In 1977, the Army's *Desert Environmental Handbook* explained that "the arctic, tropic, and desert areas of the world all have one thing in common: conditions of extreme environment." Each environment "impose[s] severe stresses on men, material, and materiel." And while those stresses might be different across space, the end result is the same: "Materials fail to function as they were designed."[19] The solution, at least the one conceived in World War II, was to determine how to adapt existing practices and technologies to the extremes, how to make the environment bend to what was at hand, even if for only a short time.

As these new centers and units opened, the defense apparatus proved able to pull men and materials into its ambit with surprising speed. Individuals with what had hitherto been niche interests in research and exploration found their knowledge suddenly coveted. The Australian Sir Hubert Wilkins offered his services to the US Army's Quartermaster Corps to assess equipment and gear for extreme environments. The fifty-four-year-old fit well with Patton's penchant for theatrics and fame. Wilkins was, according to a biographer, "the last—and one of the greatest—explorers." In 1913, he had accompanied Vilhjálmur Stefánsson, who would later join the war effort as an Arctic expert, on his Canadian Arctic Expedition. A war photographer during the Great War, Wilkins had reportedly put down his camera and led men in battle when their officers were killed. In

1928, he was knighted for his transpolar flight. He was also a noted ornithologist and won a series of medals from the National Geographic Society for his work capturing wildlife. In the 1930s, he proved that it was possible to pilot submarines below the Arctic, although his goal of sliding under the North Pole proved elusive during his life (his ashes made it there on a submarine in 1959). In the early 1940s, Wilkins availed himself to both the British and the Australian governments for wartime service. Both turned him down due to his age.[20] The Americans were less discerning, or perhaps more desperate.

Wilkins wandered across the American defense establishment just as he wandered the world; his name pops up everywhere. Like many of the other experts brought in as wartime consultants, he worked with the Army and the Army Air Force at test sites and in laboratories. He observed recruits battling against heat and sun at Patton's Desert Training Center; he visited indoor climate chambers to advise on clothing design; he traveled to Washington, DC, to attend meetings of various panels and boards that were attempting to bring order to the chaotic, last-minute effort to acquire environmental intelligence and expertise. And like many of his explorer-turned-military-consultant peers, he delighted in participating in whatever trials and experiments he could: testing sleeping bags in the snow, donning sunglasses in the desert, sampling rations designed for survival kits.

We don't know what, if anything, Patton learned from Andrews that spring. We do know that there was a lot to learn. We also know that by early April 1942, troops had started moving west toward Camp Young, the first installation in the Desert Training Center and what would become the administrative center. Accompanying them was a rotating group of experts who wanted to study the extremes and, of course, a commander who still had a lot of questions. "I would deeply appreciate your sending to me any and all information, pamphlets, and what-not, you may have on the minutia [sic] of desert fighting," Patton wrote to a colleague in the War Department.[21] He understood that the limits of his knowledge—the limits of his nation's knowledge—could be the first sign of defeat.

The first trainees—the boys who became the guinea pigs in the Army's effort to master the desert—began to arrive just weeks after Patton first identified the outlines of the training center. Most came by train, some by truck. Most traveled from somewhere back east: Delaware, Kentucky, Pennsylvania, Ohio. Few had been west of the Mississippi; fewer still had seen the desert. The GIs were told to pack their bags for training in California, so they could have been forgiven for envisioning palm trees and oranges, Hollywood signs and movie stars.

The men of the Fifty-Fourth Armored Field Artillery Battalion stepped out of their train into ankle-deep "flour dough" (a thick sand that the engineers cursed as "fluff") and looked up to see "a brilliant expanse of sun and sand and jagged rock." Sand was everywhere; it rose in a "chocking dust." The men sent to the desert learned that sand grains would fill their shoes, their pockets, their blankets, even their canteens, and even when they swore they had screwed the lids on tight. The GIs wrote home about the ways the sand pricked their faces and clogged their ears. The battalion band struck up the tune "There'll Be a Hot Time in the Old Town Tonight."[22]

Many of the troops arriving in California were surprised by the wind. Its sharp, hot blasts took their breath away. They squinted in the blinding, searing white light of the desert sun. Out there, the sky seemed higher, the light brighter, the air crisper and sharper than anything they had experienced before (fig. 1.3). The GIs cursed the heat. "It was so damn hot you could fry an egg on the palm of your hand," recalled John C. Coveney. Temperatures soared into the triple digits. The men of the Ninety-Third Infantry Division, one of the few units of African Americans in the segregated US Army, slept under wet towels to stay cool. Everywhere were reports of men burning their hands on tools that had been left too long in the sun.[23] Dog tags left scars. It seemed that everything in the desert had either a sting or a thorn—scorpions, snakes, spiders, cacti, and more. The president had spoken about exotic and strange places, but California hardly seemed one of them.

On the eve of war, the US military's experiences training in the Mojave seemed only to confirm a relatively simple view of what a "desert" could be. Deserts were wastelands, vast, empty, and changeless.

FIGURE 1.3 GIs relaxing in the Desert Training Center, somewhere in the Mojave, ca. 1943. Over one million US soldiers were trained in the desert from 1942 to 1944. Image courtesy General Patton Memorial Museum, Chiriaco Summit, California.

Not unlike the War Department's cartoon that opened *A Pocket Guide to North Africa* (see fig. P1.1) or the *Saturday Evening Post* image of North Africa (see fig. P1.2). Trainees adopted the imagery immediately, sending letters home complaining of a place that was "empty," "lifeless, and "the land that God forgot." "It is going to be hell," one GI wrote days after arriving at Camp Young. "Some of the guys are cussing California," he explained, "but you can't blame them not seeing any place else in the state than dropping them off in *no mans land*."[24]

Common associations underwrote the idea of the desert as a dangerous wasteland. Littered across elementary school primers and popular accounts were tales of overland pioneers and forty-niners who had braved the desert to reach the "promised land." The list of potential hardships was long. Some were avoidable, and others a matter of chance. You could pack poorly, hire the wrong guide, start too late in the season, use an outdated map, take an unproven shortcut. The desert was a dangerous place that one should cross quickly. The nature writer Mary Hunter Austin, in 1903, remarked that every year the Mojave took "its toll in death," leaving "sun-dried mummies" as a warning to the next group. Textbooks and Westerns described nineteenth-century encounters, both peaceful and violent, with Native Americans. Although many more people had settled in the area by the 1940s, it remained terrain that most visitors sought to cross over and get through, not linger in. As a result, it was relatively easy for Patton and the US Army to turn the desert into a military reservation. Patton explained that it was better (and easier) to bomb and run maneuvers in a place that lacked roads, streams, and houses. Modern military needs—large-scale training and testing with mechanized vehicles and planes—required a lot of open, unused space. The public seemed to agree; there is no record of anyone protesting the acquisition of twenty thousand square miles of desert land for military exercises. No one raised alarms about the creation of a training center that hosted nearly one million GIs between 1942 to 1944, or about any of the other desert lands appropriated for military needs.[25]

Almost immediately the Desert Training Center came to be called "Little Libya." Patton seems to have made the connection right away, explaining that the DTC territory he had picked out was as close to Libya in climate and geography as one could get. The War Department's desert operations manual acknowledged that "the information included . . . is based primarily on the experiences in the Western Desert of Africa," which was thought to be "applicable generally to all deserts." All the worlds' deserts, in this reading, were the same. Nothing about training in the DTC suggested otherwise.[26]

GIs began to use the moniker Little Libya as well, and rumors spread that the Mojave was even worse than the Sahara. Not that any of these men, Patton included, had ever been to Libya, but it was a comparison that carried cultural and strategic weight. The desert Southwest had long been a stand-in for the deserts of North Africa and the Middle East. The towns of Mecca and Bagdad (both in California) were just miles from Patton's base camp and named for their perceived similarities to the deserts of the Middle East. A desert in one place was like a desert in another, even half a world away. That April, as the first troops arrived in Indio, California, the last stop before the training area, the *Los Angeles Times* welcomed them to the "Southern California Sahara."[27]

Little Libya was a simple, seemingly innocuous shorthand linking training to war. Indeed, had the US military actually landed in Libya or Egypt in the fall of 1942, lack of good intelligence would not deserve mention. Patton explained that he wanted his "men to take just as rough a beating as I can give them in as near the situation they will have in North Africa."[28] But the desert is not where the North African invasion took place, and simple thinking was misleading. US forces that took part in Operation Torch landed not in a hot, dry desert, but in a wet, wintery coastal area. In this case, then, the Americans had trained for battlefields they would not see, and their experiences in the Mojave perpetuated expectations about deserts that were not to be. Put another way, had the US military been more attuned to the nuances of the desert, it could have better prepared for the desert variations presented by the war instead of rushing in on the basis of thin expectations that would prove terribly disappointing.

The journalist Ernie Pyle, while reporting on the war in North Africa, carefully explained that nothing about any part of North Africa looked "the way it does in the movies," even though it seemed that everyone thought it would.[29] That could lead only to a bad ending.

★ 2 ★
MUD, RAIN, FROST, AND WIND

North Africa is where the United States learned the costs of poor environmental intelligence. Brigadier General Donald A. Davison had the misfortune of being one of the first to experience the information shortfall. In 1942, Davison was one of the country's leading engineers trained in aviation needs. His job was to create the runways and airfields that would allow Allied aircraft to take off and land again. But as soon as he arrived in Algeria as part of Operation Torch, he knew he had a problem. Having anticipated a dry, desert environment, Davison and his teams of aviation engineers had packed the wrong stuff. As his boots sank into the muck, and his materials and planes were mired in crud, Davison admitted what many others were thinking: the mud was their main foe (fig. 2.1).[1]

In retrospect, of course, it is easy to think of mud as a small detail, an inconvenience that had to be overcome. Indeed, it was overcome. By May 1943, as the skies cleared and the Americans finally gathered the right equipment, Allied forces kicked the Axis powers out of North Africa and back across the Mediterranean to Europe. Both US and British commanders were able to plan their assault on the continent, set to begin that summer. But those in the know understood that getting to that point had taken them longer than expected. The Allies had hoped to get to Tunis within three months of landing. Instead, "great rains and appalling mud," President Roosevelt admitted

"I knows mud pretty good mate, but this sort's new to me. Maybe they imports it special for airports."

FIGURE 2.1 *Stars and Stripes* cartoon, February 1943. The cartoon pointed out what all GIs in North Africa already knew: although people back home thought they were in a hot, dry desert, they were actually wet, cold, and stuck in the mud. Bruce Bairnsfather / © 1943 2024 Stars and Stripes, All Rights Reserved.

in early 1943, "have delayed the final battles of Tunis," first by weeks, then by months.[2] The war dragged on.

Operation Torch, when written about at all, is framed as a learning exercise for the Americans, something of a tune-up for the war in Europe. When its failures and missteps are discussed, they are ascribed to a host of factors that all point toward the newness and inexperience of the American military machine: poor training, inefficient logistics, choppy communication lines, the growing pains of cooperating as part of a multinational military force.[3] All legitimate factors in military shortcomings, to be sure. But if the environment is taken seriously, new ways of seeing the war and planning after the war emerge. First, as the president recognized, the unexpected rain and mud prolonged the battle for North Africa, which delayed the push into Europe. The importance of terrain and climate were hardly novel to military strategists; these things were vital to any military campaign. But in the case of Operation Torch, the failure to properly anticipate the environmental peculiarities of where the troops were heading, combined with new needs of building for modern war, created a potent mess. A postwar report was clear on the problems: "We should have known more about North Africa before invasion." "We came to Africa," the authors admitted, expecting to find "the schoolbook land of desert and sand and hot winds," only to be met with "mud, snow, and winter weather." The environment had made fools of them all.[4]

Hints of trouble had appeared in September 1942, two months before the invasion parties were to land. On September 16, Robert Murphy was summoned to London to visit with General Dwight D. Eisenhower. Murphy—an American diplomat who served as liaison to the Vichy regime in Algiers and quietly chatted with the president in the Oval Office—traveled under the pseudonym Lieutenant Colonel MacGowan, because, as General George Marshall admitted, "nobody ever pays attention to a lieutenant colonel." Secrecy was critical because very few people knew about Operation Torch, and things needed to stay that way. Murphy was one of the few American officials who had recently been to North Africa, which is why Eisenhower—appointed supreme Allied commander for the operation—requested him.[5]

Murphy flew across the air routes that the president had told the American people about and that military engineers were constructing. From Washington, DC, up to Labrador and then over to Scotland before coming down into Hendon Aerodrome. From there he was whisked off to Ike's residence, Telegraph Cottage. The men chatted for nearly twenty-four hours. They started in the backyard, where the skies were unexpectedly clear and sun dappled the lawn. Sometimes others joined them; sometimes they chatted alone. Most of the questions Eisenhower peppered Murphy with were political: Who in North Africa could be trusted? How would the Americans know whom they could confide in? Murphy had been in Europe for over a decade and was well versed in the French political landscape and its personalities. He was confident that there were enough potential collaborators in North Africa that they would provide significant assistance to an Allied invasion there. This was welcome news to the general and the handful of advisers who came and went from the conversation.

As the sun set and the air cooled, the men stretched from their seats in the backyard, crossed a flagstone terrace, and entered the house through its french doors. A coal fire warmed the room. Over soup they continued to chat about who could be trusted, which airdromes might be readily available, and whether the Spanish would get involved in the fight. The room grew exceptionally smoky since the windows remained closed and heavily draped due to the blackout. The general and the diplomat chatted well into the night, and then again over breakfast the next morning.[6]

At some point, before Murphy's departure early in the morning, the conversation turned. Murphy recalls being asked what Northern Africa was *really* like. The men in the room seemed to have rather primitive and fuzzy ideas about the continent. Murphy assured them it was "more like California than a tropical wilderness." While the statement was technically true, it may have been misleading. Perhaps unclear as to what he meant, Eisenhower asked whether the men would need "winter underwear." Operation Torch was set to begin in November, after all. Murphy responded that, yes, indeed, winter underwear might be a good idea.[7]

Soldiers' undergarments did not usually come up in meetings about high-level strategic planning; it was an odd enough detail that Murphy noted the conversation in his memoirs. In retrospect, the exchange hinted at a rather alarming environmental ignorance on the eve of battle. The assembled military planners did not know what climatological conditions the troops would encounter in just six weeks' time. Worse, even if they had chosen to act on the new bit of environmental data, it was too late. Requisition lists had been drawn up in August from the meager information available. Patton and his fellow commanders created packing lists based on expectations rather than real data. This was partly due to the necessary secrecy surrounding the ongoing planning, and exactly where the troops would land was not always known. Patton and others planned on the basis of the assumption that they were going to a desert like the one where the British were fighting, like the desert where they were training.

There was little to complicate their expectations. Earlier that year, Frederick Wulsin, one of the men called in to consult the military on desert war, had gone to Libya to observe British forces and assess their uniforms. Wulsin's claim to desert expertise was based on the expeditions he had conducted with his wife, Janet, in the Gobi Desert during the 1920s. From 1921 to 1925, the Wulsins traversed Asia, collecting plant and animal specimens and photographing people, landscapes, and religious ceremonies along the way. The couple cited the work of Roy Chapman Andrews as their inspiration. In 1926, *National Geographic* covered their expedition, giving Janet zero credit for any of it. Her contributions went unknown until the 1960s, when her daughter began recovering the story.[8] The couple divorced in 1929, the same year Frederick earned his PhD in anthropology.

It is no surprise that of the two, it was Frederick who was asked to join the war effort. I have found no record of women being consulted for environmental expertise. But in 1942, the Quartermaster Corps appointed Frederick to lead its desert studies unit. Wulsin was asked to help find the best equipment to protect troops operating under extreme conditions. He took a trip to Egypt to observe British equipment and gear. Wulsin was apparently taken with the havelock, a flap affixed to a cap that protected the neck from the sun. Back in California, the American GIs disagreed—they did not like it.

The havelock was the least of Wulsin's misjudgments. He also returned to California with news about the conditions that US troops would likely encounter when and if they reached North Africa. He reported that severe temperature swings—like those the men experienced in the Mojave—should be expected because the desert could go from extremely hot days to frigid nights. Wet conditions, however, were not to be prioritized. In North Africa, Wulsin explained, "there was, most of the year, little rain and therefore no need for protection from it." Aviation engineering literature, too, wrote of dry conditions across the region, emphasizing potential problems in maintaining runways because of the absence of water.[9]

In any case, even if Eisenhower had gotten word back to Washington about "winter underwear," there was no time to change the packing lists. Procurement, manufacture, and transport were far too complicated to allow for that. The global reach of war—the fact that battles would be in these far-flung places, stretched out along thin supply lines—meant that information had to be gathered well in advance of deployments. The military commanders would have to make do with what they had, even if what they had was not up to the challenge.

And it wasn't. Just weeks after landing in North Africa in November 1942, the GIs realized that Tunisian nights were bitterly cold. To stay warm, the GIs across North Africa wore every article of clothing they had and looked around for more. Brigadier General Theodore Roosevelt III described his layers: "wool trousers, and shirt, then a sweater, then a lined field jacket, then my lined combat overalls, then a muffler, then my heavy short coat." He was glad he had brought his own woolen underwear. But not even that could keep him warm. He was so wet and cold that he did not change his clothes, including his undergarments, for days. "We are not well prepared," Roosevelt stated plainly.[10]

So bad were the conditions that mud became something of a punch line, a collective misery shared by soldiers across North Africa. They came expecting sand dunes and instead found sinkholes. The war reporter Ernie Pyle christened the Americans the "mud-rain-frost-and-wind boys." *Stars and Stripes* reported that "mud and mystery" were the year's November and December holiday decor. At the airdromes, crews had to use shovels and sticks to scrape mud from wheels just so that planes could taxi to the runways.[11]

Bad weather was not wholly unexpected. Military intelligence had acknowledged the possibility of winter rains.[12] Eisenhower and his staff understood that the race to Tunis was going to be a race against winter. But that acknowledgment of seasonal change in intelligence quarters did not translate clearly into operational plans. Nor, as fighting in North Africa made clear, did it help engineers account for the tasks of modern war. Modern war fought with world-spanning technologies, they learned, required environmental intelligence far in advance of battle.

Just as Roosevelt had explained in his February fireside chat, aviation transformed the world, the war, and the types of environmental intelligence required in order to deploy not only in the obvious sense that ordnance dropped from the air created its own form of destruction and terror, but also because airplanes rearranged experiences of space and time. Suddenly, men and materials could be moved halfway across the world in a few days rather than the weeks or months it had previously taken. But that acceleration of movement was not accompanied by a commensurate quickening in intelligence gathering and knowledge production. The need for runways and airfields to be quickly constructed in faraway places necessitated advanced data and reconnaissance, but little of that had been done. Indeed, aviation not only opened up the air but also the ground beneath it. Refueling stations and frontline airdromes meant that conditions of local terrain should be known well ahead of time, not at the last moment—especially because, unlike today's airplanes that can soar over the ocean without stopping, in 1942, long-distance flights required multiple stops to refuel and resupply. Take one example: in January 1943, when Roosevelt flew to Casablanca to meet with his British counterpart, Prime Minister Winston Churchill, he first took the train to Miami, where he caught a Boeing Clipper to Trinidad. In 1942, that flight took ten hours; today, it would take fewer than five. From there, the president's entourage headed to Brazil and then across the Atlantic to Gambia, before the final leg of the journey up to Casablanca.[13] It was the first time in history a US president had flown for official business. A trip that would take eight hours today took Roosevelt five full days.

Every stop along the way required special maintenance depots and resupply. This was the work that Davison and his fellow aviation engineers were tasked with completing and that the president had mentioned in his fireside chat. Each airfield needed runways, taxiways, hangars, maintenance facilities, fuel lines, and storage. The more sophisticated the planes got, the more extensive their ground needs became. World War II–era fighters and transport planes required runways at least three thousand feet long; the heavy bombers needed twice that, often more. The work was unglamorous and underappreciated but nonetheless essential. As the *Military Engineer* declared, the "preparation of airdromes" was the "dominant engineer task" of the war.[14] Without airpower, little else could be done. And without detailed environmental data—a kind of data not previously required—airpower was a fantasy.

War had turned aviation engineering from a specialized field to one of the country's most pressing needs. During World War I, airstrips had been little more than "grass plots." But modern aircraft and their importance to the war meant that the new airfields required "extensive tracts of land cleared of obstacles and provided with all-weather runways."[15] This sort of construction required ever-more-sophisticated environmental data and plans—environmental intelligence. As the country lurched into war, the Corps of Engineers scrambled to update engineering manuals and practices with the new realities.

While overseas emergency airdrome construction was central to the war effort, so too were airports at home. Within the United States, the military constructed nearly three thousand fields. Designs had "come a long way," but still, everyone knew this was just the beginning. In 1942, the design of airport and runway pavement was "still in its swaddling clothes." Given constant changes to airplanes and aviation technologies, the Corps had to "start almost from scratch."[16]

There were no good specifications for building modern military runways. There weren't even great or consistent standards for building the nation's roadways. Paving technologies were novel. Different agencies, municipalities, and companies used different formulas and standards for roadways and runways alike. War changed the calculus by pouring resources at the problem. As war raged, the Corps' aviation experts were determined to "eliminate [the] wide variation in designs" and "limit the use of unproven theoretical design methods."[17] A state of emergency provided the centralized planning that could bring order to what had been a messy maze of local and state procedures. And to create standards for paving over terrain, one had to start with the most basic material of all: what was underfoot.

If one wanted to know about dirt, Arthur Casagrande was the man to call. Born in the Austro-Hungarian empire, Casagrande had immigrated to the United States in 1926, and he spent the late 1920s touring European soil mechanics laboratories—where he often collected vials of dirt—before landing at the Harvard Engineering School. There he created his own soil library. In 1936, he organized the world's first International Conference of Soil Mechanics and Foundations Engineering. A rather large contingent of Corps of Engineers officers attended and joined discussions about the best ways to evaluate different soil types and ground conditions for construction purposes.[18]

In 1942, the chief of US military engineers hired Casagrande to bring order to the messy, incoherent system of defining soils and paving runways.[19] This was a new field of study. Never had there been a need for so many runways, built so quickly, in so many diverse places, and all at the same time. Knowing soil was the key to designing the best runways. A modern runway capable of catching larger and faster-moving planes needed a stable subbase, which consisted of specially calibrated granular materials that could serve as insulation between the ground below and the repetitive stresses of aviation above. Without the right subbase, a runway would list or buckle, warp, or crack, rendering it useless. Identifying and creating the best subbase, then, required careful understanding of the terrain. And as planes got heavier and faster, the types of subbases and pavements had to change.

Casagrande approached the problem from two directions. First, he recognized that the field of runway design lacked standards and basic protocols. This ultimately led to a problem of soil classification. Without a good understanding of specific soils around the world, how could runways be built globally? Casagrande advised the Corps to begin a rigorous evaluation of the world's soils so that different terrain types could be assessed in terms of aviation construction needs. The resulting Airfield Classification System divided soil types into major classes according to their functionality for runways.[20] The hope was that engineers around the world could arrive at a particular place, test the soil for basic properties of size and strength, compare those values against the new classification system, and then design stable and durable runways.

For the system to work, however, Casagrande insisted on his second approach to soil mechanics and engineering: training. The engineer was the linchpin that allowed US Army Air Force specifications to be imprinted across a variety of terrain types under conditions of great duress and speed. Unlike domestic construction in the United States, no one was going to have access to laboratories and modern testing equipment. Wartime engineers needed to head into the field with the ability to read the soil.

So it was that, in July 1942, Casagrande began welcoming US engineers to Cambridge for "Soil School." His first class only had twenty-four students, but over the course of the war, he trained more than four hundred engineers. Because nearly none of them had training in soil mechanics, they had to start from the beginning. In addition to classes and readings, the men participated in Tuesday field trips, either afternoons spent in the surrounding area or visits to Casagrande's lab, where he had accumulated soil from around the world. Casagrande held forth on the art of observation. He reportedly rolled a sample between his palms and across his fingernails and would inhale deeply to determine its scent; sometimes he would even taste it.[21] The trainees practiced field testing so they could learn to confidently read the ground. The most experienced engineer, Casagrande intoned, might not even need lab tests to know what was underfoot.

Davison did not train with Casagrande. His leadership was needed closer to the battlefield. But he, too, understood the importance of knowing the dirt. It partly explains his prioritization of the advanced reconnaissance of potential airfields; it was never enough to have mere documentation on a place. To build quickly and well, to smooth over runways for warplanes, the engineers would have to put their hands into the dirt and feel around.

As one of the country's most experienced aviation engineers, Davison was at the forefront of thinking about construction in a time of war. He had enough experience with runways and airplanes to know that the more data he had on hand before he showed up somewhere, the better. Since 1940, in fact, Davison had been conducting worldwide reconnaissance with an eye toward military construction needs. Like Murphy, Davison traveled in disguise, usually as a Lend-Lease agent. But his real task was to scout potential runway sites and test soils. Well before the United States entered the war in December 1941, Davison had been all over Great Britain and to Iceland, scouting the best locations for air bases, creating lists of available resources, and trying to draw up building plans for still-hypothetical runways.[22]

Davison's report from Iceland is indicative of the sort of information the modern engineer needed. Iceland, he wrote, was rugged and forbidding; the winds fierce and the infrastructure lacking. Davison took note of roads—impassible in the winter, mostly one lane, simple gravel surface—and examined the meager air facilities already there. Paving new runways would be difficult because the unfamiliar soil was little more than a spongy peat moss that seemed to suck materials into soggy holes. He took samples and sent them back to Virginia, where the Corps of Engineers had its headquarters. Davison concluded that, except for rock and sand, nearly all supplies would have to be shipped over from the United States. Still, there might not be enough time to get everything done. In October 1941, cargo ships carrying timber, Nissen huts, canvas, bags of Portland cement, and heavy machinery set out from Boston Harbor, bound for a construction zone in Iceland.[23]

Davison had not been able to visit North Africa in advance of landing there. Without boots-on-the-ground data, he did what everyone else did: he took whatever partial information he could get. And that information was wholly inadequate. He collated a mishmash of materials from all over. The British provided the Americans with data from their colonial possessions in East Africa, but for the western areas where the Americans would land, available data were meager. The Americans received old French maps from the 1920s and 1930s, although most were too large in scale to be of use for combat operations. Other Allied maps of the Sahara showed thousands of square miles as blank spaces, detail yet to be determined. By September, the wartime Strategic Intelligence Branch of the Army Corps of Engineers had amassed material from British and American construction companies, diplomats, geologists, academics, and tourists who had been to the area. A 1939 report on North African airfields drew almost entirely from Michelin guidebooks. The resulting volumes of information were detailed, though not always accurate and often incomplete. In fact, large areas of the maps remained barely sketched in, mere outlines of the unknown. On the eve of arriving in North Africa, one commander dashed off an angry letter to military authorities: "I feel that it is absolutely necessary to invite your attention to the fact that this command is going in without adequate photographic coverage of the terrain over which it is going to operate."[24]

The end point is not surprising. The aviation engineers had not been able to run the sort of tests that Casagrande had taught or to draw up engineering plans for the terrain they found themselves in. They showed up to clear runways and create airdromes with "midget bulldozers." The small machines saved space on their ships, but they were not useful in the North African mud. The engineers had left their heavy equipment back at home. "The allotted equipment," a postwar report admitted, "was too light for the assigned work." Power shovels, tractors, plows, rubber-tired and sheepsfoot rollers, scrapers and shovels with draglines, bulldozers, and grading machines all would have been useful. There was no asphalt available in North Africa; it all had to be shipped in from elsewhere, meaning that Davison had no way to improve or pave over the existing sludge.[25]

Davison and his teams worked hard to make do, of course. For one, they continued moving to new, drier places, looking for better materials and terrain from which to launch fighters and bombers into the sky. They moved further to the south and into the open desert. They adapted local materials to the needs of runway maintenance as best they could. In Telergma, Algeria, the engineers bartered with American cigarettes to use French road-building equipment.[26] Everywhere they tried to employ local labor. Davison reported that he had hired nearly three hundred local Arab workers to haul and move vast quantities of earth with shovels made from small trees.[27] The teams would get the tasks done, but that would take longer than expected.

Roosevelt heard from his generals shortly after he arrived in Morocco for the January 1943 Casablanca Conference. Eisenhower admitted that "the improvement of airfields had been one of the greatest problems." Instead of reaching Tunis that December, as initially hoped, the Allies were stuck for the winter. Eisenhower also acknowledged the lack of appropriate materials and equipment, citing the engineers' pressing need for steel mats to cover the ground. The mats, like winter clothing, had not been requisitioned. The local materials they had been able to procure instead were inadequate for the machines of modern war. "Broken stone merely sank into the mud," Eisenhower explained.[28]

In the end, men like Davison were able to engineer their way out of the mud and muck. When Davison arrived in November, there were only 9 usable airdromes; by May 1943, there were 127—although that did not stop a postwar aviation engineering retrospective from the honest assessment that their work in North Africa was "not impressive." The reasons? "Weather, lack of equipment and general confusion."[29]

Throughout the long six months in North Africa—and well after—the engineers accumulated a great deal of information and experience that they scratched into notes and memos sent back to the States. Lessons were learned and knowledge and experience filtered into postwar manuals that looked out at the world as if anticipating what was to come. Engineering pamphlets and publications, ranging from the quickly copied *Aviation Engineer Notes* to the glossy *Military Engineer*, reflected this interest. *The Air Force Engineer*, too, began publishing sometime in the middle of the war. Each of the engineering battalions crafted their own commemorative pamphlets, highlighting what they had done and where. One can imagine the copies being passed around, shipped from one place to the next, dog-eared and worn like the copies that made their way to Corps of Engineers filing cabinets. These publications made visual note of the muddy mess that had been North Africa, complete with cartoons of men with their feet mired in mud.

Davison himself wrote up lessons learned from North Africa that were printed in the *Military Engineer* and disseminated across the front lines. Engineering was an iterative process, and the men involved took their mistakes seriously.[30] They studied the problems, the soils, the way small things could escalate into disaster. They studied what went wrong and what went right. If they could find the patterns, they believed, they could create standards and manuals and maps that would make sure they would not make the same mistakes again. And they believed that knowledge about the world—a deeper, more accessible knowledge that was grounded in standards and readily shared classification systems—was a determinant of national strength and their own capabilities going forward. Given the technologies of the day and their trajectory—ever more sophisticated airplanes, missiles, satellites, and more—there would be no time to gather data in the event of conflict, no place for those who did not already have the information. The Americans would have to know in advance how to deploy to any and every point on the map. No place was too extreme, and no island too remote. Crucially, then, they believed that whoever had access to environmental intelligence would hold power.

It was under the auspices of environmental intelligence that the engineers created their own dirt collection. In the fall of 1945, the war just about over, memos went out to air bases around the world. The engineers hoped that base commanders could collect a few soil samples from wartime runways. They provided a collection box, shovel, and instructions. In addition to fifty pounds of dirt, they also wanted detailed descriptions of the local area, especially conditions near the runway.

Samples were returned from over five hundred airdromes. Some commanders heeded instructions better than others. From the

Philippines and other bases in East Asia, the engineers received cartons of over fifty pounds along with narrative descriptions. Bases in Algeria tended to provide less than fifty pounds, but enough to be significant. Still more commanders seemed not to have read the instructions at all: they sent samples of dirt in simple envelopes, which, when opened, emptied their contents ingloriously. The hope from gathering soil was to take advantage of wartime deployments and experiences to begin building a more comprehensive database of global soil conditions. To begin mapping and charting where and what engineers could expect if, in the future, they were sent back out into the world.[31] Poor planning before the war was not going to be repeated.

★ 3 ★

THE CACTUS CULT

The war in North Africa was over, but the quest for environmental intelligence had just begun. It was driven increasingly less by the manic urgency of looming battle and more by the need to gather long-term and detailed data, the kind of data that would help you stay in place. Advanced environmental intelligence—what one engineer report called "logistical intelligence"—was critical to imagining the future.

The collection of soil samples from worldwide airfields suggested the potential breadth of this sort of thinking: it was far more intimate, meticulous, and methodical than anything that had come before, and it was happening on a far grander scale. Away from the front lines a quiet shift was taking place in the minds of engineers about the types of environmental data that they needed to support a modern military and US worldwide interests. The most important information was no longer going to be focused only on battlefield maneuvers and operations, but on what we might think of today as back-end design. Being able to send the right equipment at the right time meant being able to anticipate more accurately what would be needed and where. And so even as the war was raging, military engineers understood that a more robust global American military presence would require more careful planning. They did not want to show up somewhere at a future date and find that all they had experienced during the war was lost.

But what type of data? Most obvious was the same sort of information Davison had wanted: terrain assessments, as well as intelligence on existing infrastructure, the availability of local resources and labor, water sources, and climatological information. The sort of information that could help them plan. They did not want anyone to have to complain that they had the wrong bulldozer, not when moving dirt was such a vital part of winning a war. As the engineers proclaimed, "dirt-moving" had become the "primer of offensive power."[1]

Back at the Desert Training Center in California, as the war continued, engineering continued to push and pull on the sand and dust. The engineers set down hundreds of miles of road, linking one training camp to the next, sometimes just angling roads toward nowhere so they could test out equipment and gear. They compared paving roads over sand dunes and gravel washes, dry lake beds and desert pavement. They measured the grains they found with their sieves, checked their angularity, and evaluated how scrapers might work on a desert pavement that was really just a thin veneer over the dreaded "fluff"—a material more like flour than sand. In some locations, they used giant fans to suck up dust and sand and hurl it at their machines, testing for wear and tear, for efficacy, for how long gear would work before the sand mucked it up. They took pictures and crafted articles for the *Military Engineer*.[2]

Other men came, too, to take measure of how the human body operated in arid environments. Described as a crew of "bearded and unbearded, natty and matted and tattered" experts, they studied boots, clothing, rations, and sun exposure for the Quartermaster Corps.[3] If the engineers were to desert-proof machines, these men were to desert-proof the American GIs. Or at least insulate them enough from the elements to make them efficient.

For every topic and idea the engineers and experts chose to pursue, however, there were many more that they ignored, even ones that were literally next door. All around the deserts of the American Southwest were people who could have been asked to share their expertise and experiences. These included Native American communities whose ancestors had lived in the region for hundreds, in some cases thousands, of years. The desert area that Patton chose, after all, was named after

FIGURE 3.1 The Joshua tree. The Desert Training Center had to curve around the area that, in 1936, had been set aside as Joshua Tree National Monument. Competing views of desert environments at midcentury highlighted how little understood these places were. In fact, the monument was threated by wartime training. National Park Service.

the Mojave people, who, by the 1940s, had long since been forcibly confined to reservation land near the Colorado River.[4] By midcentury, plenty of Euro-American settlers had decided to stay put in the desert. Miners, cattle ranchers, adventurers, artists, a few recluses were thrown in. Many came for a visit and stayed. The author Mary Hunter Austin was one. Born and raised in Illinois, she arrived in California in her twenties and never left. "For all the toll the desert takes of a man," she explained in her 1903 book *Land of Little Rain*, "it gives compensations, deep breaths, deep sleep, and the communion of the stars."[5]

Alternative forms of expertise were inscribed on the Desert Training Center (DTC) maps (see fig. 1.2). The little towns scattered around the area suggested permanent communities. The infrastructure that ran through signified the federal capacity to work in such terrain. The bite-size chunk taken out of the western edge of the training area was the site of Joshua Tree National Monument, created in 1936 (fig. 3.1). Clearly people had known enough about the desert to want to set some of it aside. What morsels of knowledge did they hold?

Of the most potential interest for military purposes would have been the work of botanist Forrest Shreve. Attached to the Desert Laboratory in Arizona, Shreve had just published a map of the plants of the Sonora. A reviewer exclaimed that it was as "great [a] contribution to geography as it is to botany."[6] Compared to existing Army maps of the area, it was highly detailed indeed.

Forrest Shreve was a plant man—more specifically, a desert plant man. Shreve had moved from the East Coast to the Arizona Territory in 1908 so that he and his wife, Edith, could join the research staff of the Carnegie Institution's Desert Laboratory. Opened in 1903, the lab was devoted to coming to a better understanding of the world's deserts. It was one of the few private research institutes in the United States dedicated to desert studies. Shreve never left his work there, even after Carnegie funding stopped in 1940. The facilities were transferred to the US Forest Service. In 1960, it became affiliated with the University of Arizona and is still operating today.[7]

Pictures of Shreve show a man who liked to blend into his surroundings (fig. 3.2). He wore earth tones, high-waisted canvas pants, a tan fedora, and small round-rimmed spectacles. The botanist spent as much time as possible quietly wandering the desert, measuring, observing, categorizing. Shreve had great passion for the intricacies of desert vegetation: how plants adapted to extreme conditions, why they grew in certain assemblages, what those assemblages could say about the climate, terrain, and region.

Sometime in the 1930s, Shreve began mapping the plant populations of the US deserts, starting with the Sonora. He was particularly interested in the edges of the desert, where one environment seemed to become another, where, for example, the Mojave turned into the Sonora, or the Sonora into the Chihuahua. These edges could be messy and unstable—there was no line in the sand, so how could scientists understand any shift? Plants held the answers. Desert plants had evolved with specific adaptations that could be read as proxies for environmental and climatic conditions. This meant that most deserts contained their own endemic plants, and no two deserts had exactly the same species.[8] If scientists knew their plants, they could determine their location in the world.

FIGURE 3.2 The botanist Forrest Shreve in the field (1940). From the early 1900s through the 1950s, Shreve painstakingly mapped desert vegetation in the Sonora. His maps were largely ignored during the war, but his practices and efforts were prized afterward. From *A Sense of Place: The Life and Work of Forrest Shreve* © 1988 Arizona Board of Regents. Reprinted by permission of the University of Arizona Press.

Some of the GIs sent to the Desert Training Center may have noticed small shifts—the "edges" that Shreve wrote about. In the southern sections of the DTC, in Arizona, south-facing hills were dotted with the giant saguaro, the iconic desert cactus that appeared in all movies and comics about the American West. The saguaros thinned as troops moved north, and they all but disappeared when troops hit the higher elevations near Camp Young, just outside of Indio, California. There, at the DTC's first camp and administrative center, GIs encountered playas full of creosote shrubs and, higher up, groves of the otherworldly Joshua tree.

Austin certainly noticed the changes. For decades, Austin chronicled the desert country. She spent her days wandering the desert in flowing white dresses and broad-brimmed hats. Her writings charted changes to the plant populations of the Mojave. Desert blooms, she wrote, "shame us with their cheerful adaptations to seasonal limitations." But she also recognized that they were fickle, prospering only in certain years. In some years, flowers exploded with "tropical luxuriance," but other times, nothing seemed to grow at all. Botanists later realized that the seeds of desert plants are configured so that they germinate only with a certain amount of rainfall, which washes away protective chemical compounds to allow for germination.[9] Austin loved the rhythmic vagaries of the life she encountered in the desert—the way plants turned their leaves edgewise against the sun, the way different species grew in small clusters, the way communities stubbornly survived the harshest conditions.

Plants were the desert's silent guides. The leaves of the creosote bush grow in a north-south direction to avoid the buildup of direct heat under the constant sun—hence the bush's other name, the compass plant. Or take the Joshua tree. A person standing in a grove of them can know a handful of things about their location: elevation (higher than three thousand feet), aridity (high), type of desert (rain cloud), the expected seasons of rainfall (winter), and the average diurnal temperature ranges (large in the high desert). Knowing how and where desert plants grew was in itself a guidebook all on its own, a plant-based GPS, if only one knew how to use it.

By 1943, the Army Air Force was pulling this sort of environmental information into its survival manuals. Concerned less with large maneuvers in desert areas, the service's Arctic, Desert, and Tropic Information Center (ADTIC) was intent on ensuring that any downed pilots or crews could outlast novel environmental conditions. Even after the battles for North Africa were over, the Sahara remained strategically important as long as air bases and supply routes were scattered across its edges. Indeed, many of the runways built for combat remained operational for the rest of the war. Air crews continued to soar over sand seas, sometimes disappearing into them. But the

Army Air Force had to worry about other deserts, too. "Our men are equally as likely to find themselves operating from bases in or near" deserts in "India, China, in Syria and Palestine, in Australia." It was thus a good time, the ADTIC understood, to take advantage of "lessons learned."[10]

The first lesson: "Start unlearning." Contrary to popular expectations, the ADTIC's desert pamphlet informed, not all deserts are sandy or hot. Even though "all deserts conform to the same general pattern and exhibit fundamental characteristics," the reader was to avoid the mistake of taking this to mean that "differences do not exist." On the contrary, echoing Austin's views of the desert but without the literary flair, the ADTIC explained that the world's deserts exhibit "many local differences." Take the Sahara, for example. The deeper a person goes into the desert, the drier it becomes. But "the coastal plateau areas" were more humid. In 1943, after the experience of Operation Torch, the ADTIC noted that "the relatively heavy precipitation" of the coastal areas could seriously inhibit operations. Small details, hard earned.[11]

For the airman who was downed in the desert, details were important. Desert plants could contain clues to water, if he knew what to look for. Sometimes, the leaves or pads of plants collected dew in the evenings, which could be harvested in the morning. "One flyer kept himself alive for days in this manner," the ADTIC proclaimed. Groves of plants might suggest a water source—the greener the vegetation, the greater that chance. Certain plants, too, contained water reserves; they acted like floral camels, in a way.[12]

Desert plants could be good for a lot more as well. US military personnel in the DTC, for example, encountered the creosote shrub with some regularity. Austin had called it the "immortal shrub"; engineers learned that it was certainly tenacious. It was hard to dig up and dispose of. In fact, the oldest living organism on Earth is thought to be a creosote shrub called King Clone. The shrub appears in clusters because of the peculiarity of how it propagates: it clones itself. A single creosote bush is often surrounded by a ring of genetic equals.[13]

This "immortal shrub" has many uses. Over centuries, Native

American tribes took advantage of creosote's medicinal powers. The Tohono O'odham people used the plant as treatment for snake and spider bites; they boiled the leaves and used the broth to induce vomiting. Salves from creosote were used to treat rashes and cuts. The scholar Gary Nabhan has identified at least fourteen afflictions and diseases that were treated with the shrub, sometimes called "greasewood," including colds, dandruff, body odor, distemper, and stomach issues. The shrub's resinous gum was also used to cement arrows to shafts.[14]

Not that any of this information would have been visible to the trainees arriving at the DTC in 1942. They did not need the shrub's resin to bind their metal weapons, and camp infirmaries dispensed aspirin. Nor were the troops likely aware that the desert had already been lived in and through. For centuries the Mojave people had operated a vast trade network that connected their settlements on the Colorado River to outlets on the Pacific coast, cutting through and along the desert. Small groups of Mojave men ran along these desert paths at night, guided by reflective white stones. They could travel as far as a hundred miles per day, an astonishing distance given that the GIs could barely muster a mile in one afternoon.[15]

There was one area where local knowledge would have particularly helped the trainees: hydration. "A lot of us will croak here," one GI wrote home after his first day in the desert. He had found three men dead in a tank, killed by heat and lack of water. He clearly had reason to be concerned. During the two years that the Desert Training Center was in operation, an alarming number of GIs died of thirst. The historian Matt Bischoff found that there were as many as one thousand dehydration-related casualties. Occasionally, newspapers carried stories of the fatalities. In the summer of 1943, a platoon of soldiers went missing for three days. Thirty-five of them found their way back to base and water, but three died. They had become disoriented and wandered out farther into the desert.[16]

It was not that the Army lacked water provisions. In fact, one of the reasons Patton chose the boundaries of the DTC and the location of its headquarters at Camp Young was access to water. The Colorado

River Aqueduct had been built just a few years before; it funneled water from the river to the parched communities of the Southwest. The aqueduct's system of pumps, tubes, and shafts was considered an engineering marvel. In 1955, the American Society of Civil Engineers recognized it as one of the "Seven Modern Civil Engineering Wonders"; today, it is more widely known for having helped drain the Colorado basin to a crisis point.[17] In 1942, military engineers working in the DTC added extra shunts and spigots to siphon off more water. In addition to basic water lines, they built a five-hundred-person shower in the middle of the desert. The Army paid a fee for water used—much of it wasted—but in any case, the trainees had a ready and plentiful supply.

Still, an alarming number of GIs ended up in hospital beds. The Army's Quartermaster Corps sent experts to investigate. So it was in the summer of 1942, when Hubert Wilkins and Weldon F. Heald arrived in "sizzling hot" desert country to consider what extreme heat and dryness could do to a man and his equipment.[18] The goal of the DTC was to give its approximately one million trainees rigorous practice before being thrust into battle, to harden American soldiers for combat. But those soldiers were also unwitting guinea pigs in a giant experiment to assess how men and machines responded to extremes. Men like Wilkins and Heald had come to take measure.

When they arrived in California in 1942, Heald (beardless) and Wilkins (everyone commented on his Vandyke) would not have seemed obvious choices for desert observation. Neither man was a desert expert, but given the type of expertise that was prized, their selection made sense. As a child, Heald had developed a lifelong passion for the mountains. In the 1930s, he summited peaks on four continents, ran a packtrain across the Cascades, and served as vice president of the American Alpine Club. When he could not travel, he retreated to his library, which contained a renowned collection of materials on the world's higher altitudes.[19]

Heald and Wilkins went to the desert to consider the effects of both dry heat on men and sand and sun on materials and hardware. They carried with them the equipment the Quartermaster Corps wanted tested: boots, helmet liners, new canteens. But Heald reported that studying men was far more interesting and challenging than studying their kit. Wilkins and Heald took the temperatures and pulses of men who had been out training every day. They weighed more men than "a couple of nurses in a maternity ward" would have weighed babies. With graph paper and pencil, they drew charts of men's fitness alongside climate conditions and daily activities. What relationships could they discover between heat and health? Between time of day and activity levels? Why were so many GIs in Patton's training camp landing in the infirmary?

The answer was simple: not enough water. Patton had wanted to forcibly acclimatize troops to desert heat. He rationed men to one canteen (one quart) per day, regardless of activity. The policy "did save water," admitted Heald and Wilkins, "but it expended men." In country where sweat "oozed off men like juice from a grilled chop," their bodies needed a lot more water. Heald and Wilkins did not know precisely how much water would suffice, but over time, they created even more charts and tables of how much water a man likely needed under different extreme conditions, hot and cold, wet and dry.[20]

Patton was hardly alone in his thinking about water, acclimation, and efficiency. In March 1942, the Army's hastily published *Basic Field Manual Desert Operations* recommended that restricted water consumption "become habit" for men in the desert. Not only because it would make men become more efficient in combat, but also because drinking too much water during the day would simply be "thrown off in excessive perspiration and thus wasted." At the first craving for water, the manual suggested, troops were to moisten their mouths and throats by taking only small sips from their canteens.[21]

There were plenty of people in 1942 who could have told Patton and the US Army that this strategy was wrong, although few could have explained precisely why. In Egypt, for example, the British had settled on no less than a gallon per man, per day, and more in the summer months. Only the most hardened and experienced desert enthusiasts could get by with less. Ralph Bagnold, a British military engineer and student of the desert, had spent two decades exploring the world's sandy expanses. He was considered one of the world's reigning experts on desert mobility, so much so that he had been pulled out of retirement in 1940 to set up the renowned Long Range Desert Group, which sent small cadres of men around the Sahara to harass Axis forces. Even Bagnold insisted that the bare minimum ration for his own men—the hardened desert adventurers they were—was four pints. More was always better.[22] In the desert, soldiers had to drink whenever they could.

People who lived in the Mojave could have shared that same advice. In 1903, Mary Austin warned that one should never "underestimate one's thirst."[23] Austin walked along desert footpaths and chatted with the folks who had also chosen to make the desert their home. A wide mix, she admitted, of Shoshone and Paiute people, prospectors, thieves, recluses, and romantics. Those, at least, were the ones who made it. She often encountered desiccated bodies of men and women who had died of thirst so were unable to tell their own stories. Perhaps they had ignored the signs of dehydration, entered the desert without enough information, or become distracted from its myriad dangers for one important moment. Those looking for shortcuts were the most at risk. American mythology—and likely the textbooks that DTC trainees read in grade school—was littered with tales of pioneers and forty-niners nearly kept from California's promised land by poor planning or a wrong turn, both of which could lead quickly to dehydration and starvation. Only God or a lucky break would save someone.

In the 1940s, the study of thirst and the physiology of hydration were fledgling research areas; the world of sports drinks, electrolytes, and special nutrients were still decades away.[24] The data accumulated during the war would certainly help jump-start studies, but in 1942, information about when, what, and why to drink was localized and particular, often just snippets of advice and wisdom passed from person to person. The drier the environment, the scarcer the water and the more urgent and fine-tuned that wisdom became. The adage "drink when you are thirsty" makes sense only in places that have ample and obvious supplies of water—and even that became standard medical advice only late in the twentieth century.

Minerva Hamilton Hoyt would have known all these lessons (fig. 3.3). She had spent the better part of twenty-five years getting to know

FIGURE 3.3 Conservationist Minerva Hamilton Hoyt. Hoyt was a Southern California socialite who became convinced that the desert needed saving. Throughout the 1920s and 1930s, she threw time and her considerable fortune at turning desert conservation into national policy. Image courtesy Los Angeles Times Photographic Archive, UCLA Library Special Collections.

the desert. Her conclusion was not that the desert was a wasteland but that it was in danger of being wasted, destroyed by heedless human activities. It was because of Hoyt's efforts that the Joshua Tree National Monument was set aside in 1936. In 1942, when the Army created the DTC, it had to avoid the new parkland. The Army objected to the restrictions, of course. Why should conservation land be more important than military training? For the Army, it was not—it tried to snake new roads and test sites onto park lands. The military was stymied by a scrappy gaggle of park rangers who took photos, wrote reports, and demanded that at least "one place in the desert" be kept "natural." Military maneuvers, they argued, had "very rapidly injured" much else.[25] The park lands held.

Like the desert itself, the Joshua tree was hardly an obvious choice for preservation (see fig. 3.1). "The most repulsive tree in the vegetable kingdom" is what John C. Frémont called it in 1844. The nineteenth-century explorer, mapmaker, and spokesman for western expansion did not stop there. The Joshua tree, he explained, was "stiff and ungraceful." Others went further, describing the plant as a "misshapen pirate with belt, boots, hands, and teeth stuck full of daggers." Austin did not disagree. Sixty years after Frémont's slander, the writer described Joshua trees as terrifying carcasses that haunted a moonlit terrain.[26]

Despite the tree's dubious reputation, by the 1920s, Hoyt had decided to save them. There were so few of them, and their numbers were shrinking. Joshua trees grow only under very particular conditions. They do best between above three thousand feet in sandy, well-drained soils. The plants tend to cluster in groves on high-desert playas, grouped together in "unearthly" forests that, a 1920 travel writer reported, seemed to eerily cast no shade precisely in a place where one desperately needed it. If no comfort for humans, the Joshua trees provide essential food and shelter for desert insects and animals. The right mix of rain and temperature is needed for germination, which is dependent on a single pollinator: the yucca moth. Charles Darwin exclaimed the relationship between the tree and the moth to be the "most wonderful case of fertilization ever published." He said as much because the two are codependent: the moth pollinates the tree, and the tree nourishes the moth's eggs.[27]

It is precisely the rarity of the Joshua tree's habitat and the oddities of its existence—the "wonder" of its life cycle—that makes it extraordinarily sensitive to any shifts in temperature, water, or wind. If one factor is disrupted, the trees die. Despite the fact that Joshua trees have been around since the Pleistocene, which began more than two million years ago, the National Park Service estimates that, because of climate change, at least 90 percent of the Joshua trees will be dead by 2100.[28]

But human-induced climate change was not Hoyt's concern. At midcentury, even the most ardent desert enthusiasts had not yet linked the health of a desert to the health of the planet, or vice versa. Well into the latter part of the twentieth century, in fact, deserts were considered discrete environments, specific landscapes contained by their own harsh limits. What happened in a desert stayed in the desert.

Even so, by the 1920s, Hoyt had started to notice that the desert was more than the sum of its parts. At first, the Southern California socialite had revered the desert for its tranquil emptiness and the solitude it offered, the timeless expanse of sand and sky. But in the time since her first trip to the Mojave in the 1890s, she realized that what happened in one part of the desert mattered to the other parts. Desert plants, she wrote, "are slow growths." Destruction "sweeps out of existence in a day what it has taken nature hundred, perhaps thousands of years to make." Along with the plants went the animals and the curious birds that thrived on cactus flowers.[29]

Throughout the first decades of the twentieth century, Hoyt witnessed the slow-motion destruction of the Mojave: by tourists, vandals, developers, even plant enthusiasts. They trampled, chopped down, and dug things up. The region's gardeners—including many of Hoyt's peers—raided the Mojave for exotic species: barrel cactus, spiny ocotillo, Joshua tree. By the 1920s, membership in what the *Los Angeles Times* called the "cactus cult" was so coveted that sheriffs and their deputies were called in to guard endangered desert plants. Arizona politicians threatened to arrest Angelenos who crossed the border to steal cacti. Hoyt offered financial rewards for information on desert vandals. People could make $100 for tipping off the authorities to the identity of the young men who had taken to driving into the desert, lighting a match, and cheering as a Joshua tree's gangly limbs went up in bright, varicolored flames.[30]

Hoyt devoted her considerable fortune and status to changing dominant views of the desert. In 1928, she sponsored a large desert exhibition at the New York Botanical Garden in the Bronx. The next year she sent equally elaborate exhibits to Boston and London. More than fifty species of desert plants were crated and moved, along with rocks and sand, to help re-create the Mojave. It took two refrigerator cars and one freight car to haul the materials across the country. In London, two fake coyotes were added to the scene for realism. Hoyt reportedly had delicate plants and flowers brought in each day via airplane so she could make sure that the scenes were as sumptuous as could be. She kept the plants in her hotel bathroom so they could stay fresh as long as possible. She wanted visitors to understand that the desert could be harsh and unyielding, to be sure, but also soft and familiar. There was nothing as dramatic as the unexpected dainty violet bloom of Mojave asters exploding across sandy terrain, or the promiscuous fuchsia blossoms of the beavertail cactus set in relief against the tans and browns of desert rock. The plants, according to the *Los Angeles Times*, were "ambassadors of a movement to save from despoilation the beauties of America's arid districts."[31]

Hoyt had allies, to be sure. In 1930, she created the International Deserts Conservation League. Among the organization's stated goals was to "define a desert area" in terms of water retention, and "using to some extent at least the flora as an indicator," as well as to create a map of the world's deserts and catalog the flora and fauna in such places. Shreve wrote in favor of her conservation plans. Hoyt traveled to and met with conservationists and politicians around the country. She advocated for a cross-border cactus park in the United States and Mexico. A cactus was named after her.[32]

Her work did not end there. In 1930 she presented her ideas to the International Botanical Congress in England, where she begged the audience to protect the deserts of the world. Closer to home, she lobbied conservationists and politicians in Sacramento and Washington. She spoke passionately at meetings of garden clubs and women's clubs. She sent two photo albums to President Roosevelt revealing the novelty of the desert.[33] One featured panoramic vistas of the desert; the other, color close-ups of desert plants in bloom. In 1936, the president agreed to set aside land for desert conservation, making Minerva Hamilton Hoyt the John Muir of the desert Southwest. Clearly, then, there were folks—even people in high places—thinking about deserts as something other than wasteland. That her work has largely been forgotten is testament to the enduring erasure of both deserts from environmental thinking and women from the history of conservation.

Most people at the time, though, did erroneously see the desert as wasteland. Reporters sent to cover the Army's preparations treated the desert for what most people believed it to be: good for little else. Columns of "tanks, mobile tank-destroyer guns, truckloads of motorized infantry, artillery, and little jeep cars" plowed through the sand, creating an endless cloud of dust. Trailing the main column were supply trains: "trucks of gasoline, water, food, ambulances, mobile machine shops" all stretching fifty miles down the desert valley. The military was impressive and prepared, readers were to understand, that its overcoming of the harsh environment made it so. From 1942 to 1945, nearly one million GIs cycled through an area that had housed just fourteen thousand before the war. Patton boasted that, "with time and perspiration, you can go anywhere."[34]

What "time and perspiration" did to the Mojave remains evident today. From the air, the outlines of camp areas and the grooves of tank tracks are distinct. On the ground, it is still possible to stumble across rock berms, foxholes, gun emplacements, the flattened ground of sleeping pads, and the everyday refuse of camp life, including ration cans and bottles. On occasion, hikers drop off shell casings and wartime debris at the General Patton Memorial Museum, which sits just off Interstate 5, near the Chiriaco Summit, nestled between what is now Joshua Tree National Park and the former DTC. More often they take them home as souvenirs.[35] If hikers find unexploded ordnance, they must call a local sheriff's office.

As Hoyt understood, deserts tend to preserve things. They scar easily and heal only over great periods of time. Although Joshua trees can live a thousand years, they take decades to take hold. Most young plants never make it. In 1928, the landscape architect Frederick Law Olmsted Jr., working with Hoyt, had written a report for California recommending the protection of the state's deserts. "Nowhere else," he wrote, "are casual thoughtless human changes in the landscape so irreparable."[36]

It is not particularly surprising that the US military in the 1940s would not have invited Hoyt, Shreve, or any other desert denizen to join the Quartermaster Corps as consultants. There were plenty of social and cultural barriers that would have made their doing so impossible, even if it were considered. But it is important to see how a variety of ideas were, indeed, available and that specific choices had been made along the way about which to include and which to keep out.

The choices made in the fog of war would continue to reverberate as US strategic planning looked more and more earnestly to the world's extremes. What would have happened had, say, more nuanced ideas of these environments been incorporated from the beginning? Had a Forrest Shreve or a Mary Hunter Austin been invited to share the table with the men we will turn to next—men certain that if they took the desert indoors, they might learn even more about what it took to operate there.

★ 4 ★
THE DESERT ROOM

David Bruce Dill could experience both the dry heat of the desert and the piercing cold of the Arctic in single afternoon. All he had to do was walk across the hallway that separated his lab's "hot room" from the "cold room" in the "semibasement" of Harvard University's Morgan Hall. Nothing about the spaces looked like the desert or the Arctic. They were little more than glorified closets, cramped and dark rooms that contained a treadmill and odd-looking machines and tubes for measuring how a man responded to environmental stress (fig. 4.1). There was no sand or snow, at least not yet. But it was in rooms like these—climate chambers—that wartime research teams increasingly turned to grapple with the corporal challenges of extreme environments.

While the war provided ample real-world opportunities to study such places, scientists realized that bottling those up and putting them indoors made measurements far easier and, hopefully, transferable. Bringing the experiences inside also made such places more accessible. Part of the reason desert studies had languished is that not many scientists wanted to spend a lot of time in a place so uncomfortable and remote. Laboratories provided a sense that things could be assessed and quantified, even in a world gone mad. Where so many variables seemed uncertain, where every environment was a possible battlefield, and thus every environmental condition had to be readied for, it was good to have standards, to have a basis for

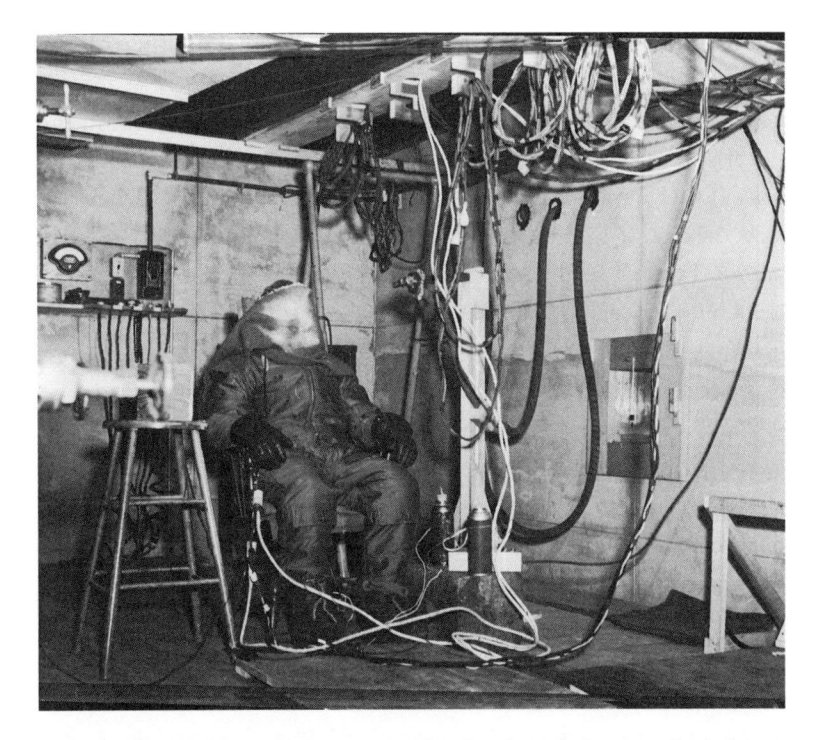

FIGURE 4.1 Harvard Fatigue Laboratory test chamber. From the late 1920s through 1946, the lab was the premier location for physiological research. Tests were aimed at maximizing the efficiency of men and equipment. Personal (Photo—F. McVay) 1944, from the Harvard Fatigue Laboratory records, 1916–1952 (inclusive), M-CE03, Harvard Medical Library in the Francis A. Countway Library of Medicine.

comparison. Of course, indoor studies could not totally replace outdoor conditions. Dill understood the need for both. His teams at the Harvard Fatigue Laboratory, where he had worked since the 1920s, used lab work as the first pass; field testing was when theory was put into action and tested. He could then go back to the lab to refine. Indoor simulations were a shortcut to managing environmental uncertainty.

Bringing things inside was a wartime example of what historians have described as the systematization of the modern world, the use of numerical indicators to parse the world into recognizable and moldable scientific units, displacing local and cultural knowledge.[1] Instead of observing men during desert exercises and trying to account for changes in temperature, solar radiation, time of day, or what they ate for breakfast, in climate chambers the researchers could choose their variables. They could crank up the heat in the desert room and then alter humidity levels, or perhaps change the type of sock a man wore to see how his feet responded. The working hypothesis was that if they turned things into numbers, they could gather the most accurate data. This was true not only for human subjects but for mechanical ones as well. Why not test tires and gears, engines and eventually entire airplanes, in chambers where everything could be closely accounted for?[2]

Ways to measure man and material abounded during the war years. Women were never included or considered in any of these studies. Rations for different climates could be determined on the basis of calories and weight. Windchill emerged as a way to consider the combined effects of cold and wind on man (a concept designed by Paul Siple but that remained that classified until after the war). Clothing allowances were reconsidered given a new measurement called a "clo." Everywhere units of measure were being crafted and deployed, just as soils were being classified. Charts were made, graphs drawn up, and data pulled from simulated places, then applied to the real world. What the military really wanted to know was how to make soldiers and their gear as efficient as possible. It didn't matter if the soldiers complained, as long as they could function. The goal was to weather-proof—or environment-proof—the fighting man.

Dill understood the stakes right away. As the world tumbled into World War II, he saw that the Harvard Fatigue Lab's niche research could meet the US Army's needs. Indeed, if military commanders had questions, Dill could provide some of the answers. And he didn't wait to be asked. Back in 1940, he had dashed off a memo to his colleagues recommending that the laboratory pivot to military programming. After all, the lab's subjects were generally "athletic young men comparable to soldiers," and the lab's tests evaluated which factors "limited the performance of these men," be it "fatigue, altitude, heat, cold, diet, etc."[3]

Dill was already something of a legend in the admittedly small world of physiological experimentation. He believed firmly in the need to evaluate and experiment on real people. "It is not enough to make neat studies of frogs' nerve-muscle preparations, of swimming rats and of panting dogs," Dill proclaimed, "man himself must be the subject." As if to prove the point, Dill was often the first volunteer for his experiments. For example, he was the first member of the 40-40-40 Club, an honor that Dill's own lab bestowed on those who successfully completed a forty-mile endurance walk in under twelve hours while in a climate chamber set to negative forty degrees Fahrenheit and mimicking oxygen levels at forty thousand feet. Dill also held the lab's record for the most time spent naked in that same climate chamber: twenty minutes. His son-in-law came in second.[4]

Most of the subjects brought to the lab were not forced to endure such extremes, although their experiences would be uncomfortable enough. The volunteer soldiers—whether soldiers can be considered to have volunteered for such things is another question—were stripped down to their underwear, had their vitals taken, and then sent to walk on a treadmill while the heat was cranked up to over 110 degrees. Dehumidifiers ran round the clock to keep the air dry. The scientists sometimes blasted their desert room up to 160 degrees and measured what happened to body temperature, heart rate, blood oxygen levels, and general sense of fatigue. They changed the rations each man ate and tested various diets on acclimatization. They experimented with different shoe fabrics and the heft of a backpack. They found that only one of seventeen enlisted men brought in could complete a full session of marching tests: three rounds of fifty minutes on and ten minutes off on the treadmill while carrying pack and rifle in a room set at higher than 102 degrees. Most were simply too exhausted to keep going. Some developed blisters and cramps. Even those who managed to finish felt dizzy when they stopped moving. Subjects were encouraged to lay down to alleviate the symptoms of heat exhaustion. If the researchers were dismayed at the condition of young men, they were heartened at how quickly they could get the men in better shape: with seven days of training—for only about an hour or two a day—most could complete the task. After twenty-three days, they were all were in decent shape.[5] Now those were results that they might be able to use.

Dill served as something of a fulcrum for turning academic research to government wartime needs. "My chief role," Dill scrawled in his customary blue fountain pen, "was to break down the barrier between civilian and scientific resources and the military . . . not only were direct contractual relations undertaken, but in addition some members of the staff were appointed as expert consultants to the Armed Forces."[6] Applying his interests and work was both patriotic and good business. Before the war, the lab had worked mostly with companies on large construction projects in difficult places, such as the Hoover Dam. There, too, the goal was to find ways to make men most efficient under environmental duress, certainly not to wonder whether it might be bad practice to conduct hard labor under difficult conditions. Which inputs could be used to insulate man from extreme heat and aridity—rations, water, salt, gear—and function as well as they did in a more temperate climate?

Dill thought it would be best to have a few lab employees become commissioned officers within the military branches. That way they could "serve the double purpose of keeping the Services in close contact with civilian work and of keeping the Laboratory informed of current problems in the Services." In 1941, Dill himself took a commission with the Army at the aeromedical lab at Wright Field in Ohio. Other Harvard Fatigue Lab researchers followed suit, joining the Quartermaster Corps, the Army Armored Medical Center at Fort Knox, the Naval Medical Research Institute in Bethesda, the newly opened Climatic Research Laboratory in Lawrence, Massachusetts, and more. They managed to cover a remarkable breadth of centers involved in environmental research. Throughout the war, letters and memos flowed through and across these men and their various organizations. At the center of many of these conversations was the Harvard Fatigue Lab, under the wartime leadership of William Forbes. Dill and Forbes kept in constant communication. Their weekly correspondence serves as a blueprint for the tight links that evolved between various individuals, labs, battlefields, and research programs regarding extreme places.[7]

Sir Hubert Wilkins made the rounds of indoor and outdoor test sites. His personal preference was the out-of-doors, but he accepted the need for more controllable places to study the extremes. After all, most men, unlike Wilkins, did not actually choose to be there. The former Australian explorer had been hired by the Quartermaster Corps to advise the military on clothing and gear. A map of his travels tracked pretty well what the country was up to: he visited training areas in Alaska and swooped down to climate chambers in Ohio and Texas. In 1943, he dropped into the basement offices of the Harvard Fatigue Lab, something William Forbes commented on with great excitement. Wilkins was a legend in the world of explorers and physiologists; his approval was like intellectual gold. He also made his way to the Mojave, where he spent a few weeks in the Desert Training Center examining whether the Quartermaster Corps' equipment could help soldiers function in dry heat and trying to figure out why men were dying.

While Wilkins and his colleague Weldon Heald identified the reason so many men were collapsing, they still could not precisely say how much water or salt any given man needed—not to mention that one of the biggest problems of taking measurements in the desert was the reliability of the subjects. Given troop rotations and training schedules, it was not always possible to get the same units and men to show up each day, and it was harder still to believe them. Did they really do the work they said they had done? Did they actually limit themselves to the water allotted to them, or had they found a way to sneak more? A controlled lab space with controlled subjects might have yielded better, clearer results.

In both lab and field studies, human beings were seen as measurable mechanical bodies whose functions could be tracked and charted. The physiologist Ancel Keys wrote explicitly of the "human machine." Keys had been at the Harvard Fatigue Lab before starting his own center at the University of Minnesota. He is best known for the starvation studies he carried out on conscientious objectors during the war. For nearly one year, Keys kept "volunteers" alive on just a few hundred calories a day. But this was hardly the only extreme experiment Keys ran at his lab. He also studied the effects of various vitamins (and their withholding) on men's performance, of exposure to excessive moisture over long periods of time, of adaptations to heat and cold, and of bed rest on the human physique. He kept six conscientious objectors in bed for one month.[8]

All these studies were directed toward a similar goal: to find the right mix of inputs to enable the machine to function efficiently. To be able to place soldiers in any global environment and be confident of their success. This could mean maximum efficiency during work or the bare minimum to keep a man alive. Comfort and quality were not measures of progress. Rations, for example, had to sustain a soldier so he could perform basic duties, not so he would enjoy the meal (though problematically some rations were so bad that men stopped eating them). The scientists and their "volunteers" were part of the system that was being created to conquer the world's environments. Efficiency was measured through blood samples, urine tests, heart monitors, and more. The files of these wartime labs are full of hand-drawn charts that painstakingly plot results of testing temperature against weight and chemical analysis of urine samples against humidity. The men in the labs wanted to know whether caloric intake needed to be adapted for different climatic extremes. In ferreting out answers, not only did laboratory data seem more replicable and demonstrative, but it was also more readily pulled out and broadly applied.

But the world was awfully big, which raised questions about whether every single place had to be studied or whether patterns could be discerned. Could the same rations work in the Arctic as in the desert? Were old wives' tales about the importance of vitamin C (ascorbic acid) borne out? Was salt a factor in dehydration? Did hair color correlate to heat sensitivity? What about skin color? Did the weight of a man correlate to how much water he had to consume? "Big men need more water than little men," a wartime conference concluded. The anthropologist Frederick Wulsin was more specific still: "Men who are plump at home and get thin" due to training, he explained, "are better than those who start thin."[9]

Everywhere in the documentary record are statistics that separate subjects by race, ethnicity, age, and education, although rarely do such distinctions seem to matter. Indeed, although the men of the segregated Ninety-Third Infantry Division, sent to the DTC in the summer of 1943, were subject to various desert studies, it was wrongly assumed that Black soldiers were not capable of performing at the same level as their white counterparts. The beliefs in links between race, acclimatization, and environmental determinism remained strong. Many of the men researching extreme environmental stress and acclimatization of the white, male soldier during the war had earlier conducted work examining how racial difference might explain varying capacities to withstand certain conditions. For example, Dill had taken part in a study on how Black sharecroppers in Mississippi dealt with heat and humidity. Were nonwhite people better able to withstand the extremes?[10]

The biases inherent in purportedly objective studies were encoded into the very measurements used to track them, which meant they would be replicated in any solution. Take the idea of the "clo." As an Army report later explained, the clo was specifically defined in 1941 "to be thought of in familiar terms," so that "generals, admirals, congressional or parliamentary committees" would be able to understand what it meant. One clo was equal to the "thermal insulation in a business suit as worn in Philadelphia, New Haven, or Toronto." It was a definition with a point of view. Clothing was measured for the northern white businessman in a suit, not workers in coveralls or women in nylons, and certainly not Black sharecroppers in Mississippi. Across the research establishment men were filtered by race and class, revealing white scientists' assumptions that different races would perform differently. The fallacy appears over and over in the reports, even though at no point do the accumulated data support it.[11]

The invention of the clo also tracked trends in the science of thinking about human comfort in extreme environments. Clothing and specialized equipment were meant to help the human body remain in balance with the elements around it, but the question was: Is clothing an extension of the man or part of the environment?[12] These sorts of questions gained traction during the war not only because of the vast geography of conflict but also because war in the air also meant new types of acclimatization. Human bodies would be spending a lot of time in places they had not encountered before, such as at higher and higher altitudes for longer periods of time. Airplanes created all sorts of novel environments, sometimes combined in odd and alarming ways. Added to this was the reality that airplanes could go down anywhere in the world—in deserts, the Arctic, the tropics—and the need for research into human adaptability and, more importantly, survival was crucial. Environmental intelligence had to be gathered and measured against the "average" man.

The US military provided an unprecedented stream of men for various research programs tied to national preparedness and security. There were a host of other medical experiments conducted on troops, many that were far more invasive and damaging than the ones mentioned here, such as mustard gas and radiation experiments. Despite their many differences and racial presumptions, these programs all exhibited blithe disregard for the ethics of human subject testing that would later prevail in military and academic research circles. In the 1940s, there were few legal and moral limits on using human subjects in testing, and consent was a highly elastic concept. Troops in the DTC, for example, were told to show up for weight checks and urine samples, not asked whether they wanted to participate. Signing up for the military meant signing your bodily rights away. Not until after the Nuremberg trials that indicted Nazi doctors in extermination camp atrocities were the ethical dimensions of such practices more widely addressed. But that did not stop the experiments. For example, in the 1950s, the US Army's Arctic Aeromedical Laboratory conducted radiation experiments on indigenous populations in Alaska. Scientists there were convinced that they could somehow illuminate the ways that indigenous people had what they believed was an innate ability to adapt to cold, and they justified their experiments in the name of scientific expertise about the Arctic.[13] As strategic planning during and after the war opened up "environment" as a category to be evaluated, scientists and researchers rushed in to provide answers with seemingly few reservations about what could and should be done.

Within the growing military bureaucracy, the question for outside experts like Heald, Wilkins, Dill and others, was how to insert their expertise into programs and what that might lead to. The question of how was more immediate and rewarding. Suddenly, conversations held in wood-paneled rooms of the Explorers Club in New York had national and international importance. The best way to navigate in a desert, the right socks for the Arctic, which local food supplied the most power—these all had become conversations that pertained to national security. In normal times, questions about hydration, vitamin intake, relative body weight, and more, would have languished in grant proposals and small centers tucked away in university basements, which is precisely where they had been for the previous twenty years. War meant that obscure knowledge had new utility. These were heady days for people interested in how man and machine functioned at extremes.

Given the possibilities and opportunities, it was easy to put off considerations of outcomes for a while, usually just long enough that the choice no longer mattered. There was a war to be won and a lot of intelligence was required. So Heald and Wilkins joined the steady stream of men asking how they could apply their expertise to the war. For those who did take a moment to think through the potential outcomes, the future would have seemed unlikely: their individual expertise churned and crunched through studies and analysis, emerging on the other end as a one-size-fits-all solution to the problem of the extremes. In fact, after the war, the Army Air Force's Arctic, Desert, and Tropic Information Center acknowledged that, for the modern military man, the knowledge of explorers was not particularly useful. A pilot would not have desert experience, for example, and did not require any to draw on. Instead, he just needed the right gear.[14]

So it was that as the experts and explorers entered government service, they had to come to grips with the reality that wartime necessity and military utility reframed once obscure fields in ways that few could anticipate. At first, much of their work revolved around individual survival and efficiency. It is what many of the experts and explorers themselves knew. Men such as Wilkins wanted individual soldiers to be comfortable. In their studies, soldiers were treated as somewhat clumsy explorers in military attire. But what might be great for an individual was completely impractical for equipping an army.

The focus of research shifted from the individual and explorer to the group and to machines and gear, which required a shift not only in scale but also in imagination. Winning war was about maintaining equipment and logistics, not about individual comfort and survival. In the long run, airmen and soldiers were not going to be trained to be explorers. A better strategy was to insulate the human from the world, to create machines and equipment that would protect the human body, no matter how average (or even subpar). Even in extreme environments where great care went into planning for specific items, nuance drifted off on the winds of war.

Take the example of Frederick Wulsin and his metallic suit. The explorer's brief trip to North Africa to observe British gear and equipment in early 1942 was just one small part of the task he was given. As head of the Quartermaster Corps' newly created Desert Warfare unit, the bespectacled Wulsin spent the early years of the war collecting information. He took advantage of the overlapping network of explorers and adventurers who converged in New York throughout the early half of the twentieth century. He canvassed his colleagues and friends at the Explorers Club, consulting those who had been to North Africa or had spent time in desert country.[15] Partial to a brown fedora and high boots, Wulsin visited the Desert Training Center in the summer of 1942 to get a better sense of how newly designed footwear and canteens functioned.

Wulsin dispensed with any previous designs and went all in on ultramodern desert wear. In 1943, he began experimenting with a "thermal armor" made of metallic material, meant to reflect the sun's rays and thus keep the wearer cool. At night, the suit would keep heat radiating inward to combat the desert's chill. While the costume demonstrated Wulsin's knowledge of the diurnal temperature swings that characterized most deserts, it was completely impractical for the Army. For one thing, Wulsin expected to use either aluminum or copper for the suit. Aluminum was discarded quickly as a possibility. The Office of the Surgeon General had already argued that aluminum was a poor helmet liner because it could get into head wounds and complicate healing. It was also expensive. At the same time, the Army noted, a reflective suit would simply serve as a gleaming target, giving away positions and welcoming enemy fire.[16]

Wulsin was not alone in aiming for optimal rather than practical military apparel. The year before, the polar explorer Vilhjálmur Stefánsson had insisted that the best clothing for the Arctic was made of fur and animal hide, as practiced by people native to the region. But to equip tens of thousands of men with fur was unreasonable, although enough thought was given to the idea that furriers lobbied for it and Native tribes were queried about supply.[17] Not only would the Army's needs quickly outstrip the supply of hide; there were not enough people to craft such clothing and no facility capable of mass producing it. Different textiles, perhaps even synthetics, were needed.

The idiosyncrasies and bright ideas of world explorers—no matter how effective they were in administering their own expeditions—were not terribly useful when it came to clothing an army. As the US Army Quartermaster Corps acknowledged, any specialized gear had to be "domesticated" and reconsidered for "manufacture by American production lines." Military gear had to be efficient. Its production had to fit into US business practices and into factories that were prepared to turn textiles into clothing.[18] The imperatives of war meant that the heroic individualism of the explorer or the extraordinarily effective and time-tested materials of local people had to give way to the efficiency of the crowd.

An idea came to the young physiologist as he passed a shop window in Cambridge: Did they even need human volunteers? What if they could make a man that could be sent into test chambers, kept at negative forty degrees for days on end? Across the city, mannequins were used to display the latest fashions. Could the Harvard Fatigue Lab use a "thermal manikin"—a human-shaped model for medical and clinical scenarios—to display the power of materials?

This was Hardwood "Woody" Belding's Frankenstein moment. He was studying wartime physiological needs at the Harvard Fatigue Lab, but his particular expertise was in how to protect the average male body from cold. Belding studied the insulating properties of sleeping bags and clothing, electrically heated suits for aviators, and more. But men were poor test subjects. The men who came to the lab day after day often passed out, and very few ever proved as tough as the lab's scientists. Worse, they tended to be unreliable narrators of their own experiences. What, then, could be better than a huge doll?[19]

Belding cobbled together his early thermal manikin from a stovepipe and sheet metal, stitching it together and filling it with a heating unit and fan (fig. 4.2). It was a torso without arms or head, described by some as "portly." In 1943, Belding worked with General Electric to create a full-bodied, electroplated, copper-skinned manikin based on measurements taken in the lab from male Army recruits. The perfect proportions, cast in copper. The manikin was heated with various materials and clothes draped across it. Calculations were taken of how much heat escaped. This led to the measurement of the "clo" values that were attached to the materials and synthetics that were increasingly making up military garb. The third derivation of the copper man was given a name: Chauncey Smythe Depew III. Chauncey was just one way that environmental testing was moving increasingly toward abstraction and simulation. The solution for a global military, one with bases in the Arctic, the desert, and all places in between, was going to be properly insulating men from those variations with the best materials and machines. They would environment-proof the materials, not the man.

The Air Force's Eglin Field climate chamber was the ultimate representation of environmental planning. Renamed the McKinley Climatic Laboratory in 1975 (after the man who created the initial

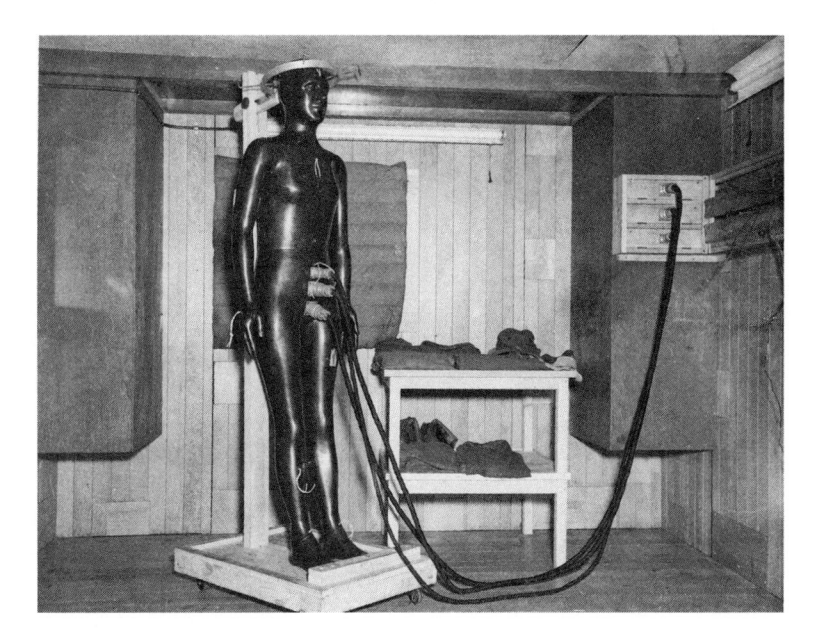

FIGURE 4.2 The copper man. During the war, the Harvard Fatigue Lab began using copper manikins to test physiological responses to extremes. Human subjects were considered too fickle. Clothing (Figures) 1947 Photograph & Fatigue Lab, from the Harvard Fatigue Laboratory records, 1916–1952 (inclusive), M-CE03, Harvard Medical Library in the Francis A. Countway Library of Medicine.

test site), it was a giant, indoor hangar that could simulate a range of climatic events for giant aircraft. The main chamber was 252 feet by 201 feet and 70 feet tall; temperatures could be brought up to 165 degrees Fahrenheit and down to negative 80. Winds could gust to sixty knots. It was possible to manufacture snow and ice, hail and sandstorms—anything you wanted to hurl at the military machines.[20]

The work of Belding and his colleagues from the Harvard Fatigue Lab, now spread out across military institutes and universities, set the groundwork for and anticipated the concerns of scientists trying to send men to the Moon in the late 1950s and 1960s. In all those places, the goal was first to re-create the temperate environment—to mimic what it felt like to be somewhere unremarkable—and then deploy that technology to the extremes. If they could protect the human from the environment, did the human response matter at all?

Or how could they maximize the efficiency of the entire system by managing the various organic and inorganic inputs? As Paul Siple, a polar explorer and later Army consultant, explained in 1947, in environmental studies, "personal opinion must be ruled out in favor of a scientific approach."[21] Was there room anymore for the Wulsins and Wilkins of the world?

★ 5 ★

EXTREME WISH LISTS

What will happen when the war ends? It was a question that nagged at the men involved in researching environmental extremes. To be sure, every one of them wanted the war to be over. But by late 1944, the eventual cessation of hostilities raised two potential problems. The first was what would happen to their generous funding. War had altered the trajectory of their research programs. William Forbes, wartime head of the Harvard Fatigue Laboratory, dourly predicted that defense funding would "drop to a low ebb."[1]

But the question of funding highlighted a larger, potentially devastating possibility: What if the US defense establishment simply turned away from the work that had already been done? What would happen to the knowledge already accrued? Or the programs started but not finished? The task for the men involved in extreme environments was about a lot more than money; it was about expertise and consequence. How could they ensure that their niche fields remained relevant?

The last few months of 1944 were particularly anxious days in the basement offices of the Harvard Fatigue Lab, where these questions circulated constantly. There was no doubt that wartime contracts had elevated the lab's research areas to national importance. After all, what organization other than the US military would fund studies of electric underwear for B-17 crews or would provide dozens of test subjects for studies of rations of pemmican, a high-calorie, lightweight amalgam of pulverized dried meat, melted fat, and berries that explorers favored. One of the varieties that the Harvard Fatigue Lab studied was so unpalatable that even a dog stopped eating it after a few days.[2]

William Forbes, David Bruce Dill, and others exchanged notes and memos that highlighted the need for rebranding. In late 1945, for example, Forbes recommended that the lab reach out to various industry groups, including in food processing, textiles, footwear, handwear, and clothing, that shared a "particular attention to climatic extremes." The lab created a brief informational film to highlight its contributions to exercise physiology. Forbes made a pitch to the *Harvard Alumni Magazine*. "What are the problems which the laboratory will deal with in times of peace?" the article queried. "Man will continue to need both food and shelter, and therefore the study of nutrition, clothing, and acclimatization is of fundamental importance," not just to the US Armed Services but also to industries and the intrepid American citizen, perhaps more willing than ever to move out into the extreme environments of the world.[3]

Closer to the halls of power, Lieutenant General Eugene Reybold was also concerned, albeit for different reasons. As chief of the Corps of Engineers, Reybold knew that his organization would not be abolished. The military engineers would always have jobs to do. The question, then, was not whether they would work but where. Would they remain deployed overseas, or would most of their activities return home, where their prewar activities had largely been contained?

Reybold hoped for the former. Indeed, he warned that retraction was the wrong path. The engineers had acquired unparalleled international experience during the war, Reybold explained as he stepped down as chief. They had built equipment that could work across environmental conditions; they had created new tools and machines; they had amassed a trove of information about the world's environments, including the soil samples being collected as the war ended. Yet while he praised the accomplishments of the work already done, Reybold cautioned against complacency. Drawing on tropes of exploration and national mission, Reybold linked wartime engineering work to a long history of the "discovery" of the "new world."

Military engineers, he said, had "made a new exploration of the world—exceeding even that of the Seventeenth Century." He went on: "Out of War's Pandora's Box of Evils," he promised, is "the one great hope of building an environment" not for the aims of war, but "for the causes of peace." The key for the future was to maintain a wartime level of interest, to resist the pressure to abandon projects and forgo extreme environments whose relevance was less obvious beyond war. "We must keep building," Reybold urged.[4]

Continuing to build meant continuing to push the envelope of scientific knowledge and method. There was so much more that had to be learned and mastered. Engineering documents and publications suggested that the military engineer's job in the postwar period was to keep account of all that had been done, to measure and catalog as much of the world as possible, and to maintain an archive and database of as much detailed information as could be gleaned from everywhere, all at once. This expertise would enable US interests to continue flowing out around the world, in the building of defense-related facilities, to be sure, but also in the infrastructure of economic and corporate power.

Reybold and the military engineers insisted that postwar reconstruction would be key to keeping the peace. War had left the industrial powers of Europe and Asia in tatters. Much as FDR had called for the construction of shipping lanes and air routes to connect the world in a web of wartime logistics, US postwar aims required ever more robust and plentiful paths that would enable trade and security. "Both survival and progress will be measured by what we build," the Corps insisted. The engineers' new role was spelled out in places like Greece and Turkey, where the United States committed to helping rebuild war-ravaged countries. Truman Doctrine aid was aimed at ensuring that those countries were politically aligned and economically connected to the West, not the Soviet Union. Corps-led projects became conduits for US companies to enter markets as lead contractor. In fact, civilian-military collaboration was considered vital to promoting US interests around the world.[5]

Although a great deal of postwar work was in Europe, first through the Truman Doctrine and then the Marshall Plan, the extremes also came into sustained view. "The old days, when wars were limited to temperate zones are no more," the *Military Engineer* explained. The engineer was going to be "required to serve in the most rigorous climates." He had to be "prepared to operate in these inhospitable areas . . . both polar and desert environments."[6] The war had merely been the warm-up.

In retrospect, it is easy to see that many of the worries about postwar retrenchment were unfounded. Institutional support for research into extreme environments remained steady, if not robust. As Forbes and Reybold both recognized, the United States was probably not going to return to its prewar strategic posture, although what its global presence might look like remained uncertain. The Corps of Engineers reported spending $127 million on equipment that was sent to overseas facilities in 1948 and another $200 million on maintaining Army posts at home and abroad.[7] The question in the mid- to late 1940s, then, was not whether the US defense apparatus would stay engaged overseas, but how, where, and when. It was within that margin of uncertainty that extreme environment experts and other interested parties had to maneuver.

The end point is no mystery here. Interest in extreme environments roughly tracked the trajectory of defense funding for the environmental and Earth sciences, which both continued to grow. The Cold War amplified strategic interest in, and therefore funding for, those fields. This was not only because the whole world suddenly seemed a potential battlefield, but also because the effectiveness of modern weapons required knowledge about things such as atmosphere, magnetism, radiation, and more.[8] Work in extreme environments was just a small sliver of this, but its continued prominence in the postwar world brought commensurate funding and influence.

But paying attention to the beginning and the end risks obscuring this story's contingency: how this happened and what direction the interest in extreme environments would take. There is no natural path of scientific study or a quest for information, no inevitable research direction. Wartime strategic needs clearly pitched inquiries into desert knowledge toward operational information and expertise at the expense of more place-based knowledge and detail. To simply jump to the end point means that we would miss the middle, which is where decisions were made about what to fund, who to fund, and how to do so. While extreme environments were often on the fringe of public discussions about spending, they were increasingly central to how the world was becoming known. The key question here is thus not whether funding happened, but how that funding shaped the nature of what was known.[9]

Work on extreme environments took a ragged path out of the war, propelled along by the lingering laments of insufficient prewar preparations and driven by eager men with specific, sometimes extraordinary agendas. Why had the military planned so poorly for North Africa? How specific must environmental information be to be useful? Could the same runway be paved on all terrain types? Was it possible to create a global soil map? Was it possible to change the weather?

Early signs of what was to come were not encouraging. In late 1944, the Army closed the Desert Training Center, which meant that much of the Army, including the Corps of Engineers, no longer had a desert-testing site. The quartermaster general began to disassemble the environmental boards that had managed special equipment for the world's deserts, the Arctic, and the tropics. Extraneous workers were released, and expertise was consolidated into a few key locations. The Army Air Force's Arctic, Desert, and Tropic Information Center (ADTIC) was shuttered with little fanfare. Without a war to fight, what would be the purpose of keeping them open?[10] The Harvard Fatigue Lab, too, ended its two-decade reign, its researchers and scientists fanned out to the various military centers still able to fund their type of work.

The closure in 1947 of the federal Office of Scientific Research and Development also concerned extreme environment experts. For much of the war, the office had overseen research and development for defense purposes, bringing together private business, university research teams, and military interests in a stunning display of collaborative productivity. The scientist Vannevar Bush had been appointed by the president in 1941 to lead the office. Backed by a seemingly endless stream of funds—some of which appeared at the training centers, physiology labs, and quartermaster development centers—the office was able to steer research and development toward critical military ends. Best known are its big science projects, such as the creation of the atomic bomb and the invention of radar. But many other smaller, less obvious projects benefited as well. This was how Ancel Keys, for example, was able to carry out his starvation experiments and how new equipment for war was designed and tested.

The marriage of business, research, and military funding was meant to be temporary. Bush explicitly intended the Office of Scientific Research and Development to last only as long as the war required that level of coordination. He assumed that all parties would want to return to the prewar arrangement of arm's-length collaboration and communication. But he was wrong. Not only did private and university researchers covet government—particularly military—funding, the size and speed of postwar research projects and needs seemed to demand a sort of "district attorney," as the historian Ronald Doel has called the office. There was a need for a federal body that could manage and encourage cooperation between the physical sciences and defense needs. The Joint Research and Development Board (JRDB) was the result. Though never as powerful as the wartime boards, the JRDB recognized that "events of the past ten years have demonstrated that the problem of national security must be faced on a global basis." A coordinated response was required, and Bush returned to lead it. In late 1947, the *Joint* was dropped from the JRDB when the US Air Force was established as its own service. The National Security Act kept the RDB around as a coordinating body but, as we will see, changed some of its roles. The entire project of the RDB was disbanded in 1953, but not before the apparatus had helped influence the warp and weft of extreme environmental research.[11]

Most of the panels of the RDB were technical, but there was also an umbrella panel called the Committee on Geographical Exploration that temporarily housed interest in environmental extremes. According to the official announcement, the committee was created for the "continuing study of means, programs and plans for the observation and recording of geographic information in regions not normally accessible." The goal of Cold War strategic thinking was to acquire and maintain "intimate knowledge of, and contact with, all parts of the world," particularly the "arctic, desert and tropical areas."[12]

The emerging Cold War was central to the continuation of extreme environmental programs. As tensions rose with the Soviet Union and as various US policymakers began to sound alarms about preparedness, defense budgets stabilized and grew. The story of Cold War spending has been told elsewhere. Important here is that the type of weapons prized by Cold War strategists and budgets—missiles, airplanes, reconnaissance devices—dovetailed with extreme environmental research and a need to know the whole planet. In putting the whole Earth under surveillance, the Cold War ended up highlighting those areas that remained little known.[13]

Still, the problem of how to know the whole world remained vexing. Even with new globe-spanning technologies, access to places and environments previously inaccessible, and a wartime arsenal of data, the questions of how best to gather more information was open ended. Thus, the RDB's panels. These panels were advisory boards. They met a handful of times a year; their members drawn from related government and military agencies. They had no money and no material power, but the weight of their collective expertise could influence decisions. Made up of equal parts civilian and military representatives, with plenty of corporate executives thrown in, the committees accumulated data, made recommendations for areas in need of more thinking and planning, and then guided where that work could and should happen.

Plenty of familiar faces populated the various RDB panels. The Committee on Geographical Exploration was a holdover from the war: men such as Hubert Wilkins (the Australian who became a US consultant), Ancel Keyes (who ran the starvation experiments), and a smattering of others who had worked with the Harvard Fatigue Lab all continued to meet and mingle their ideas. Wartime concerns were also reflected in the three subpanels that made up the Committee on Geographical Exploration: the Panel on Expeditions, the

Panel on Geographic Environments, and the Panel on Exploratory Physiology.[14]

The makeup of the committees was important to the conversations that took place in the Pentagon, where the RDB groups often met. The men invited were enthusiasts, not skeptics. They all arrived convinced of the need for more programming, not less. As a result, skepticism about strategic environment research or critiques of military programs were in short supply. There were few constraints put on the types of fantastic programs presented and promoted, from efforts to change the weather to the creation of bomb-throwing bats. The minutes from the panel meetings are remarkably congenial. The only arguments were about which service or organization would conduct which research program, not whether a particular research program was needed. For example, everyone seemed to agree that ice runways were vital to strategic interests, but the Navy, Air Force, and Corps of Engineers all believed they should be in charge of them. Interservice rivalries were everywhere apparent. But all shared a strategic vision of environmental intelligence that emphasized military operations and mobility at the expense of the more place-based knowledge that had found some purchase during the war. It was a narrowness in thinking that would drive work on and access to the extremes for years to come.

Paul Siple emerged as a leading proponent of extreme environmental research. While his true love was the cold, he also threw himself into desert operations. As a young Eagle Scout, Siple had been chosen to accompany Richard Byrd on his Antarctic expeditions in the late 1920s. From then on, Americans became accustomed to seeing Siple, replete with toothy grin and fifty-nine scout badges, splattered across the nation's daily newspapers as he became the poster boy for polar exploration. In 1939, he received his PhD in geography from Clark University in Massachusetts. His dissertation, "Adaptations of the Explorer to the Climate of Antarctic," made him an ideal fit for US military needs during the war. By that point, Siple had honed his nationalistic polar career, inserting his interests into seemingly all aspects of the country's cold-regions practices.[15]

The consensus across the panels in 1947 was that work on extreme environments remained woefully underfunded, even after a global conflagration that highlighted the need for precisely the information such work would generate. The defense establishment had not yet created a central repository for the reams of environmental data that had been collected during the war. Files from the Desert Training Center had gone missing, and other basic data about North Africa was lost. When Hoyt Lemons of the quartermaster's office went to London to see which materials he could scrounge up on British operations in the Sahara, he found that it would be nearly impossible to collect much. Documents were scattered across different agencies and files.[16] Ongoing efforts to acquire detailed data about the world's deserts, it seemed, had barely moved beyond prewar inertia.

The committee described work on "not normally accessible" environments as akin to the nation's commitment to mapping: poor. US cartography, the committee wrote, is "characterized by well laid and ambitious plans with an attendant lack of funds for the accomplishment of the work." The same was true with information on deserts and the Arctic. A lot of important work and knowledge was "locked in field notes, aerial photographs and other records." Even when data existed, there were too few experts to evaluate it.[17] A strong central agenda was needed to make sense of it all and set a future course. More specialists had to be trained. In keeping with the wartime push toward laboratory and scientific work, the panels recommended rigorous scientific investigation of wartime reports and experiences. The function of these panels was thus part of the postwar push for efficiency, standardization, and administrative order; to wrangle a rather scattered and slapdash wartime grab bag of programs into something coherent and useful.

In early March 1948, Vilhjálmur Stefánsson wrote to Vannevar Bush with a few concerns. A famed Arctic explorer—named the "eminent authority on the Arctic" by the Quartermaster Corps—who had consulted the US government during the war, Stefánsson was worried that the US Army had not quite figured out how to clothe troops for extreme climates. The boys in Alaska, he noted, were wearing stiff and unwieldy coats with too many flaps and buckles. Their clothing weighed nearly twenty-eight pounds and had so much padding that the troops "waddled instead of walk[ed]." Stefánsson had a potential solution, which he outlined in the letter to Bush and in a series of speeches and comments that winter. The Army, he assured his audiences, needed to pay more attention to native "Eskimo" dress. He recalled his own travels in the Arctic winter, when he wore a suit of caribou fur that weighed under nine pounds, leaving him warm and agile.[18]

The actual contents of Stefánsson's letter are less important than what the ensuing controversy reveals about emerging plans for extreme environments. Tensions remained in research circles over which sources of knowledge were best: the experiences of an explorer or the experimental expertise of laboratory scientists and quantifiable research. Where Stefánsson prized tried-and-true caribou coats, the Quartermaster Corps focused on testing materials that might mimic the benefits of caribou-style clothing. In one case, the Army explored whether vests made of "sponge rubber balls" might somehow emulate the air-flow properties of animal skins.

This clash was increasingly obvious in military institutions, where the mass manufacturing of the war had turned into perpetual mass preparedness afterward. Indeed, the very continuation of the apparatus of the RDB signaled that big thinking and big solutions were required for the problems of the postwar era. How long would the military have to entertain the pet projects of heroic individuals?

From March through May 1948, a series of memos and letters shuttled among the committee members, either refuting Stefánsson or trying to defuse the volatile situation. His critics labeled him a "crank" and a relic of a dying generation. He was ridiculed for his obsession with "going native." Lemons declared that Stefánsson was "unenlightened, archaic, and antagonistic" to the needs of the US military. Siple, as an explorer of younger generation, made it personal: "Taken as a whole, it is my impression that [Stefánsson's] letter begins on the 'crank' class, written by a man whose ego is inflated beyond his merit or usefulness, and who now feels he has an ax to grind." Lemons tried to quantify Stefánsson's errors by using wartime inventions. For example, he reprinted a chart that showed the clo values for caribou skin (7.7) and a "1-inch thickness of laminated fiberglass" (5.3), which, Lemons explained, showed that "for its weight and volume," the skins were "far inferior as an insulator."[19]

Others defended the sixty-nine-year-old. Wilkins (still sporting his Vandyke), who had crossed paths with Stefánsson both out in the world and during the war, insisted that Stefánsson's comments were being taken out of context. The man liked to cause a stir, Wilkins admitted, but he was not wrong. Perhaps the US Army would do well to pay a little more attention to some of the properties of Inuit dress and adaptation to the cold. Indeed, even Siple had to concede that the Army was trying, however limply, to do just that.[20]

The structure of newly constituted RDB committees and panels, however, highlighted the fact that vests of sponge rubber balls were to be victorious over caribou coats (at least metaphorically). *Exploration* and *explorer* were banished from the new RDB panel titles. The earlier focus on "expeditions" with a military flavor was erased. In an effort to distance itself from the old approach, the new Panel on Desert and Tropical Environments declared that the "present method" of working with "private expeditions" was "haphazard and wasteful."[21] To accumulate good and comprehensive data, the US defense establishment was going to have to assert its authority more firmly over the sort of fieldwork that was done. And to be sure, fieldwork had to be done; it was the only way to acquire necessary details about the world's extremes. But this work had to be closely coordinated with "office and laboratory research" so it could be quickly translated to military applications.

Within the new Committee of Geophysics and Geography, eleven technical panels were organized to "cover the whole field of earth sciences." The Geographical Exploration Committee was no more, nor were that committee's three subpanels on expeditions, geographic environments, and exploratory physiology. In fact, physiology and exploration were nowhere to be found. A late 1947 memo to Bush from the Quartermaster Corps raised concerns about shifts in funding priorities, noting that the meager allocation to the Environmental

Protection Section of the Corps would starve all military agencies of their work on "several thousand of individual items of clothing, food, and equipment" that the "individual soldier" used to combat "environmental factors."[22] There is no indication that Bush responded.

The new RDB panels reflected a more technoscientific plan centered on the physical sciences, not the biological sciences. This included new panels on the atmosphere, cartography and geodesy, geology, hydrology, geomagnetism and electricity, oceanography, seismology, and soil mechanics. The only geography-specific panels were organized around the extremes: the Panel on Arctic Environments and the Panel on Desert and Tropical Environments.

The interests of the RDB panels also reflected a shift toward quantifiable science. The postwar shift to bureaucratization, classification, and replication were all evident in the ways that Cold War science and technologies were promoted. Part of this was a shift to lab work, but it was also a shift toward thinking about regions and climates rather than places, thus the new panel names. Localized work was never totally elided, and there was always tension in this process, but efforts to streamline and create shortcuts to knowing a vast world were powerful.[23]

The RDB's new geophysics and geography committees immediately got to work reviewing proposals and projects underway across the country, in military centers, corporate spaces, and university labs. Wartime collaboration was assumed in postwar planning. By the end of 1948, the panels had evaluated 1,500 projects across public and private agencies. The next year that number increased to over 1,900.[24] From the various proposals, the RDB created "unsolved problems" lists, which it then used to set priorities across its research areas.

Behind those large numbers were research agendas that ranged from the futuristic to the utterly ridiculous. Indeed, reading through panel minutes and agenda items is to glimpse giant wish lists for the extremes. The Panel on the Atmosphere wanted to study machines to modify weather and to eliminate smoke and fog on a battlefield. They demanded more funding for high-level wind forecasting to help with aviation and for basic weather forecasting. They praised projects that used "mathematical terms" to describe Earth's atmospheric conditions so that better forecasts could be made not by men out in the field, but by computers using punch cards. Weather forecasting in the deserts, the Panel on Desert and Tropical Environments admitted, was abysmal, not only because there was no long-term data, but also because there were no facilities on the ground to do the collecting. Weather collection stations were needed. It was not enough to know that a desert could be hot and dry—the Air Force in particular wanted detailed and intimate data to forecast all around the world.[25]

Climate was just one area of interest. Studies central to engineering needs regarding construction and mobility, too, were prevalent. A proposal circulated for the study of "practical methods" to erase the "tell-tale tracks of men" on snow and sand. Could a tool be developed to remotely test the stress deformation and strength of any given soil or rock? New vehicles were needed for moving across extreme terrain. Could mud and sand be quickly solidified to enhance mobility? Could a catalog be created that would help military personnel respond faster to terrain, whether it be sandy, snowy, icy, or muddy? Engineers noted that this was important because heavier, faster equipment led to new ways of thinking about mobility and "trafficability." "Earth itself is no longer terra firma," Reybold declared, referring to the new ways men could now interact with it. All materials had to be understood.[26]

The Panel on Cartography and Geodesy asked for navigation and global positioning systems. Pilots needed to know precisely where they were at all times. Other military leaders and panel members suggested tools that could calculate ground conditions entirely from the air. The Panel on Geographic Research Techniques wanted more detailed data on mineral and fuel resources around the world, as well as terrain features that might be considered vulnerable to environmental shocks, such as landslides and flooding. Studies of the origins of sea ice and nature of the Arctic Ocean pack ice were also on the wish list. In the most extreme cases, RDB panels discussed ways to weaponize nature itself. As the historian Jacob Darwin Hamblin has shown, Pentagon planners wanted to change the weather, initiate earthquakes and tsunamis, and study the utility of releasing pathogens on unsuspecting populations.[27] They wanted to make their own extremes.

There were wish lists aimed at opening the world up to US interests beyond military needs. It was assumed, after all, that US corporate power would flow along those routes, sometimes helping set them down. How could private companies be brought into these conversations and—crucially—help push them further along?

If many of these technologies and capabilities sound familiar today, it is because parts of the wish list were fulfilled, designed, and developed within US military channels. But the lists also highlight how research programs were being channeled into particular fields. As a result, even as the US defense apparatus expanded knowledge about the whole Earth, moving toward global awareness of environments and conditions, there was no effort to reconceive of what the "globe" meant. The strategic vision was narrow, rooted in the expansion and maintenance of US military power.

Despite the sense that anything was possible, not every wish could be realized. After all, as a memo to the head of the RDB explained, the Defense Department would "never budget sufficient funds to accomplish all of the projects." The RDB panels and committees were thus charged with making choices about what was needed and what was not. Within those parameters, what was the best way to get to know the extreme environments of the world? As the RDB's Panel on Desert and Tropical Environments reminded, they needed information that would "provide maximum efficiency in all phases of military operations."[28] Successful proposals could be linked directly to national security needs, which were something of a moving target. Research programs to aid the design and performance of advanced weapons systems were prized, such as missile guidance, high-altitude flying, navigation, and communications.

What did this mean for deserts? A 1949 Army report stated that "little has been done in the last two years," regarding desert operations. As Hubert Wilkins reminded his colleagues in the RDB, experience and information were dangerously "inadequate." If the military wanted to be able to maneuver over dunes, deploy to hot and dusty places, and learn to best hide tanks in the desert, it needed to study the desert. Certain that everyone else in the world had more desert knowledge than they did, members of the RDB's desert panel asked that the Air Force's newly reconstituted Arctic, Desert, and Tropic Information Center (ADTIC) conduct a review of desert research activities around the world being carried out under nonmilitary auspices. The ADTIC complied and created a file cabinet of three-inch-by-eight-inch notecards full of data, 1,780 cards in all that represented "scientific research institutions in 55 countries." This included one facility in Afghanistan focused on meteorology; eighty-two in India focused on a range of issues, including agriculture, forestry, and geology; one in Saudi Arabia, where there was a tropical desert lab; four in Egypt for agriculture, medicine, and meteorology; seven in Morocco for medicine and agriculture; and ten in South Africa focused on geology, chemistry, building, and medicine.[29]

Most damning was the fact that US programs in desert studies could be counted on one hand, housed at a few military research centers and universities, but there was little else. The RDB thus recommended that the Corps of Engineers establish its own desert testing facility. A number of possibilities were evaluated. A research center in Saudi Arabia in conjunction with the Aramco oil company was considered briefly. Better, however, would be desert test sites at home. A number of locations in the former Desert Training Center were considered before settling on Yuma, Arizona. There, at the Yuma Test Station, desert construction and operations could, once again, be conducted.[30]

The opening of the Yuma Test Station was a sign of the US military's renewed interest in the world's extreme environments, motivated not only by meetings of the Research and Development Board's various panels but also by war. The moment when North Korean forces crossed the thirty-eighth parallel on June 25, 1950, the US military perceived potential weakness everywhere. Korea was hardly considered an extreme environment—although the first winter of the war was brutal—yet the action focused the strategic gaze on all the nooks and crannies that the military had failed to account for. Cold War tensions had been percolating for some time, of course. The fall of

China to communism in 1949 and the Soviets' earlier-than-expected detonation of an atomic bomb fed US fears of an expansionist communist bloc. Any remaining reservations about building up the country's military capabilities and need for constant vigilance evaporated. The Cold War was on.

Fighting in Korea did not involve desert operations, but it did throw into sharp relief the need for advanced preparedness. If simply because it reminded military planners how much they had not yet learned. Eager to be useful, Siple went to East Asia in January 1951 to test out winter gear. He reported the unwelcome but predictable news that "environmental information was not adequate." While the clothing was generally good—he was photographed wearing a winter Army parka and smiling his typical grin—he declared that some of the rest of the equipment was inappropriate for the hard, frozen ground. Shovels snapped in half as men tried to dig foxholes. The hats were warm enough, he reported, but the earflaps meant that men could not always hear well. Worst of all, the men themselves often failed to take basic precautions. More training and "indoctrination" were needed if US troops were going to be sent out to new environments.[31] Planning for the extremes had not gone far enough.

War in Korea reaccentuated the dangers perceived to lurk everywhere. Just as important as knowing how a shovel would behave on frozen ground was how a shovel might behave in sandy, rocky terrain—particularly because it seemed likely that the Cold War, even more than World War II, was going to be fought on the margins, in precisely the environments that remained unknown to Americans. The extreme battlefields of World War II had not been anomalies; rather, they served to introduce US military planners to a global geography that required firmer integration into all aspects of preparation. This was particularly obvious as the wishy-washiness that had swirled around what sort of global role the United States would assume cohered into a set of plans that meant engineers would be busy building for years to come and soldiers would be deployed in these places on a semipermanent basis. In 1950, the US Air Force alone requested that the Corps construct dozens of new air bases for them, many in the middle of the extremes the engineers were still learning about. Reybold, though no longer chief engineer, certainly would have been pleased.

By the time the engineers began packing their bags and hiring private companies to aid in their work, the contours of what had to be known were becoming clear. Environmental intelligence, still in short supply, needed to be gathered for construction and operational purposes. How precisely to get that done was not always obvious; what was clear was that anyone who wanted to work in these types of environments might have a patron, if they were willing to channel their work into finding the right kind of answers.

★ 6 ★

ENGINEERING DESERTS

The world is running out of sand. At least, that is what twenty-first century headlines declare. Although it is more accurate—if less sensational—to say that the world is running out of some kinds of sand. Sand is foundational to modern life, but not all types of sand are. Beach and river sands are key ingredients in the concrete used to build cities and roads. Computer chips are made from silica sands, as are beer bottles and windows. Even some toothpaste has sand in it.[1]

Scarcity was not on the minds of US military engineers in the 1950s and 1960s as they encountered sandy places in new ways. Instead, the men who showed up in North Africa to build giant air bases wondered why there was so much of it. In Libya, the one thing nearly everyone in Tripoli could agree on was the overwhelming amount of sand. It was everywhere. British military officers, charged with administering postwar Libya before the country's independence in 1952, reported that blowing sands "overwhelm[ed] houses and cultivated land." US military personnel marveled at how quickly the runways and roads were rendered invisible. Drifts as tall as men formed overnight. Judy Radeker, of Kentucky, explained that the sudden arrival of the Ghibli—a hot, dry, and dusty desert wind—could "ruin the best of days." Her husband was stationed at Wheelus Air Base, and they spent years in Libya. When home on holiday in the United States, she reported to the *Louisville Courier-Journal* that the winds could be counted on a few times a month, rocketing the temperature up to 115 degrees and depositing "sand and dirt a foot high inside windows." Sand, she said, "makes housekeeping difficult."[2]

It made a lot else difficult as well, which is why sand received so much scrutiny from the US military. The Cold War turned sand and sandy places from mere curiosities into strategic priorities. This was both because defense planners worried about needing to operate across the world's strategic deserts and because the Pentagon developed so many military facilities in them. The Desert Training Center might have been closed at the end of the war, but by the late 1940s, the southwestern part of the United States was host to dozens of other test ranges and military bases. Cheap land, clear skies, and isolation seemed to invite military cantonment. As the historian Richard White wrote, "For military planners all the old liabilities of the West suddenly became virtues."[3] Age-old tropes of the desert as wasteland fit easily into Cold War needs for militarized spaces.

The US military exported the idea that deserts were expendable. In the early 1950s, the Air Force asked the Corps of Engineers to create bases in Saudi Arabia, Morocco, and Libya (fig. 6.1). Strategically, the locations made sense; technologically they were necessary. The military required bases within striking distance of targets behind the Iron Curtain, but bombers did not yet have the range to be stationed any further away. These bases would contain and deter the Soviets. At the same time, the importance of oil to the global economy made the Middle East and Mediterranean increasingly vital.

The desert locations were telling as well. Given the paucity of resources and people in many of these places, the US military hoped that their bases would be less visible to populations who might object to housing foreign troops. Much like in the United States, it was considered wise to operate military bases in sparsely populated areas, where they had a chance of being ignored, if not welcomed. In Morocco, French colonial authorities insisted that the five proposed US air bases all be located far from population centers.[4] The message was that US Cold War military bases might be flash points for the rising independence movements that threatened ongoing European colonial rule.

Wheelus Air Force Base in Libya was the ultimate example of the

FIGURE 6.1 F-86 Sabres roar off the runway, likely at a US air base in Morocco, ca. 1955. In the 1950s the engineers constructed dozens of facilities in the world's "strategic deserts." At each site, military engineers served as intermediaries between strategic plans hatched in the United States and local conditions. Courtesy of Headquarters, US Army Corps of Engineers, Alexandria, VA.

military finding utility in the desert. The base became vital to US interests not because it housed bombers, but because it became the gateway to the Sahara. By 1954, the United States had established not only a massive air base and a string of radar sites in the country but also a twenty-three-thousand-acre bombing range ninety miles from Tripoli called El Uotia, where pilots from around the world could practice bombing and strafing, firing, and dropping bombs on mock targets painted on the desert floor. One pilot remembered that "there were hundreds of miles of desert to spare for a gunnery range, and for recreation." The whole area, according to another, "was just sand . . . it was nothing, but it was perfect for the task of catching simulated nuclear weapons slung off of fast-moving jets." While pilots soared overhead, engineers worked the sand, building mock targets (even a mini-Kremlin was created) and smoothing an emergency runway out of sparse chaparral. So active was the practice range that a plane took off from Wheelus every forty-five seconds, pausing only to honor the call to prayer that sounded from Tripoli's minarets. During a 1961 visit to the base, the aviation writer Humphrey Wynn reported that the Americans dropped seventy-five thousand bombs on the Sahara every year.[5] Military activities spread out from there. New missile test ranges were created, radar sites built, and desert maneuver areas established. The Sahara—like parts of the American desert—became a military playground (fig. 6.2). Such was the case until 1969, when the US military was kicked out of Libya following the Libyan Revolution.

Given the importance of deserts to strategic thinking and military preparations, learning about sand and sandy places was a priority for engineers. This was true in Libya, where a base turned into a bombing range, just as it was true in Morocco, where the US Air Force built five new facilities, and in Saudi Arabia, where engineers were designing and building military and civilian airports. At each location,

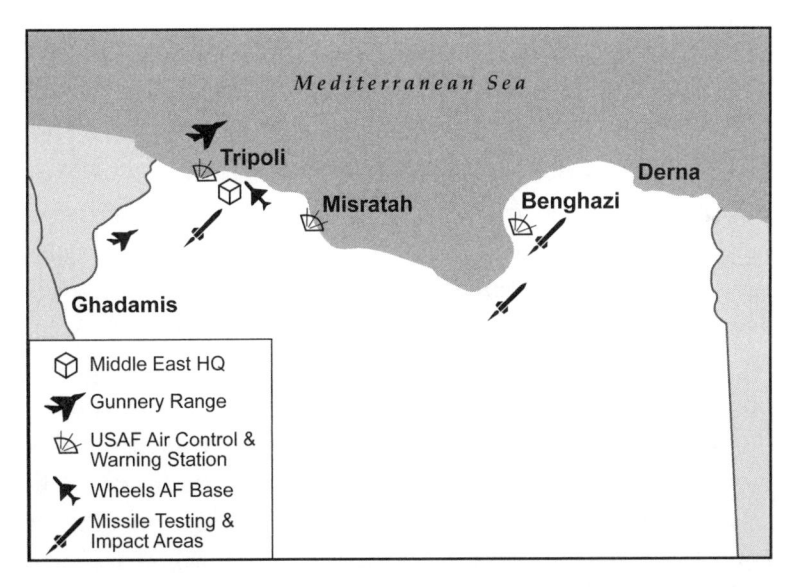

FIGURE 6.2 Military facilities across Libya, ca. 1960. When the US engineers set down bases around the world, additional facilities often followed. In Libya, Wheelus Air Base was first, followed by missile-testing sites, radar facilities, and a giant testing range, El Uotia. Some locations are estimates based on best available information. Graphic by Heather Dart Designs.

the engineers had to be incredibly attentive to place; they became mediators between strategic priorities and local peculiarities. They had to manage contractors and subcontractors, laborers and local regulations, the acquisition of property and—ultimately—the environments. As the engineers moved between high political plans and specific strategic needs, they looked for continuities, ways that mastering the peculiar material foundations in one desert might help them think about another. Working across the Corps' own network, military engineers met and pooled ideas. They looked for shortcuts that might make the entire planet—suddenly all of which was of such vital concern—a little more manageable. All the while they filtered data and information, shaping the contours of what they believed should be known. Though largely overlooked in the historical record, military engineers gathered ever more environmental intelligence and—importantly—further linked that data to strategic ends. Local projects fed into a database of global knowledge that enabled planetary thinking.

The US military may have preferred to think of deserts as empty wastelands, but everywhere the engineers went, they encountered evidence to the contrary. Lieutenant Colonel Paul D. Troxler, of the Corps of Engineers, discovered this immediately when he arrived in Tripoli in 1950. He had been sent to take command of the Corps' newly created Middle East division, which would oversee new military construction in Morocco and Libya. Troxler's list of projects was long; his budget, substantial. The Air Force anticipated spending more than $60 million in the coming years to expand runways, upgrade facilities, and build additional military sites in Libya alone.[6]

As Troxler made his way to North Africa, traveling along the military's air routes, he carried with him sheaths of paper: memos, contracts, reports, charts, blueprints. For months already—since the Air Force had asked the Corps to construct its new North African and Middle Eastern air bases—Troxler had been planning. As lead engineer for the Libyan sites, his job was to bring the various people and resources together to make a base possible. The first step was to hire US-based design firms to help outline the project; at the same time, he would put out a call for bids to American construction companies. Without a wartime army at its disposal, the Corps would not actually do the building and heavy lifting but would oversee the work being done by US-based firms. In the case of Libya, the contract went to a joint venture called Crow-Steers-Shepherd.[7] Different conglomerates were hired for the jobs in Morocco and Saudi Arabia.

Troxler lamented the difficulty of getting US firms engaged in international work. Not only did they have an abundance of work to do within the United States, but also the barriers to entering foreign markets were high, and the rewards—at that point—seemed uncertain. The Corps would help change that. The carrot was something called a cost-plus-fixed-fee (CPFF) contract. The Corps used these instruments widely in the 1950s (and would later be castigated for them, given how costly they became). Basically, a CPFF allowed a firm to enter a contract with almost no risk because the US government paid all costs and a set fee. Regardless of how quickly the work was done, how well it was done, or how expensive it became, the American taxpayer footed the bill.[8]

It was a mechanism by which the US military facilitated the movement of US business interests out into the world. Indeed, short-term military construction seems to have paid handsome dividends for the companies involved which often received follow-on contracts, local connections, and footholds in overseas markets previously inaccessible to them. These private companies would also have the benefit of receiving Corps research reports aimed at a better understanding the global extremes.[9]

It is easy to imagine that Troxler looked out the window as his plane descended into Tripoli in November 1950. Troxler had been to Libya before, during the war. In 1944, he was stationed at what was then called Mellaha airfield to help maintain the wartime airdrome. Troxler left with most of the GIs as the war ended, but a skeleton crew of personnel remained behind. Their job of maintaining the status quo changed in 1947 as Cold War tensions led to an expansion of US overseas facilities. The base went from housing a dozen men in 1947, to fifty in 1948. By 1950, the number was tenfold that. Troxler arrived just in time to turn an old wartime base into a place that not only was habitable for US airmen and their families but also could house the machines of modern war.

In 1951, Troxler's first and most important task was to expand the Wheelus runway. The need for careful survey and soil sampling remained. Runways would have to be longer and stronger; Cold War bases were to be built for ten or twenty horizons, not as temporary wartime structures that might be used for just a few months. No one wanted a new bomber to be ruined because of runway failure.[10] Troxler would need even more data than his predecessors.

Out of the window, Troxler would have seen the relatively short and primitive runways and taxiways he was asked to expand. But he also would have gotten a bird's-eye view of the environment and the people who lived there. Along the coast, a thin stretch of green hugged the sea as far as the eye could see; this was the small coastal belt where the vast majority of Libyans settled. Behind the narrow strip—only a few kilometers in some places—was the sand, a yellow glare against the cloudless sky. The desert made up 95 percent of the country, stretching 650 miles to the southern border. Caravans measured their crossings not in hours or days but in weeks and months. Ambassador Henry Villard marveled at the "limitless empty hinterland."[11]

Along the edge of the sand was the city of Tripoli, which rose up and out, stretching toward the Mediterranean. It was an oasis surrounded by sand. In his memoirs, Villard noted that "feather-duster palms" and minarets pointed toward the blue sky. As an engineer, Troxler may have been more interested in Tripoli's harbor, which was flanked by two protective breakwaters meant to facilitate shipping. One had suffered serious wartime damage. From his plane above, Troxler may have worried a little. Already 770 tons of equipment was steaming across the Atlantic from the United States toward Tripoli. Using the engineer network, Troxler had nearly seventy machines being sent from construction zones in surrounding areas. In a few months, fifty thousand bags of concrete were to arrive from Yugoslavia. Troxler knew that the port would see a lot more activity in the coming months and years.[12]

The uncomfortable reality for Troxler and his peers was that even as the Americans continued to write about Libya being a "huge, barren slice of Sahara," Wheelus was not an airfield out in the middle of the desert; it was in the middle of a city, pressed right up along the Mediterranean. Although this meant that the American families who eventually came to live year-round at Wheelus—such as Judy Radeker—would have access to a lovely beach, it also signified that Wheelus Air Force Base was a plot of prime real estate.[13]

Indeed, as Troxler's plane came in for landing, he may have noticed the white, limestone houses scattered near the base, or the earthen walls that formed barriers between fields of barley, date palm, and melon. The people who lived near Wheelus—what they called Mellaha—grew a variety of things. In the summer they cultivated small crops of maize, pumpkin, watermelon, eggplant, and pepper. In the winter, farmers turned to barley and aniseed. Near the coast and the edges of Mellaha, the farms tended to be small, bounded by beaten-earth walls topped with local plants such as prickly pear and thorny capers, some so thick they looked like hedges. Large gateways led into farms where oranges grew. Date palms, mulberries, and pomegranates crowded in patches near wells and water sources. Olive groves with trees hundreds of years old filled the valleys. On good years, when the rains hit or exceeded the average of two hundred millimeters or more (the minimum for farming without irrigation), farmers had surpluses to sell and trade. In 1949, the British authorities still administering Libya reported an above-average harvest of wheat as well as of new potatoes, peas, and tangerines.[14]

This local reality would soon be an intimate part of the engineer's experience. One of Troxler's jobs was to acquire land. Local patterns of ownership immediately befuddled the engineers. It turned out that property did not mean the same thing in Tripoli that it meant back in Toledo. In Libya, it was possible for a man to own just a small percentage of any given plot of land or a single date palm on a larger family plot. Contracts would have to be written up for each item. Engineers' maps and charts are full of multiple names, plots of a few acres broken up into smaller units. Working with the local British administrator, the engineers set about to buy up as many parcels as possible. But not everyone wanted to sell. Local landowners complained. We know this because they voiced concerns to the British administrator in the area and to local Libyan authorities, who brought their complaints to Ambassador Villard. In many cases, landowners ended up with 50 percent more than was initially offered to them. They also demanded that local boards be established to determine property values and help with resettlement costs. Building an air base was going to take a lot more than environmental intelligence.[15]

The conflict over property and payment is an important reminder of the violence at the center of US military bases. Lands were taken and people disposed; fences were thrown up where none had previously been, bisecting communities and routes. US officials were not oblivious to the awkwardness of this relationship or to its potentially destabilizing outcomes. Diplomats stationed in places where bases were constructed issued warnings about the effect on local people and politics. No matter how poor a place might be, they insisted, or how much money the US military could throw at it, housing foreign troops was always going to be contentious. Although such reminders did little to stymie the expansion of US power at midcentury, there were certainly long-term consequences. In Morocco, for example, where the United States had spent hundreds of millions of dollars on new air bases, the newly independent government asked the military to leave in 1956. In Libya, the United States was kicked out in 1978.

If someone in the mid-twentieth century wanted to know about sand and deserts, there were few places to turn. The study of sand, like the study of desert plants and deserts more broadly, had never been a thriving field. Work in these environments remained parochial. French colonial authorities in North Africa, for example, might study sand and desert agricultural practices, but they were not likely to share what they had found. Other people interested in deserts—like the individuals who wrote about and researched in the Mojave and Sonora—might acquire a lot of information and anecdotes but had

FIGURE 6.3 Ralph Bagnold, known as the "Sandman." Throughout the mid-twentieth century, Bagnold studied and wrote about sand and dune formation. He gained most of his experience with the British military, but by the 1950s, he was working in the private sector and as a consultant for the US government. Churchill Archives Centre, the Papers of Brigadier Ralph Alger Bagnold, BGND E.20.

limited ways of publishing them.[16] It was modern construction and operational needs at midcentury that forced a new reckoning with the stuff: flying sophisticated planes, building giant air bases, supplying troops with the luxuries of home—all created new requirements for knowing sand and desert environments.

There is one name that appears over and over again in midcentury writings about sand and sandy places (in fact his name still appears): Ralph A. Bagnold, the "Sandman," who had been studying sand for a long time (fig. 6.3). As a British military engineer in the 1920s and 1930s, Bagnold was able to travel across the world's deserts along the sinews of British imperial control, writing about what he saw. He drove from Cairo into Transjordan on one expedition; on another, across the Sinai and into southern Palestine. In 1929, he drove from India to Egypt and then on into the great sand sea of the Sahara.[17] All the while he scribbled notes in cramped cursive. His handwriting looks rushed, as if he were trying to get everything down in an instant. Later his corpus became vital to Western studies of desert environments, including those in the United States.

Stories of Bagnold and his desert exploits appeared in US publications. In November 1930, for example, the *New York Times* explained that Bagnold had been the first Western man to reach an oasis in the Libyan desert long rumored to have been the "lost oasis" of Zerzura. Bagnold insisted he had not found the mythical desert Shangri-la, and privately he doubted its existence, but that did not stop him from participating in the Zerzura Dinner Club for men who shared his passion for desert adventure. In the interwar period he kept up a correspondence with two giants of desert exploration: T. E. Shaw, also known as "Lawrence of Arabia," and Count László Almásy, later the basis for the protagonist in Michael Ondaatje's novel *The English Patient* (1992). Zerzura was less important as a place, of course, than it was as an idea: a quest that kept people searching and exploring. Bagnold was less romantic in his own travels. He wanted to chart little-known territory. To reach the edges of British imperial maps, where known features of Egypt faded away into a blankness stippled vaguely to indicate sand, ending with the simple "limit of sand-dunes unknown."[18]

On his expeditions, Bagnold practiced and perfected many of the tools and strategies that would later become standard issue for the British and US militaries. This included deflating tires to travel over sand, a sun compass he invented for navigation in the desert (that was also used by US engineers in Greenland), and water rations for men in extreme desert heat. Indeed, so novel was his expertise that in 1939 the British military called him back into service to arrange the Long Range Desert Group, which comprised small roving teams of men who traveled through the desert harassing Axis troops.[19]

But it was Bagnold's scientific pursuits that became of particular interest to US engineering centers and military projects in the 1950s. With a degree in civil engineering, Bagnold had become fascinated with how the desert worked. How and why did desert sands form ripples, dunes, ridges, or flat sand seas? What were the mechanics of how the wind, sand, and terrain combined to create these otherworldly formations? It was not random, Bagnold insisted. The desert was not just an odd conglomeration of sandy particles that sometimes formed into dunes and sometimes did not. "Instead of finding chaos and disorder," Bagnold wrote, one is "amazed at the simplicity of form, an exactitude of repetition and a geometric order." Sand formed into dunes that "move inexorably, in regular formation, over the surface of the country, growing, retaining their shape, even breeding." The desert was not a barren wasteland; it was alive with little-understood processes. Sand grains, for example, seemed to self-accumulate, gathering in hills and dunes rather than spreading out evenly across the ground. Dunes "seemed to behave like living things," he wrote, which made the desert "vaguely disturbing to the imaginative mind."[20]

Bagnold certainly had an imaginative mind. He was committed to better understanding what he saw. He spent a great deal of his time out in the desert collecting qualitative data. He scribbled down notes, made abundant field observations, and wrote widely cited and praised articles. But to really understand how dunes formed and move, to get a clearer sense of the mechanics of the desert environment, he believed he had to take the desert inside. There was no such thing as Earth-orbiting cameras or devices to take in the whole desert or sand sea at once. There was no way to observe the movements of great dunes in one sweep. Bagnold instead planned to mimic it on a small scale. He needed to carefully observe and chart how grains behaved. And so, like the wartime engineers, he bottled samples of sand and took them home to London, where he built a wind tunnel and began to experiment. He would observe grains in his lab and then go out to the desert to test his results. Out of his 1930s experiments and fieldwork, he drafted what is still considered a seminal text on the topic, *The Physics of Blowing Sand and Desert Dunes* (1941). Much of what we know today about the behavior of sand starts in Bagnold's wind tunnel.

Bagnold was interested in the size and shape of sand grains. From his own travels, it was obvious that not all grains were equal, and that difference might account for some of the properties of different desert systems. The composition of sand from Death Valley is distinct from the sand of the Sahara, which is different from the sand of the Mississippi River's edge; each grain is a token of an area's underlying geology. Just as Shreve's studies of botany could help place one in a particular desert ecosystem, sand grains are signifiers of a particular place. "The birth of a sand grain," the geologist Michael Welland wrote, "signifies the death of a mountain."[21] It is a never-ending process, however, and sand is not the end point. Sand grains continue to rub against each other, grinding and refining. They get rounder and smaller over time.

The size and shape of sand grains are important to how they behave. As Bagnold discovered, dune formation is determined by the local type of sand and wind conditions. Different grain sizes lead to different sizes and shapes of dunes and hills. This might be practiced in a kitchen by comparing the piles formed by regular table salt and sea salt and sugar, all of which are granular materials that behave like sand. A tipping point will eventually be reached: one extra grain will lead to an avalanche down the side, an angle of repose.[22] But every type of grain has its own capacity for piling and its own angle of repose. The grains matter.

The most common sand is that which might be found on a beach, rough pellets of rock and shell. The grains themselves can be many things: quartz, bits of seashell or coral, or even lava. Namibian sand is speckled with tiny diamonds. But sand is categorized by its size and granularity, not its content or location. Any sediment with a diameter between 4 and .00625 millimeters is considered sand. Anything smaller than .00625 millimeters is considered silt or clay; anything larger than 4 millimeters enters the gravel or boulder categories.[23] A standard engineering sieve kit would help engineers figure out what sort of grains they were dealing with.

During his wind tunnel experiments, Bagnold discovered the physics of how sand grains behave. For example, sand grains bounce when picked up by the wind. When they landed, they collided with other grains, sending those grains bouncing and colliding. Bagnold called this "saltation," a process that could cascade into a sandstorm as millions of grains bounced and rebounded, collided and scattered. Sand was not only picked up and flung about by the wind, Bagnold intuited; it propelled itself. That explained something curious he witnessed out in the Sahara: a real sandstorm (as opposed to a dust storm) picked grains up only a few feet off the ground. He could stand in a sandstorm and have his legs pelted by grains while his head remained above it all.

By the 1950s, Bagnold was one of the few men writing in English who had done dedicated, meticulous research into the physics of sand. He showed that in studying the desert, one could determine a lot about the underlying geomorphology (the relief features of a given area), about wind, and even about how to build things in and near the sand. So important was his work to understanding desert environments that in the 1950s he was hired by Shell Oil to evaluate desert sand formations for possible oil deposits. He left after his first year, uninspired by the work, and returned to the world of eolian processes, studying how particles move in fluids. But for the remainder of his life, he was asked to consult various oil companies and governments on how they could cope with encroaching sands.[24] Eventually the National Aeronautics and Space Administration came calling. In the 1970s, the space agency needed help interpreting pictures of Mars. NASA staff wanted to know if Bagnold could use the shape of the dunes to tell them about the underlying geology of the Red Planet.

Bagnold may have sought out sandstorms to test his theories, but the military engineers genuinely hoped to avoid them. A sandstorm was just about the only thing that could shut down an air base. At Wheelus, planes could be grounded for days. Visibility was poor and ground operations dangerous. Pilots could get disoriented; equipment could stall. The sand did more than wreak havoc on operations; it also ruined equipment and the built environment. Whirling sands were like sandpaper on exposed surfaces, scraping, rubbing, and eroding. Sand accumulated in engines, leading to excessive wear and pitting.

The same was true for runways and aprons. Sand dunes formed on the pavement. The sand was an everyday menace. The soil of the base did not act quite like soils in other places. For one thing, it moved—it moved from below, which made erosion and runway collapse more frequent than expected. More visibly, it moved over land, as drifting dunes but also as swirling masses of tiny particles that could sting the face and hands. Pilots complained that it got in their ears. Even the mess-hall operators expressed dismay. It was impossible to "fix" things "tight enough to keep out the sand."[25]

Back at the recently opened Yuma Test Station in Arizona, engineering teams studied the effects of sand on military machinery and materials. Sand itself was not the problem, they admitted; it was sand picked up by wind that led to trouble.

In addition to trying to figure out the damage done by sand and storms, the engineers also wanted better ways of predicting such storms. They tried to account for the lack of data that existed before the war by keeping better accounts after. It is hard to predict the future if you don't have a sense of what has been. Engineers and related agencies thus began building data sets about dust storms and climatic conditions in the world's deserts. They found that in the US Southwest, no more than two sandstorms occurred per year, so that is what they could expect. Most of North Africa had a similar pattern, though the engineers noted something of great interest: areas fought over and trained on during the war had greater incidence of dust storms. Military activities led to the "destruction of protective vegetation, the pulverizing of the surface sediment, and the crumbling of surface crust." A study of dust storms in Egypt, for example, showed that the average of three to four per year before the war was dwarfed by the forty to fifty such storms that occurred during the war when tanks tore up the desert. Human activities, a 1963 report admitted, would hasten and exacerbate desert storms.[26]

Around the world, the military engineers continued to mediate between local environments and strategic needs; between questions they had and reports and studies being done to help answer them. (Many of those reports, incidentally, cited Bagnold; some also referenced Shreve). Stymying blowing and drifting stands was one obvious need. Although the US deserts lacked the sorts of vast dune fields found in the Sahara, dunes in Death Valley were studied for their migration patterns. Knowing prevailing winds was vital to understanding when and where you would lose materials to drifts. Seasonal patterns were also important if you wanted to know a desert environment.

In Arizona and North Africa engineers experimented with new and local methods to try to keep dunes from erasing their roads and runways. In some cases, sand was simply removed, shoveled up and sent somewhere else. Perhaps useful for a time, this was an awfully labor-intensive way to manage things. "Oiling" and paving could also be useful, an engineering report explained. In California, oil was occasionally dumped on a sandy area to keep the grains from encroaching on roads. Troxler also turned to local practices. Libyans bound the sand with vegetation to stabilize the dunes that seemed to be creeping toward town. Planting trees could help limit airborne particles. Fences strategically placed could limit sand from ruining crops. In 1954, the Wheelus team used local practices to plant trees as part of a $50,000 erosion control program. But they planted everything at the same time rather than at intervals as suggested by local practices, and they lost it all.[27]

As the American "experts" were learning, the best solutions seemed to be small and local. Nothing seemed suitable for keeping sand off massive military installations. The answer, the engineers in Yuma concluded, was "intelligent site selection." If you wanted to build things in desert areas, it was best to avoid areas where sand would be most invasive.[28] It was not always possible—no matter how modern and technically sophisticated the gear—to overcome the environment.

To address the problem of sand and desert climates, engineers working in the world's deserts began to create connections across them. Right away the teams working in Saudi Arabia and Libya began meeting and passing memos. They held conferences and discussed how to best manage their global projects. What were the best strategies for avoiding flies and bugs? The engineers shared ideas for setting concrete in extreme heat (avoid the daylight) and how to anticipate local labor difficulties. Given the sparse natural resources in the desert, the engineers shared information on where they had accessed supplies of needed material. A key item—they noted with some irony—was the need for sand. The grains of the Middle East and North African deserts were often too round to make strong concrete. They imported beachlike sand from Yugoslavia. Small details that had enormous implications for the type of work the men had to do.[29]

In environments that lacked familiar construction materials, like trees, creative solutions had to be found. In Libya, the engineers realized that it would be costly and wasteful to build the traditional wood-frame barracks and family houses expected on US bases. Instead, Troxler's team adopted the practice of using masonry, tile, and stone, as seen in the small houses scattered around Tripoli. They calculated savings of nearly $100,000 per building compared to US-style methods. Through the engineering network, the specifications for that sort of construction found their way to American projects in Saudi Arabia and Morocco.[30] The military engineers were taking locally constituted knowledge based on environmental peculiarities and folding it into their Western engineering practices.

The connections went largely unremarked on in the documentary record, but the borrowing and sharing of ideas and resources ran deep. The regional network of desert engineers shared ideas, held frequent conferences, and attempted to write new regulations and protocols for military construction in desert climates. The affinities were so obvious, in fact, that the Corps of Engineers reorganized in 1952—creating the Mediterranean District, which combined the former East Atlantic District (just Morocco) and the Middle East District (Libya and Saudi Arabia). Iran and Pakistan would later join that division.[31] Environmental affinities became institutional organizing priorities.

This was part of a bureaucratic reorganization of the Corps, but it also was motivated by an emerging sensibility about the problems of military construction in extremely arid environments. Back in the United States, Corps districts were already organized around environmental resources, namely watersheds. By the 1950s, the Corps began exporting this concept to the world: environmental factors would be as important as political ones in parceling out the planet.

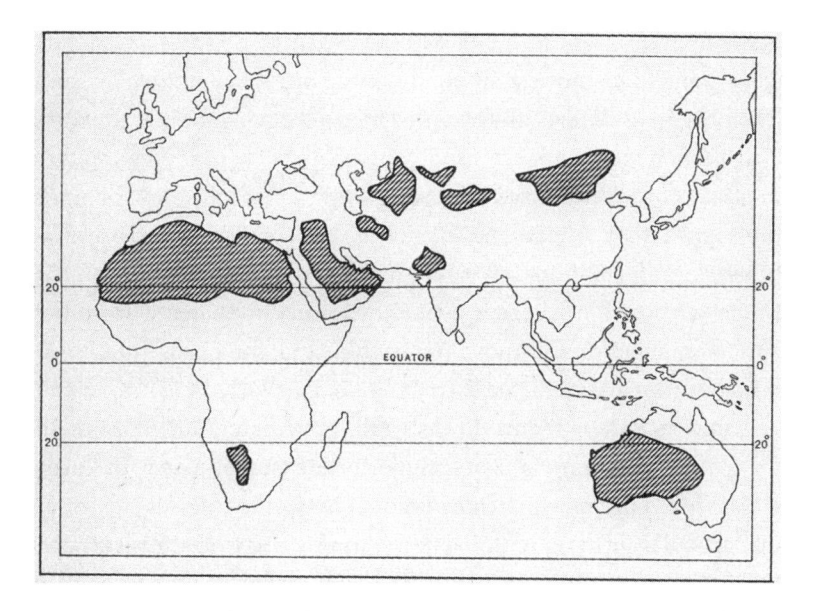

FIGURE 6.4 Strategic deserts of the world. By 1950, maps like this began appearing in defense materials, including in RDB files and the pages of the *Military Engineer*. The map reflects new ideas about Earth that divided the globe by environmental zone rather than geopolitical borders. Image originally appeared in *Military Engineer* 43, no. 295 (September–October 1951). Reprinted with permission of the Society of American Military Engineers.

There was purpose behind dividing the world into environmental regions. Engineers wanted to be able to more easily share ideas and resources, move materials and people with less friction, and think expansively across desert regions. The materiality of the environments stitched their construction work together in ways far more meaningful than expected. Taken to its logical extreme, the engineer's view on the ground would look something like a map that appeared in the *Military Engineer*, a world defined by environmental similarities, not geopolitical ones (fig. 6.4).[32]

It was more than a rhetorical shift. The engineers *intended* to work regionally; they actively sought ways to avoid national borders and distinctions. They framed "environment" as a category that would eradicate geopolitical boundaries through shared material conditions. They refused inspection from Libyan harbor officials and demanded the right to travel without passports.[33] After all, if they were going to be moving air conditioning units, cement trucks, and tractors across deserts, they wanted to avoid paying for that transfer.

Although engineers are usually invisible background characters in stories of US military power and imperial reach, their behind-the-scenes work was crucial to creating the sort of networks and libraries of information that would make the continued projecting of US power possible. There is a vivid historical irony here. Given the anticolonial nationalist movements of the era, the Corps' interest in a sort of transnationalism is striking—particularly in Libya, which declared its independence in 1952 and immediately sought to build the institutional trappings of a modern state (sometimes with Corps aid). It is a reminder that transnational networks can work to shore up particular forms of national power as much as to diminish others. Using the shared objective of creating military landscapes, engineers mobilized networks that obliterated state boundaries by following environmental ones.

★ 7 ★
ANALOG DESERTS

Sir Hubert Wilkins drew the world in Technicolor blocks. With blues and yellows, reds and hatch-marked green, he mapped shapes that adhered to no known geopolitical borders. Instead, he was following environmental ones. Throughout the early 1950s, Wilkins created scores of images like the one in figure 7.1, with the continents traced out in bold graphite, sometimes on thin tissue paper, other times on firmer cardstock. He colored the prominent environmental features in bright tones: red for desert, blue for "tundra and alpine," and so on. Some of the maps got more specific, using green, red, or white in relief to show areas of seasonal vegetation patterns across the planet. The maps are beautiful and utterly modern. They are slightly disorienting in that Wilkins centered his maps on the top of the world. It takes a while to find the familiar shapes of continents, and there are no national borders. With the Cold War geopolitical context, the maps seem almost subversive.

Subversion was not, however, Wilkins's mission. Rather, he drew his multicolored maps for a very specific military task: to create camouflage patterns for US troops sent to the extremes. In 1950, just as Pentagon planners were mapping out a vast global military infrastructure, the US Army Quartermaster Corps rehired Wilkins because it needed new uniforms for both the desert and the Arctic. Wilkins would spend two years traveling the world collecting information. He stopped at Wheelus Air Base in Libya to sample soils and

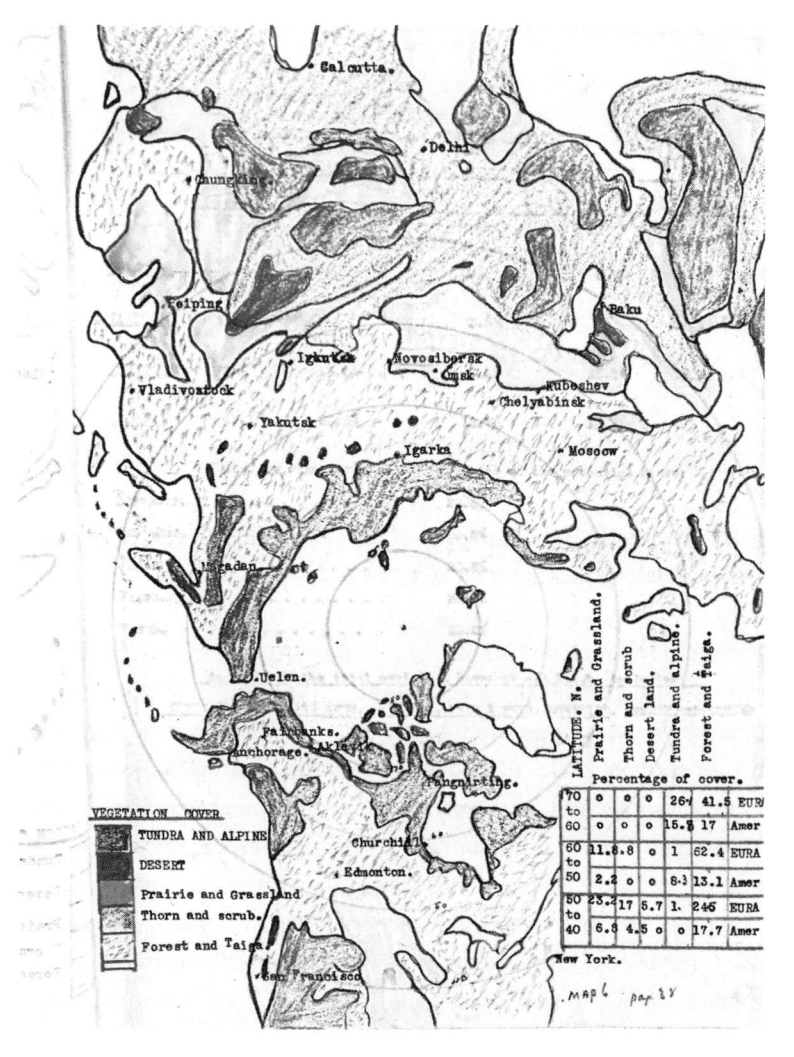

FIGURE 7.1 Global environments. Throughout the early 1950s, Sir Hubert Wilkins created hand-drawn maps that divided the world by environmental features rather than geopolitical borders. Many of these maps were in color, but this black-and-white version is still representative of a key characteristic of Wilkins's maps: they are disorienting. Viewers might find it difficult to know precisely what they are looking at because there are no geopolitical borders and the map is centered on the North Pole. Sir George Hubert Wilkins Papers, SPEC.PA.56.0006, Byrd Polar and Climate Research Center Archival Program, Ohio State University.

interview engineers before moving on to Morocco. Along the way he took photos and films, drew up charts, and sampled groundcover. He wrote lengthy reports about terrain, vegetation, the people and places he encountered, as well as climatic details.

Wilkins's archive is an eclectic jumble of reports, handwritten notes, maps, charts, and other drawings that provide a stunning snapshot of the world's hardest-to-reach places at midcentury. These were places that US policymakers and diplomats had very little experience in, but that were all suddenly of great strategic importance. Rifling through Wilkins's papers, then, raises interesting questions about all the ways his access could have been put to use. United Nations agencies, for example, may have benefited from careful reconnaissance of agricultural lands. Local communities could have used his aerial surveillance to track changes to the desert dunes. But because Wilkins's project was military, his remarkable globe-spanning reconnaissance missions were secret. The environmental intelligence he collected to help the military more easily and simply parse the world was not shared but ferreted away in files at the Corps of Engineer and the Quartermaster Corps.

Wilkins's goal was to study as many places as possible so that he could mix local data to create one representative pattern, a single palette that might reflect all the deserts of the world or all Arctic areas. Ultimately, this is where strategic environmental planning was headed: trying to identify certain characteristics about an environment that could be abstracted into broad and recognizable categories. Just as Troxler and the construction engineers were sharing materials across desert environments, trying to create global desert databases, other military agencies were looking for shortcuts to grapple with difficult places. "Desert" and "Arctic" were easier categories to deal with, for example, than "Sonora," "Mojave," "tundra," and "boreal forest," particularly when the US military was looking at deploying thousands of men for long periods of time. Creating the average palette that was a perfect mix of all the desert colors in the world was just one example of how the strategic needs funneled environmental data into generalities, synthesized the local into the regional. It was a narrow vision and frankly an ugly one that repeatedly blurred the particulars of place in service of global security and control. While its efficacy was questionable, it was a powerful of way of framing environments precisely as the United States defense apparatus was coming to terms with the diversity and complexity of the whole Earth.

Fittingly, Wilkins began his desert reconnaissance at the Corps of Engineers recently opened Yuma Test Station in Arizona. From there, he spent the summer of 1951 touring the deserts of the United States, looking for the most representative sites and samples of sand and stone. For eight days in June, he and a small team, including a geologist, a geographer, and a physicist, traveled some 2,500 miles across California, Arizona, Nevada, and Utah. They traipsed across Patton's Desert Training Center, where tank tracks were still evident. Wilkins gathered twenty-three soil samples and took dozens of photos and films. He also took to producing dyes and paints in the colors of each area so he could create swatches that matched the terrain (fig 7.2). For the Arctic, he compared subtle tints of white that few others would be able to distinguish: "bright," "winter," "star," "antique," and "lily," with matte and smoke thrown in. For the desert, he included khaki, Yuma tan, sand with a hint of moss, and pinkish gray.[1]

Wilkins dyed his desert palette onto firm canvas fabrics and stapled those onto a poster board. The next summer he took his enlarged palette to the deserts the military had identified as the most strategically important. The title of his report, "A Journey through Hot Dry Areas in Pakistan, Saudi Arabia, Iraq, Sudan, Lebanon, Egypt, Libya, Algeria, Morocco," demonstrated the geographic breadth of the investigation and also the geography of US military power that had been consolidated since the war.[2]

For three months, Wilkins moved across these places using the stunning logistic network the United States had set down and was continuing to refine. Wilkins relied on the Military Air Transport Service to skip from air base to air base. He took reconnaissance flights with air crews or bounced along desert terrain with engineering teams willing to drive him out to sand dunes and rocky ridges. He sometimes lodged at air bases or nearby hotels. He usually got up before dawn to get ahead of the desert heat. He sought meetings

FIGURE 7.2 Sir Hubert Wilkins in the desert with the color swatches he created while trying to design an ideal desert camouflage pattern for the US Army. Wilkins spent years traveling around the world, collecting samples, taking photos, and trying to find the best mix of tints, hues, and pigments to represent not individual places but one global desert. Sir George Hubert Wilkins Papers, SPEC.PA.56.0006, Byrd Polar and Climate Research Center Archival Program, Ohio State University.

with American construction firms and their engineers who were busy erecting bases so that he could gather intel from them as well. He met local and foreign workers hired to help with the work, and he marveled at the American goods—like tennis shoes—flowing about the world along military supply lines. But his real interest was in the materials he encountered. What type of soil was at the base? Was there any more data about the runways? In Libya, he learned that base engineers had not yet had time to do much reconnaissance beyond base boundaries. He was happy to do it for them. He jotted down that the base engineers at Wheelus were good at determining the bearing stability of the ground; but they "had little knowledge of the general soil cover or its relation to overall environment."[3]

Wilkins and his companions scooped soil samples so they could best match colors to fabrics. They took photos and shot short films. The Australian wrote detailed descriptions of every place he visited. He wrote about local clothing and labor practices as a way to glean knowledge for military applications. In Saudi Arabia, for example, he interviewed oil company employees about what it was like to work in a place where the temperature was often 119 degrees in the shade. The employees had ideas about the best sort of tents for the hot desert—low with large flies, and with outsized corrugated pins to hold the fabric on the sand. The Americans workers did not do so well in the heat, he admitted. They tended to rotate quickly out of desert labor programs. In extreme cases, they could succumb to periods of "desert crazy," which entailed profuse sweating and hallucination.[4]

After interviewing Americans he met in the world's deserts, Wilkins returned to his interest in the interplay of colors—the ways a splash of vegetation could shift one's perception of the sand, how intense solar radiation would bleach out colors to pale shades of tan—and how desert plants related to the environment.[5]

Wilkins's inquiries were deep, far deeper than anyone else was able to go. It was, after all, the one job he had to do. But placed within broader conversations about knowing the world, we can see his data set as just one part of an effort to amass information so that it could be combined, processed, evaluated. Wilkins might have lingered on the small details, like the hue of a desert leaf, but there were two things about his work that would prove most useful. First was the art of abstracting, using as broad a data set as possible to home in on the similarities and patterns that might lead to simpler outcomes. Second, Wilkins's reports contained details of use for operational planning. He wrote not only about color but also about the softness of sand dunes, the size of boulders, and the prevalence of low shrubs and gangly trees.

The photo of Wilkins standing in a desert with color swatches painted on large panels (see fig. 7.2) is one of hundreds of such images. He carried swatches and fabric samples around the world, folding them up as he moved about, setting them down at various times of day, under a variety of conditions. One can imagine him with poster boards tucked under his arm as he moved on and off planes, jeeps, and boats, scratching notes down that he would later type up and edit. Wilkins tended to carry small, spiral-bound notebooks that he filled with contact information and notes to himself. The notebook he kept when he went up to Greenland later that decade, for example, was filled with ways to get in touch with various engineers and commanders who were spending time in the icy desert.[6]

Wilkins even had diagrams of how to best set up the poster boards, how far they should be from the observer and the optimal angle in relation to the sun. The goal was not to create camouflage patterns for every location at different times of day but to acquire as much as detail as possible so one panel could represent everywhere as much as possible. He explained with some disdain that the British Empire used a desert uniform—designed in 1902 in Africa—that was "not satisfactory." It was a poor match for "desert areas in the US."[7] His goal was to use "scientific methods" to determine the best pattern, not the vague preference the British seemed beholden to. He was doing for camouflage what the Harvard Fatigue Lab did for the "average" soldier: creating a standard uniform that anyone could don.

If Wilkins could truly combine his color wheels into a simple three- or four-part camouflage pattern to simulate all the strategic deserts of the world, it would be in keeping with how the military engineers and the Army were planning for the world's extreme environments. Across newly opened research centers, the Corps of Engineers was studying wartime lessons. This happened at the Yuma Test Station as well as in the then recently created Snow, Ice, and Permafrost Research Establishment (SIPRE). More established Corps of Engineers centers angled their work toward environmental intelligence, including the Waterways Experimental Station and the Frost Effects Laboratory. In these places, engineers honed soil classification systems and tried to amass more knowledge about international terrain and ground cover. The *Military Engineer* continued to publish stories about how to deal with sand and deserts. The effects could range from merely bothersome problems, like fogging on hot metal, to potentially devastating ones. High temperatures could lead to overheating; alkaline substances could corrode nearly all equipment. Sand was more intriguing still: it was nuisance. Blowing sands could destroy windshields, clog filters, and muck up gears.[8]

With its new training center, the Corps of Engineers wanted to figure out how to match Yuma to other deserts, not just to make uniforms, but to test equipment and men. Although they could certainly do some of this overseas and in different places, it made financial and logistic sense to try and center research in one place. Wilkins supported this forethought. He was certain that the Yuma area had "conditions representative of desert areas in most of the world deserts," so perhaps it could be used for evaluating and testing not just desert hues but also desert materials and equipment. Of course, wartime interest in using the Desert Training Center as a stand-in for Libya was a huge miss. But Cold War engineers decided that it had not been a problem of thinking in analogs; it was a problem of poorly conceived analogs. Major General Patton had not been totally wrong to call the DTC "Little Libya," although it would have been more accurate to have just selected a small part of the Sonora for that comparison. Indeed, because "previous investigations were carried out under the pressure of war with the requirement of immediate answers based on incomplete basic analysis," there was an opportunity to finally get environmental intelligence for military purposes right.[9]

It was hoped that the Yuma Test Station could become a giant outdoor simulation for all the world's deserts. Few people seemed to question this hypothesis, so engineering teams moved ahead with figuring out how to best make the global connections. How much information on specific deserts was needed to make useful generalizations? Like Wilkins's pursuit of the ideal color palette, the goal of engineering analog studies was to find a series of features that could somehow be pulled out of their respective environments, turned into numbers, swirled together in analytical reports, set down in columns and charts, and then reapplied to the similar places. Details would be funneled into abstractions that could be broken back out to fit other places.

But first, the deserts had to be mapped and known. Reconnaissance missions and reports proliferated, most aimed at the Sonora near the Yuma Test Station. The Desert Ground Surface Study was commissioned to "determine and describe the types of surfaces of desert areas, classifying them, on a new basis, into militarily significant groups."[10] Those groups fell into four baskets: terrain features, mobility, ground cover, and vegetation. The team consisted of geologists, botanists, climatologists, military engineers, and a soil expert. Across the seasons, they walked and drove, took soil samples and ran tests, evaluated plant species, and created highly detailed maps of the area.

To people who did not closely know or study deserts, the inclusion of vegetation would have been somewhat unexpected, although the engineers who had toiled away during and after the war understood its implications. Plants were guideposts. The maps that Forrest Shreve had been creating during and before the war on plant populations were suddenly in high demand. In the 1950s, Shreve's work was being used and cited all over the place. Engineering reconnaissance of the Sonora in the 1950s relied on Shreve's studies, a recognition of the ways that plant assemblages provided a key to understanding the rest of the environment.[11] The flora-based "GPS" that desert denizens had long recognized was becoming part of modern military wayfinding.

It would be a mistake to read any of this as attention to place or to local environments. Quite the opposite was true. Abstracting places into the category of "desert"—with all the wasteland associations that still clung to that word—not only did great violence by erasing anything local or particular about a place but also justified the violence of militarism precisely because the label "desert" defined a place as useless. After all, the US government quite easily appropriated vast swaths of Nevada's desert to test nuclear weapons—928 of them—between 1951 and 1992. That is just the most visible form of violence meted out.[12] Still today, the deserts remain militarized. In the twenty-first century, Superfund sites dot the desert landscape, and huge swaths of the country's deserts remain off-limits as a result of ongoing training and testing.

The Yuma analog project was just the beginning. Well before satellites could take detailed pictures of any place at any time, data produced by engineering field teams all fed into what was becoming a much larger military project to create analog maps of the world's deserts, called the Military Evaluation of Geographic Areas program. Engineers were assigned the duty of creating analog maps of all extreme environments, but the desert was the most intensely undertaken. The plan was to attempt to apply mathematical principles to terrain features so that particular places in the Sonora could be compared with desert areas from around the world.[13]

If analogous mapping could work, the men in charge of it argued, then things like Wilkins's highly detailed files would not be needed, assuming that the right mix of factors could be discovered. Although, of course, finding those factors would require precisely the sort of work Wilkins was doing. One of the analog project reports was clear on the limits of the individual: the "qualitative, or classical, approach to geomorphic description" relied on individuals wandering around, writing things down in a way that other people had to believe. You had to believe, for example, that Wilkins saw colors in the same way that everyone else did, or that a geologist had some intuitive grasp of landforms that superseded what a nonexpert could see. But what if said individuals were unreliable? What if Wilkins saw pink instead of purple? Or what if the author was a bad writer and conveyed the wrong thing? The individual observer was "patently unsuited for objectively comparing one landscape with another and developing terrain analogs."[14]

Instead, terrain features would be broken into numerical values that could be added and compared. "Quantitative terrain descriptions," one report explained, suggested "objectivity" and "rigor." Numerical data that could be turned into contour line drawing maps with bright colors and simple keys (fig. 7.3), like a geographic parallel to the thermal manikin being used in climate chambers to simulate the male body. Scientific shortcuts to mastery of environments as yet minimally known.[15]

Using a color-block style similar to what Wilkins drew by hand, the engineering intelligence units that created the analog maps charted

ANALOG DESERTS ★ 133

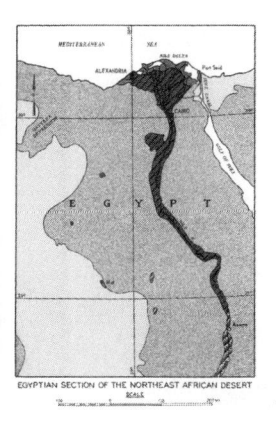

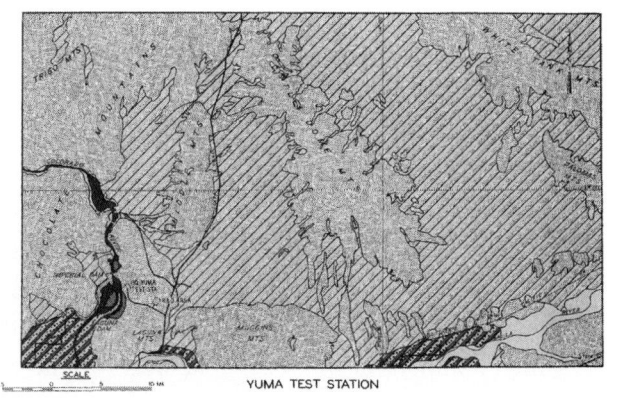

FIGURE 7.3 Analog desert maps. In the 1950s, the Army created maps to try to extrapolate conditions in the world's deserts back onto the Yuma Test Station in Arizona. By defining deserts through a set of characteristics—including soil type, vegetation, slope, and terrain—it was hoped that similar desert areas could be identified for training and testing purposes. These images show vegetation type analogs for Egypt. This was paired with an image of Yuma to highlight any similarities (there were not many). From *Handbook: A Technique for Preparing Desert Terrain Analogs* (Vicksburg, MS: Waterways Experiment Station, 1959).

the world's desert in terms of how similar they were to Yuma. In the map of Egypt, the thin strip along the Nile (red and green striped) was only "partially analogous" to terrain found way down south at Yuma. The large segment on either side of the Nile (yellow and dotted in the color version) was most similar to some of the mountainous terrain in Arizona. This particular map was for soil types; other maps were created for "ground factors," "vegetation," and "geometry of form factors" (including slope and relief), as well as "aggregate general terrain."[16] The Yuma area, not surprisingly, was the most carefully mapped and studied because it was the point of comparison for all other world deserts. That was the way that the military could then parcel out the Sonora, creating mini-Gobi landscape or Saharan wadis that were more accurate and reasonable training grounds than hastily crafted wartime analogs.

In trying to mimic the world's deserts, the engineers ended up learning far more about deserts closer to home than had been intended. No less than three field surveys were carried out in Yuma by 1955, along with aerial missions and literature reviews.[17] When the survey teams revisited a site, they would note changes. Only after all the data was collected and charted could the numerical values be crafted that would ultimately be compared with other deserts, none of which had come under the same level of scrutiny as the Sonora. Yuma became intensely studied so that the rest of the world could perhaps be better known.

In the end, the mapping project told the engineers more about their own processes than the deserts of the world. Even in the Military Evaluation of Geographic Areas reports, they acknowledged the limitations of the work, that local variances might still matter. And when engineers went out to build air bases, pave roads, build barracks in hot and dusty places, the analog maps were irrelevant. As Troxler learned in Libya, no amount of machinery or logistic power could overcome the significance of the smallest things. Place ultimately mattered. And in the desert, the sand mattered more than anything.

The real trouble with analog mapping, the military admitted, was that it tried to avoid the natural world. It was an effort to make the world what someone wanted to study, not how it actually was, kind of like Eglin's giant climate hangar built to test airplanes and engines. At the Florida facility, engineers could throw sand and salt at an engine, turn the humidity up and down, change the wind speed and direction, and so on. It was good for testing specific inputs and measuring output, but what if they needed the whole picture? Out in the real world, a Military Evaluation of Geographic Areas report noted, "nature operates with all of her attributes simultaneously." There was no way to control for one thing or the other. You could select out for four chief characteristics but who was to say those four would always be the most important? All it took was one day when something else happened for the entire system to fall apart. Environments were an amalgamation of "several things acting in concert." A "simple but devastating" reality that would ultimately make the analogs irrelevant.[18]

Strategic and military efforts to catalog environmental features to generate simple categories that could be responded to—albeit categories carefully constructed—ended up concluding at the place desert conservationists like Minerva Hamilton Hoyt and Forrest Shreve would have started: it's hard to predict the natural world. Especially when the "nature" in question was so extreme. Knowing that two deserts matched closely along two characteristics (average slope and average rainfall, for example) meant almost nothing when added to other factors or when—as could easily happen—the expected average was ignored.

New means of aerial surveillance would add to the sense of futility. The analog program emerged out of a particular moment where the whole world became of strategic interest, but the country lacked the technology to see the whole world or even just large swaths of it consistently. To study a desert meant trapsing across it, not flying high overhead; to understand terrain meant mapping it from the ground. High-altitude photography changed that. Satellites would eventually beam back more data about deserts than engineer teams had been able to acquire from airplanes or ground reconnaissance. Earth-orbiting technologies could take in entire sand seas, not just the dune fields that a man might cover in an afternoon. Men orbiting in space, too, could capture bigger pictures and descriptions of the desert.

This was the intent of the 1970s Earth Observations and Photography Experiment, which was part of the Apollo-Soyuz Test Project. Apollo-Soyuz was the first time that US and Soviet spacecraft docked together. Farouk El-Baz, a geologist working with NASA, was bullish on the possibility of astronauts being able to take the first systematic photographs of the southwestern part of Egypt's Western Desert. World War II–era photos existed for the coast, where fighting had been most intense. But until the new photos from space (in the 1970s), there had been no way to see the whole desert. As El-Baz admitted, "Photographs taken from Earth orbit provide a very practical means of studying remote, inaccessible, and unexplored desert regions."[19] With detailed pictures from above, analog maps seemed crude and ill suited to modern cartography and environmental intelligence gathering.

Earth-orbiting technologies clearly did not encourage a return to the local and particular. On the contrary, the big picture remained the norm. Seeing the world in desert hues and color blocks suggested the efficacy of big environmental categories (now seen not abstracted). Strategic planners wondered whether they could read Earth's topography for some sort of clue to what was underneath and to make connections about operations and utility. They would not need to learn about local details at all if they could see the big picture from far above.

El-Baz worried about this very problem. Although he praised the potential of the globe-spanning view, he explained that the individual was vital because the human eye could "often see more color and textural variations than will be recorded on the most sensitive photographic film." Human observers could note things like "color and textural variations"; they could attend to different times of day, shadows, angles, and more. As a result, the astronauts involved in the Earth Observations and Photography Experiment, or EOPE, missions went through sixty hours of training on how to take the best images and describe what they were seeing. For the EOPE program, the selected astronauts sat through classroom lectures on basic geological formations, and then participated in "flyovers," which gave the crew practical experience in observing, snapping pictures, and narrating what they were seeing on the basis of their classroom work. The astronauts were trained to use a handful of tools that they carried with them, including a "World Map Package" and a Munsell color wheel that featured shades of gray, beige, red, and yellow. Even more local expertise was required. The data accumulated from space was then compared with fieldwork being done across the Egyptian desert by teams that took samples, used the same Munsell color wheel, and tried to confirm or complicate what space-sensing was able to show.[20]

Even though largely ineffective, the US military's way of trying to define the world's environments mattered. It mattered because it channeled knowledge about environmental categories that bolstered the projection of American power. Classifying sand and soil types, combining color palettes into desert hues, creating hatch-marked maps that showed similarities, not actual terrain features—these were all part of the same process of trying to imagine the whole Earth in new ways, of grasping for new paradigms and patterns that might make even more possible. In establishing what it believed were the most important categories, the military could try to control the information about them. For soil studies, for example, they were interested in how to move wheeled vehicles across various terrains, not learning about the strange organisms that lived there or what habitats were being destroyed. Strategic needs drove environmental intelligence so that the whole Earth could come to be seen in the simplest terms. It would also make the slippage between the extremes easier, especially as engineering reports and manuals consistently explained that the materiality of arid places was one of the most challenging of global conditions.

Environmental expertise allowed the US military to make claims to knowledge and places. Here was a new vision of US globalism, one that enabled the rapid spread of military bases and facilities everywhere. It was not born out of a sense of environmental stewardship or some academic interest in gathering data about how the world works. It was about projecting power by accumulating and acting on rarefied knowledge that no one else could accumulate in such quantities across such vast terrain. In some ways, it was the Cold War extension of what President Roosevelt had described in his February 1942 fireside chat, when he called for the American people to accept the need that the US military had to create and maintain shipping lanes and air routes that would link the world's allies and battlefields, ensuring the free flow of arms, goods, and information. It is how the Allies would win the war.

Ten years later, the Cold War US government believed the same lines still had to be secured, once again through the US military. Engineering teams moved out across the world to build and fortify bases to ensure the continued free flow of goods and information so vital to US economic and strategic interests. While Cold War globalism is often envisioned in terms of aviation—planes, aerial views, movement—it was just as intimately rooted in places and material realities. The engineers who had to pave runways over desert sands and erect barracks without wood understood that the acquisition of environment intelligence in some places would facilitate the projection of

US power everywhere. US power relied on the continued expansion of global trade, open sea and air routes, and ever-increasing access to foreign markets. And that power depended on how well the engineers knew the material world: the sand, to be sure, but also the snow and ice and permafrost.

II. SNOW

PRELUDE

Two months after he arrived in northwestern Greenland, Henri Bader was stranded on top of the ice. The July morning had started normally enough. Bader and a US Air Force pilot had set off from Thule Air Base's newly paved runway in a twin-engine SA-16 Albatross and landed at a site somewhere in the "north central region of the Greenland Ice Cap," just over two hundred miles east of the base. Bader, a Swiss glaciologist, was chief scientist for the Army Corps of Engineers' new Snow, Ice, and Permafrost Research Establishment (SIPRE). He had been summoned to Greenland by the Air Force to locate potential radar sites that would protect Thule—and his adopted country—from surprise attack.

Bader and his young colleague and assistant, Carl Benson, had spent the previous few weeks skipping about the ice sheet in cargo planes fitted with skis, touching down to dig holes, take samples, and get a sense of each place (fig. P2.1). Benson had recently graduated from the University of Minnesota, where he earned a degree in geology. The two men had apparently unending workdays, because the sun never set in July; it just circled the horizon in an endless loop.[1]

Benson did most of the digging, as Bader had injured his back in a fall into a crevasse while doing fieldwork in Alaska. Once a nine-foot-deep hole was excavated with simple shovels, the pair would read the icy layers, looking for signs of annual snow accumulation and potential signs of seasonal melt. These were both important pieces

FIGURE P2.1 Explorer on the Greenland ice cap. Sir George Hubert Wilkins Papers, SPEC.PA.56.0006, Byrd Polar and Climate Research Center Archival Program, Ohio State University.

of information you wanted to know before trying to build something on top of the ice, but both were things that, until that moment, no one had bothered to measure.

So much remained uncharted about the icy realm on top of the world's largest island. In 1952, no one knew for certain how much snow fell each year, how deep the ice cap was, whether the ice was growing or shrinking, how cold it got in the winter, or really much about how the ice cap had formed or changed over time. Bader was interested in all those things. And while he was sent to Greenland that summer with a specific task—to find the best location to implant radar sites—he and Benson were also making other plans. They were thinking about how Greenland could become a new living laboratory for studies of ice and snow.

On the morning in question, Bader headed off without Benson, who stayed on base. The sound of hammering and sawing, the growling of trucks and hollers of men would have followed Bader to the airstrip. Construction season in the Arctic—a brief summer sandwiched between fleeting spring and fall—was intense and twenty-four hours per day, creating a constant din. Thule's construction was in its second season, and the base would be officially turned over to the Air Force by the end of the year.

From the base, Bader and the pilot headed due east over the ice, crossing over the "marginal" zone where crevasses exposed the bright blue of the ice below. Occasional summer-only lakes shimmered in depressions, cyan blue set in sharp relief against the glaring white. It was a sign of the glacier's complex plumbing system, another feature that would become the subject of scientific investigation over time. In the winter, this frigid landscape would transform into bleak tundra as the bright blues faded to dreary grays.

The aircraft climbed. The ice sheet rose steadily from the coast, peaking at 9,800 feet above sea level. Bader and the pilot were not going to the top, but they were headed 220 miles inland to a place that the Air Force had suggested was geographically ideal for a radar site, at about 6,560 feet above sea level. The Air Force chose the site, called Site II, not from their own terrain investigations, but because it was the optimal strategic distance from Thule and other radar sites.[2]

Most things that morning went to plan. After they landed, Bader evaluated the area by digging and examining. It was sunny and clear, warm enough to allow them to remove their heavy parkas (quartermaster general approved) to dig and work. After Bader's assessment, the two men got back in the plane, started the engines, and discovered that the warmth of the summer sun had trapped them. The wet snow claimed a sticky grip on the skis of the SA-16, which could not shake free of the surface. The men were stranded. They radioed the air base.

That no rescue could be immediately mounted spoke to how little the Americans knew about Greenland, despite their impressive base-building operation. By the summer of 1952, there were plenty of machines and men at Thule; a bakery was operating at full capacity, a laundry kept socks clean and paired. Benson was probably enjoying a meal at the giant mess hall that served hot food on aluminum trays. The year before, the anthropologist Jean Malaurie reported having seen a large pink confection carried to a table as dessert, puffy and wispy and so very unexpected.[3]

Notwithstanding all that had been brought to and left at Thule, there was no clear or safe way to free the men stuck 220 miles away. Clearly landing another plane was too risky. An over-ice operation was impossible. Not only did the Americans lack the right equipment or skills to move over the terrain, but summer conditions on the edge of the ice—where Bader saw the crevasses and lakes—were unstable. What looked like solid ice might prove to be a thin ice bridge liable to collapse under the weight of a person or a sledge. Crevasses yawned open like stretch marks as the ice shifted. Men had been swallowed whole. Lieutenant Max Demorest and his sled disappeared in a Greenland crevasse in 1942. In 1951, two men on a French expedition plunged a hundred feet to their death in a gap no one had anticipated. Their vehicle was wedged down between the icy cliffs and eventually crushed as the crevasse closed in again.[4] As the local Inughuit people understood, and as past explorers had discovered, there were certain times of year when it was possible to cross the ice, and others when it was to be avoided. July was to be avoided.

At any other time in human history, Bader's odds of survival would have been low. But the US military was changing the odds. Even though the men would not be rescued for two weeks, the Air Force could ensure that they did not starve or freeze to death (not uncommon fates in the history of travel to ice caps). Benson organized supply runs. The Air Force dropped tents, food, extra clothing, and communication equipment. The men could be comfortable enough while the military swung toward finding a way to get them home.

Part 2 takes us to the top of the world and to the top of the ice where Bader was temporarily stranded. Throughout the 1950s and into the 1960s, the US defense apparatus turned its focus to the High Arctic, Greenland in particular. If another war came (and there were plenty of men who were certain war was coming), it would likely be fought over the North Pole, it was thought. As the polar aviator and military consultant Bernt Balchen explained, the far north had gone "from being one of the coldest areas" to "one of the 'hot spots' of today and of the future."[5]

As a result, US defense installations spread out from Alaska, across northern Canada, into Greenland, and on to Iceland. As was the case in the world's deserts, the construction of facilities in these places required more sophisticated understandings of the material conditions found there. Part 2 traces efforts in the Arctic as military engineers moved from building Thule Air Base to establishing a research camp, called Camp Tuto, at the edge of the ice cap, and then on to a series of fantastical construction projects on top of and inside of the ice itself. The engineers dug and drilled, scooped and measured, all the while collecting data on the nature of snow and ice and permafrost, always expanding their databases about the material foundations of the world.

★ 8 ★
TOP OF THE WORLD

If there was one place on Earth that could make building an air base in the desert seem easy, it was the High Arctic. That is precisely where Colonel Morton Solomon of the US Army Corps of Engineers was sent in the middle of February 1951. It was bitterly cold, negative thirty-five degrees Fahrenheit. According to the American and Danish weathermen stationed at Thule, Greenland, it was one of the coldest Februarys on record. They admitted, though, that their record was rather thin. The weather station had been in operation only since 1946, and that year they classified as "unseasonably warm."[1]

It was no coincidence that Solomon was set to arrive in Greenland around the same time that Troxler was putting plans in place for Wheelus Air Base in Libya. Both were part of the Pentagon's Cold War buildup, an aggressive drive to contain the Soviet Union through a network of bases. The goal was to deter Soviet incursions by threatening immediate retaliation. While the Americans justified their bases as defensive, that was hardly how the deployment of men and weapons would have seemed to those on the other side of the Iron Curtain—or, for that matter, in Greenland. As Danish authorities explained, hosting a major air base meant that the island itself had become a target in the event of war.

It was not easy to get to Thule. From Washington, DC, Solomon flew first to Goose Bay, Labrador, then over the Davis Strait to Sondrestrom Air Base in Greenland (a wartime facility built by the engineers) before a C-124 took him to the site. Weather conditions often stymied each part of this route, meaning that any one of them might see delays, in turn delaying the next leg and then the next. Solomon's view from the window of his C-124 was of a vast world of ice and snow, dark and vague, tinted blue in the polar twilight. Above the Arctic Circle, the Sun sinks below the horizon for the last time in late November and does not reemerge until early February.[2] Solomon's trip to Thule coincided with the grimmest, darkest season, during which men were rumored to lose their minds with no glimpse of the sun.

Greenland is the largest island on Earth. It is also one of the world's most sparsely populated places. In 1951, only twenty-three thousand people lived across its 2.2 million square kilometers, with most inhabitants concentrated in the southern tip. In 1951, as today, the majority of Greenlanders were Inuit, the rest Danish and European settlers. In the early nineteenth century, Greenland became a Danish colony. That is why up in the area that the foreigners called "Thule," a Danish administrator was stationed to work with the local population, the northernmost group of Inuit, often referred to as the Inughuit. Their ancestors had inhabited the area for 4,500 years.[3]

Nearly 85 percent of Greenland is covered in a giant ice sheet. Engineering documents often described the sheet as three times the size of Texas. No one lives on it. By 1951, few people—Inuit or foreigner—had ever dared cross it. Because of the ice, the island has been called the "white desert," a "bleak and barren" land, "lifeless," the "icy waste."[4] From the air, the ice has been compared to a sheet of homemade paper, buckling and waving with imperfections, or an endless white curtain that wrinkles and folds in on itself. In the 1930s, a polar explorer and naval officer declared the ice cap the "globe's greatest curiosity," monotonous and terrifying at the very least.[5] The explorer wrote that being on top of Greenland was "like standing on the surface of the dead moon, a million years devoid of life, and waiting for a single vagrant meteor to break the spell." That the lunar surface would have seemed a more familiar point of comparison for readers in the 1930s is a telling reminder of how little was, in fact, known about the ice.

Solomon's plane flew over a narrow strip of land that separated Greenland's giant ice cap from the sea. That winter, Solomon was not interested in the ice—not yet at least—but in a small peninsula in northwestern Greenland that he had been told contained a broad, flat valley; it was one of the few places that far north where an airstrip could possibly be constructed.

As Solomon's plane descended toward Thule, there was little to see. The year before, the French anthropologist Jean Malaurie had arrived in the area to begin a yearlong expedition. He hired local Inughuit guides to help him on his travels along the northwestern coast. His aim was to investigate the people who had managed to survive so far north. In his account, Malaurie declared the Thule area "severe, hostile, and pitiless." Solomon may have shared the sentiment. He wondered why on earth the US Air Force had decided such a place was worthy of modern military presence. When he landed, he would have seen a handful of sodden, earthen Inughuit houses, a few wooden structures built by explorers and Danish colonial officials, and the small Danish-American weather station. Could such a place really be turned into a gleaming modern base, capable of billeting thousands of men and supporting military operations?[6]

There were many ways to answer the question, Why Thule? Some stretched back hundreds of thousands of years and are rooted in the environment itself; others are more immediately recognizable. For the US Air Force, the most basic answer was geography (fig. 8.1). Greenland's location made it an important potential barrier and warning station in the event of war with the Soviet Union. The Air Force planned to use Thule as a forward base for bombers, a way station for top-secret reconnaissance flights, and a recovery station for squadrons of airplanes flying over the top of the world. By 1958, it had also become a storage center for nuclear warheads.[7] Greenland was to function as a giant, icy aircraft carrier. So vital was Greenland to early Cold War defense planners that in the mid-1940s the Joint Chiefs of Staff had offered to buy the island from the Danes.

The inclusion of Thule in the expanding infrastructure of the US military signaled how truly global Americans' ideas about the world had

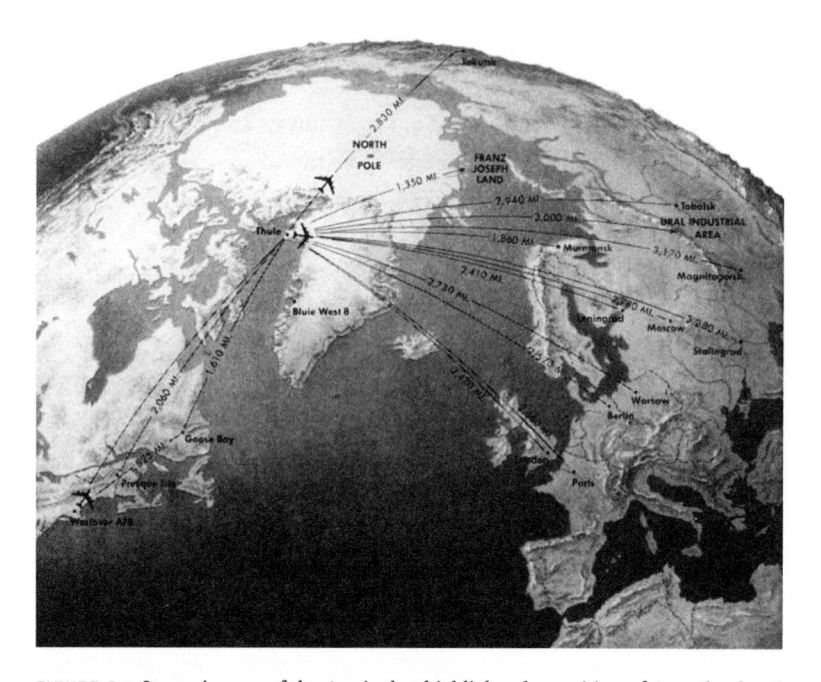

FIGURE 8.1 Strategic map of the Arctic that highlights the position of Greenland and Thule Air Base for US Cold War aims. In the 1950s, when Thule was constructed, bombers had to be stationed close enough to the Soviet Union to be able to reach targets behind the Iron Curtain and return to base before running out of fuel. This map appeared in "Birth of a Base," *Life*, September 22, 1952, p.130; Greenland Map © Estate of James Lewicki.

become. Although Greenland had hosted wartime facilities, the permanence of Thule Air Base highlighted the new nature of US control: a commitment to logistic networks and material knowledge. Greenland mattered both because of its strategic geography and because of its capacity to provide environmental intelligence that few else in the world seemed to have.

While Solomon made his way up to Thule, General Lewis Pick, the chief engineer, had his doubts that the job could be done. There were, he later admitted, so many "imponderables" when it came to building anything—let alone a modern air base—eight hundred miles from the North Pole. The Americans had never done anything like it, nor, did they suspect, had anyone else. The potential exception was, of course, the Russians. And that was part of the problem.[8]

As Cold War tensions grew, the gap between Soviet and American expertise regarding the Arctic seemed to threaten the foundations of peace and security, at least to the small cadre of men who had spent their lives studying such places. Paul Siple, for example, used his position on the Research and Development Board (RDB) to warn that the United States would lose the next war if it did not adequately account for the far north. As a consultant for the US Army in the 1940s and 1950s, Siple would become interested in all extreme environments as they pertained to military preparedness. But his true passion was the polar areas, and so he threw his energies into US planning for the Arctic Circle with abandon.

Siple's colleagues in the RDB agreed with his concerns. "We are lagging far behind a potential enemy," the minutes of a 1948 meeting recorded. Men were fond of spouting the words of Henry "Hap" Arnold, father of the modern Air Force, who claimed that World War III, should it come, would be centered at the North Pole. The Army's research director explained that, because the Arctic was the shortest distance between the Soviet Union and North America, it had taken on "new military significance," especially since the "Soviet Union [was] expending great energy learning how to overcome the natural barriers of the polar regions."[9]

Siple grasped for optimism. If there was a silver lining to the grim projections, it was that the United States clearly had the potential to overcome its weaknesses. Wartime devotion to mastering the world's environments had shown how much progress could be made (even if incomplete) when the nation turned its attention to the right places and problems. That the country lacked expertise in 1950 did not mean it was incapable of acquiring it in the coming decade. Mastering the Arctic was a matter of will. "In ten years," Siple predicted, "if we attack the arctic problems with vigor, it is possible for our forces to be superior to those of the enemy and to be able to use the arctic against his forces."[10] *If* was the pivotal part of that sentence, and it was what Siple and his peers devoted their energies to ensuring.

They had a long way to go. The US government had shown little interest in the Arctic until the war demanded its begrudged attention. But as was the case with the deserts, wartime work had been short-lived and temporary. How could the right lessons be learned in the

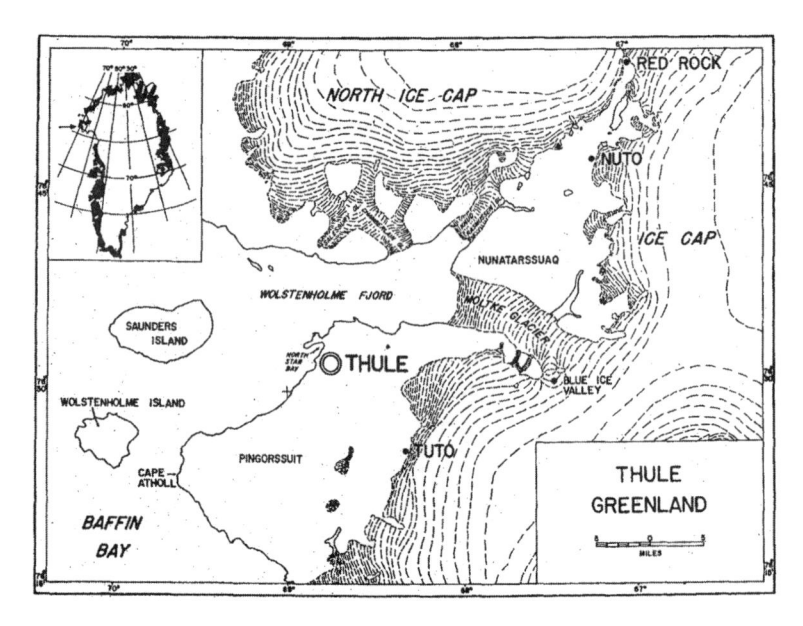

FIGURE 8.2 Map of the Thule area, Greenland, 1950s. The air base is marked by a circle. The joint U.S-Danish weather station was just to the north of the air base, and the local settlement Uummannaq was to the north of the inlet labeled "North Star Bay." From, "Permafrost Tunnel, Camp TUTO, Greenland," SIPRE Technical Report No. 73 (1960), US Army Engineer Research and Development Center.

years after? Solomon would have a deeper—but hardly abundant—bench of expertise and centers to draw on as he thought about constructing an air base inside the Arctic Circle. In 1950, for example, the Corps of Engineers' Snow, Ice, and Permafrost Research Establishment (SIPRE) was formed. SIPRE engineers would soon swarm Thule and Greenland, using the icy island as a site for learning how to master the cold. Indeed, Thule quickly became a central node in how information about the Arctic was generated and disseminated (fig. 8.2). Greenland provided an opportunity to examine so many Arctic processes in a place that was sparsely populated and always cold. Within just a few years, Greenland was a huge, outdoor laboratory, the place where all cold-climate research teams wanted in. But first Thule had to be constructed. And it was to that job that Solomon turned.

The selection of Thule was not just about coordinates on a map;

the US military is more parasitic in its acquisition of properties than an attention to raw numbers and distance to targets would suggest. As in Libya, the military built where something else already existed. If it was going to invest resources and put down the tracks and pathways to such a godforsaken place, then it had to be a place military engineers knew they could work with. US global expansion would be etched on top of places others had been; US strategic needs would thus connect these known places in novel ways.

Ocean currents, ice bridges, and wind patterns were part of that something else. In Greenland, the Americans tended to focus on the land and the ice as they contemplated their own military needs, but Thule came into their line of sight because of the sea. For millennia, the coast had been shaped by an area of ice-free water called the North Water polynya. The first Americans who came to scout the area were not totally ignorant of the phenomenon. In March 1951, for example, a crew flew over the coast and reported open water "up to quite a high point." When he arrived at Thule the year before, Malaurie had explained that "North Star bay opens onto the 'free waters of the north.'" This Arctic anomaly had been particularly important to nineteenth-century whalers, in part because it was the location of the summer whale migration, but also because the whalers could overwinter in the area. It was even more important to the Inughuit, who called it Pikialasorsuaq (the "great upswelling").[11] Once scientists started studying it, they would later describe the North Water polynya as the most biologically productive ecosystem in the Arctic Circle.

A polynya is like an oasis in the middle of a frozen sea. The polynya in Baffin Bay is formed as the West Greenland current brings warm, salty water from the Atlantic up along the coast. The warmer water rises in the northern part of the bay. At the same time, an ice bridge at Smith Sound keeps pack ice from flowing out of the Arctic Ocean and down into the warm water, a scenario that would significantly cool temperatures, like adding ice cubes to a glass of already-cold water, and lead to more freezing. Winds also keep ice from forming. As a result, each year a plankton bloom in early spring feeds a large cod population, a keystone species. Marine mammals such as walruses, seals, bowhead whales, polar bears, narwhals, and belugas feed on the fish. The open waters create an important habitat for migratory birds, which then nest on the cliffs and ledges of nearby landmasses. All these features are essential for human habitation along the surrounding coast of Greenland and Ellesmere Island.[12]

Because the polynya facilitated humans' presence, Inuit communities thrived farther north than most Europeans and Americans would have imagined possible. For centuries, the polynya has provided sustenance, transportation, and connection. The ice bridge at Smith Sound, for example, has been a way for people to travel freely from Canada to northwestern Greenland (Malaurie reported taking the trip in 1951). More recent documentation by the Inuit Pikialasorsuaq Commission, created in 2013 to grapple with transformations to the polynya due to human activities and climate change, emphasizes the food resources that the polynya provides and its cultural and social significance.[13]

US and European explorers came to the shores of northwestern Greenland specifically because of the Inughuit populations who lived there. In 1891, the American Robert Peary traveled with a small team to Inglefield Gulf (north of Thule), where he built a house and began launching expeditions. First, he wanted to map the northern edges of the island (at the time unknown), and later he sought the North Pole. Peary was explicit about his interest working with the locals. His "Peary system" was organized around adopting Inughuit practices and methods, including dog sledges and fur boots. "It is only reasonable to assume," the explorer explained, "that these people, having lived for generations under the severe conditions of the Arctic lands, have evolved the best methods of meeting the needs of their daily existence." Peary would adopt their practices and employ their men, although, as scholars have shown, this in no way should indicate a sense of reciprocity. Peary exploited the area's resources and people for his own gain, including deceptively taking six Greenlanders to the United States.[14]

Knud Rasmussen was next. A Danish Greenlander who had spent part of his childhood in Greenland, Rasmussen followed Peary north to meet the Inughuit. Rasmussen returned again and again.

He returned not only to embark on the numerous Thule expeditions that made him famous but also to establish a trading post that provided funds for his continued Arctic travels and to provide the local community consistent access to the Western resources it had become accustomed to through Peary, things like matches, wood, and metal parts for weapons. Rasmussen was credited with having given the area its Western name: Thule, a reference to what the Greeks and Romans referred to as Ultima Thule, the last of the lands, the White Island, the farthest north. At eight hundred miles from the North Pole, he wasn't far off. In any case, Rasmussen spent decades trading European items for fox pelts and walrus tusks he could sell in Europe to raise funds for his own adventures. The trading post and proximity to year-round goods meant that the Inughuit village at Thule (called Uummannaq) grew in size and importance.

It was Rasmussen who first alerted the Americans to the idea that Thule might be a good spot to land planes, although he certainly did not have a giant air base in mind. In 1926, Rasmussen visited New York City to raise interest and money for his travels. New York in the 1920s was the place to be for a polar explorer in search of camaraderie; the men came and swapped stories in smoky, dark rooms at the Explorers Club or joined the lecture circuit to raise money for their next great trip. In the United States, newspapers and magazines covered the travels of many of the men who became involved in World War II–related research into extreme environments. Richard Byrd's flight across the North Pole, for instance, was lauded in 1926 (though subsequently discredited). Just a few years later, the Norwegian aviator Bernt Balchen, who had supported Byrd's exploits, soared over the South Pole in a modified Ford 4-AT Trimotor. Hubert Wilkins (whom we last met in the desert) made history flying over the Arctic in 1928. And then there was Paul Siple, the Eagle Scout chosen to go to Antarctica who would become the polar everyman.

Rasmussen spent his time in New York like his fellow polar explorers, trading tales. At one point he met Balchen, one of the rising stars in polar travel. While Rasmussen worked with Native people to skate across the Arctic in sledges, Balchen flew over the ice in airplanes. Their combined interests led to a revelation: the Danish Greenlandic explorer informed the Norwegian aviator that there was one good place on the coast of Greenland to land airplanes: Thule.[15] Balchen seems to have tucked this detail away in his pocket, forgotten until he was asked to help win the war.

Fast-forward to 1941. That year, General Henry "Hap" Arnold of the Army Air Forces invited Balchen to join the service to locate and create weather stations and emergency airstrips on Greenland. After Germany occupied Denmark in April 1940, the Americans took over the defense of the island through a security agreement with the Danish ambassador, still in Washington, DC. The agreement granted nearly unlimited rights to establish military facilities. Balchen had to figure out where. The air war had brought the island firmly into the United States' strategic orbit. Planes needed places to stop and refuel; pilots might need emergency landing strips; airmen and commanders in Europe needed better forecasts.[16]

In planning weather stations and emergency landing strips, Balchen drew on the information and contacts that swirled through the networks of polar explorers. Thule made it onto the list of facilities that the Americans would build, one of nearly a dozen, though during the war, Thule was small and less important than the others.[17] The facilities served their wartime purpose and then, much to the horror of aviation and meteorological enthusiasts, were largely shuttered. The US government did not want to staff or pay for all the far-flung outposts, and the government in Copenhagen had little interest in allowing the Americans so much control over the Danish colony.

If interest in specific weather stations was fickle, the need for weather data and forecasting was increasingly obvious. In 1946, a joint US-Danish weather station was quietly reopened at Thule. It turned out that everyone wanted to know the weather, but only the Americans could pay for it. Glommed on top of the wartime weather strip, the new Thule station was a little bit larger and suggested a greater sense of permanence that its predecessor. Technically operated by equal numbers of US and Danish civilian personnel, there were regularly more Americans—many in military uniform—at the site, upgrading the facilities and resupplying the station each year.

The importance of Greenland to weather forecasting was well established. In 1927, the *New York Times* declared Greenland a "weather breeder." During the interwar period, meteorologists were certain that learning more about Greenland might help them forecast weather conditions across the Atlantic and even all the way to Europe. Explorers had described conditions that suggested that the island and its giant ice cap created its own weather that influenced patterns much farther afield. Wartime forecasting and stations had shown what could be done if enough resources were devoted to them. Aviation only amplified the need. "The invention of the airplane has not only made possible the penetration of the arctic," explained a Massachusetts Institute of Technology study linking technology and weather; it has also "greatly increased the importance of meteorological research in these isolated portions of the globe." As a report on Greenland later noted, "Since operations are so dependent on the weather, and in view of the general rapid movement of weather from west to east, it is imperative that every effort be made to obtain all weather information possible, however meager, from stations to the westward of operating areas."[18]

It is difficult for us to grasp today the significance of weather stations in the 1940s and 1950s. Advocates for them wanted dozens, perhaps hundreds, scattered across the Arctic. Today, satellites provide the data that create the models and forecasts we can look up in an instant (at four in the morning on Tuesday, August 16, 2022, it was forty-eight degrees Fahrenheit with a 1 percent chance of precipitation and winds moving easterly at four miles per hour). But in the 1940s, all that data came from the information men collected manually from small weather stations scattered around the world. Without those stations, they could not collect the data.

Here is what that looked like: men deployed to far-flung stations (sometimes just a couple of men, sometimes a dozen) would collect data and send it back to central locations, where men with pencils and maps charted out what was happening in the air above. Every six hours, the men stationed at Thule recorded surface observations. Twice a day, they sent pilot balloons into the sky to get more detailed atmospheric data. The balloons carried radiosonde machines that could take measurements at various intervals as the balloons rose, sending the data back down to the men on the ground. When the balloon popped (usually at around sixty-five thousand feet), the radiosonde machine parachuted back down so it could be reused.[19] In addition to ground-level temperature, humidity, and pressure data, the weathermen could thus also provide systematic and reliable data on the upper air conditions. This information was wired to a central collection station where data sets from around the Arctic could lead to maps and charts and forecasts. Small data points that were individually meaningless became, through accumulation, useful and important. The process was not really all that dissimilar from the way that environmental information was being collected around the world, each point funneling into a greater whole.

The creation and maintenance of weather stations helped deepen the grooves of the emerging strategic infrastructure while extending US interests—military and civilian—out even further into the world. Collection facilities were also part of the global monitoring regime that the United States was putting in place to measure things like nuclear fallout. The concern was not with the environments per se but with taking account of how weapons and materials might influence the course of war. Regardless of the rationale, the construction of these outposts—for weather collection, atmospheric monitoring, military deployments—all became nodes through which information could be gathered.[20] Getting things to and from these remote outposts required unparalleled logics. It was exceedingly challenging to get supplies to the far North (fig. 8.3).

Indeed, up at Thule, shipping and supply had long been the explorer's greatest obstacle. There was just a short window each summer when ships could be set into North Star Bay; Solomon and his peers were narrowing in on mid-July through August as the optimal window but were hoping for a little earlier as well. There was no way to reach Thule by land. Air transport remained minimal, at least until more and better equipment was available at the site. In the summer, the area became a quagmire of mud and mush. In the winter, it was frozen solid.

Because of its existing, albeit primitive runway, structures left

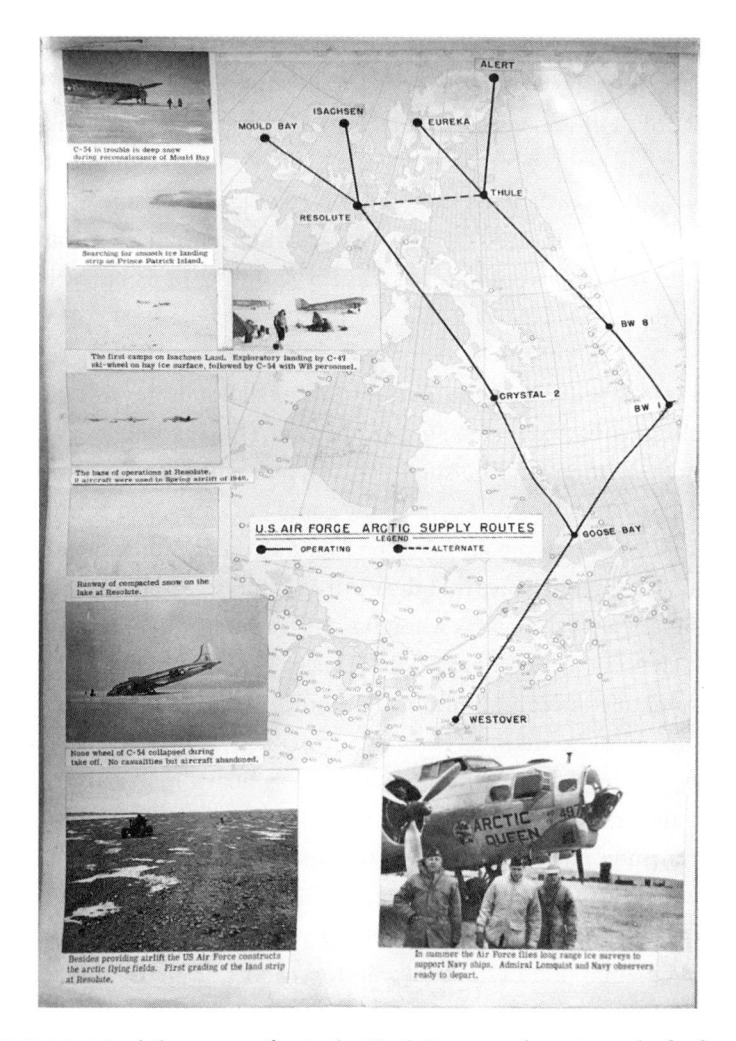

FIGURE 8.3 Logistic routes to the Arctic. US Air Force supply routes to the far-flung Arctic weather stations (including Thule), ca. 1946–1948. The US military was one of the few organizations that could consistently get materials to these locations. US Weather Bureau Arctic Operations Project, 1946–1948, album, Records Group 27, National Archives, College Park.

over from the war, and the knowledge that supply ships could set in once a year, Thule became something of a central supply node for weather stations built in the Canadian Arctic, including stations at Eureka, Alert, and Resolute. The Thule runway was improved and expanded; more barracks were constructed. Despite some political wrangling over who was going to staff the stations and who had sovereignty over the island, the Americans were in charge by the simple reality of their logistic capability. It was later acknowledged that US airmen and military engineers dressed in civilian clothing, not uniform, when in Greenland to make their roles less apparent. So, while technically not military facilities, the weather stations were entirely reliant on the network that was already stunning in its breadth and sophistication.[21]

Being stationed at a remote Arctic weather station is never easy. The problems the weathermen experienced at Thule would be instructive for Solomon and the survey teams, as well as for Arctic operations in years to come. Even if they could get ships into North Star Bay and a few cargo planes loaded and sent off to remote locations, there were myriad ways that the environment constrained what was possible. There were days when the weathermen could barely leave their shelters, which negated their tasks. Multiday radio blackouts were not uncommon when sunspots interfered with communication technologies. Sometimes resupply drops were not able to make it, which is why the small outposts tended to hoard materials. Out in the extremes—even if the US military knew where they were—the weathermen needed redundancy as insurance. The men at the Eureka weather station had four hundred days of supplies, just in case.[22]

At Thule, the supply officer complained about the odd inventory that was amassing in the station's storerooms. In great abundance were liverwurst, canned beef, navy beans, plum pudding, and A.1. Sauce. Too rare were chocolate bars and butter.[23] In 1949, the dozen men at Thule had only two water goblets to use after the rest had shattered. They were drinking out of jars. Products the men were accustomed to at home acted strangely in the extreme climate: evaporated milk separated and would not reconstitute. Eggs did not thaw well. The reports also noted an abundance of sleeping cots. While news of the facility in the United States proclaimed that the men were being well supplied with all they would want, the reality on the ground suggested some kinks and bumps in the logistics. The operation was impressive, to be sure, but hardly the well-oiled machine that was said to deliver their mail and special foods each month.

Solomon spent a week at the weather station during his first scouting mission for Thule in February 1951. He slept on one of the surplus cots and took meals with the weathermen.[24] He certainly queried his hosts about the local environment, about how cold it got, and what visibility was like. There was so much he needed to know. When would the ice on the bay melt enough for the naval ships to arrive? Solomon was interested in the best route from the beach to the runway (which depended on what had to be carried), what sort of ground he could expect in the spring (maddeningly capricious permafrost), and whether there was a ready supply of drinking water (yes, the glacier made sure of that).

The wind tore down off the ice sheet in hundred-knot gusts. The mercury plunged to forty below, the kind of cold that made it hard to breathe and that can freeze eyelashes shut. Frostbite can set in on exposed skin in fewer than five minutes, which meant that many of the tasks he was sent to do—drilling holes into the ground to test the soil, crafting rough sketches of the terrain, pacing back and forth to measure area—became health hazards. For safety, he needed heavy gloves, but he also had to be able to operate instruments. "Working with mitts," a construction worker later admitted, "was like eating grapes with boxing gloves." As many of the men would learn, touching metal with bare skin risked "metal burn."[25] Solomon would not be able to carry out the sort of field tests that wartime engineers had mastered: touching, sampling, and perhaps even tasting the soil. How could he get the soil samples that Troxler was amassing in Libya and sending back to the United States for study? The answer was that he could not.

Solomon traveled with a team of fifteen, a group that included military engineers, Air Force liaisons, and representatives of the private construction firms hired to manage operations (a group called North Atlantic Construction). Like the men in North Africa, both engineering teams had to plot ten-thousand-foot runways and map out areas for modern barracks and waste pits. They had to contemplate which area resources—such as timber, rock, and sand—might be used in construction or, as Solomon was discovering rather dramatically, to identify the environmental barriers that construction crews were going to encounter. Although Solomon was fortunate to have time to survey Thule before construction started, that time was limited. The Air Force wanted an operational base built by that fall.[26]

It had been not even two months since the head of the Air Force had requested that the Corps build an air base in Greenland. The urgency meant that while Solomon and the crew surveyed the ground, back in New York, men were already drawing up blueprints, ordering supplies, and placing advertisements for workers who would be hired in Minnesota and sent north by spring. US construction firms were hired before Solomon had returned from his mission. As was the case for the North African construction programs, the firms were enticed with cost-plus-fixed-fee contracts (basically a guaranteed profit) to enter into what had been to that point a little understood environment. The risks were high, but the payback significant, as they would receive follow-on contracts for work in Greenland and other cold-climate places for decades to come.[27]

When Solomon returned to Washington, DC, after his one-week mission, he had more questions than answers. His questions moved from climate to terrain, from weather patterns to soil analysis. How could they pave a runway on permafrost? How deep was Greenland's permafrost? When would North Star Bay be clear of ice and ready for shipments? Were there any local resources available for construction? Would it ever be possible for men to get work done without freezing their fingers? How much did it snow?

In the absence of good firsthand data, Solomon spent the winter combing through any information he could find about building in extreme cold. Engineering libraries were full of World War II manuals. The Corps of Engineers had built the Alcan Highway ("the Burma Road of the North") in Alaska and Canada, the Canol oil pipeline (considered an engineering feat until it was declared a failure), and a string of air and sea installations all the way out to the tip of the Aleutians. It had also been involved in constructing the weather stations and airstrips in Greenland. The Corps of Engineers had a manual describing that work, titled simply "Historical Monograph Greenland." Published in 1946, the study was intended to be a "brief non-technical history" of wartime Greenland work.[28] Although the study contained no information on Thule (highlighting its ephemeral wartime importance), it provided ample data on the main air bases built at Søndre Strømfjord, Narsarasuak, and Ikatek. The report reaffirmed that Greenland lacked basic construction resources, labor, and transportation networks.

It was a start, but Solomon understood the limitations. Wartime manuals provided a sense of the types of problems his men would encounter, but as Troxler was grappling with in Libya, the temporary and expedient nature of wartime work made it a less-than-ideal model for the 1950s. At Thule, the military engineers had to build a small town capable of supporting five thousand men and their machines for decades. The airplanes were heavier and larger; the machinery more advanced, yet more delicate. Airmen would need a degree of comfort and a sense of place that were not required in a time of conflict: things like plumbing, rec rooms, cinemas, libraries, and more. Because everything from screws to toilet paper was going to have to be imported from the United States, they would need a lot of storage. As Solomon looked around the world for examples, he might have wondered whether he was about to build the first American air base without a baseball diamond.[29]

As soon as Solomon returned to the United States, he began planning for supplies to be requisitioned. It was going to take time to get things ordered and sent to shipping depots so they would be ready for packing and departure in the early summer. The most important wartime lesson was that it was impossible to be too prepared. Indeed, in the years since those wartime debacles, the engineers had spent a great deal of time planning for future deployments and accumulating information about building anywhere and everywhere. It was in that context that Lieutenant General Lewis Pick, the chief of engineers, not only agreed to the rather improbable job of building in Greenland but also understood that it was an opportunity to acquire more expertise about polar environments.

In just a brief amount of time, the blink of an eye in geological terms, Thule had morphed from a seasonal Inughuit settlement to a trading post and then a weather station. The air base would be built on top of what came before, but it would also change what was possible going forward. Thule was just the start. In the network of defense installations set down about the world—aimed at mastering the world's environments—Thule was going to be a big one.

★ 9 ★
ARCTICA

June 11, 1951, was the eighth meeting of the Research and Development Board's (RDB) Panel on Arctic Environments. The twenty or so assembled men huddled around a Pentagon conference table with glass ashtrays strategically placed. They would spend most of the day holed up in Room 3E1074, reviewing the country's Arctic programs and knowledge.[1]

The agenda was long, although it barely touched on what was going on up in Greenland. To be sure, the civilian and military members of the committee knew that Thule Air Base was being planned, and they probably even knew that, in that very week, more than one hundred ships were being packed full of equipment, food, and tools for the long journey north. The men at the table would have been particularly interested in what sort of data the Greenland project might procure about the Arctic more generally. When did the ice pack melt? How far did it extend? Were there ways to predict when it would recede enough for safe shipping? Yet, curiously, Greenland did not make it onto the agenda that month; there was simply too much else to discuss.

The logic of what brought the men to the Pentagon that morning barely needed mention; it hung over everything they did. The threat of the Russian bear looming over the United States, ominously clambering over the North Pole, obsessed the men on the Panel on Arctic Environments (fig. 9.1). "Operations in Arctic and sub-Arctic

FIGURE 9.1 Wilkins with map of the North Pole. World maps centered on the Arctic became common in the 1940s and 1950s as the Cold War turned defense planning to the north. Sir George Hubert Wilkins Papers, SPEC.PA.56.0006, Byrd Polar and Climate Research Center Archival Program, Ohio State University.

environments are nothing new to the Soviets," the panel's technical capabilities report had opened, emphasizing Soviet abilities as a way to highlight US inadequacies.[2]

The June 11 meeting was intended to set out what the RDB (and therefore the country) did not yet know and to establish a course for how it might be discovered. The meetings of the panel were no doubt long and tedious. The first task was always to read and approve the minutes of the last meeting; each meeting ended with a discussion of remaining "unsolved problems." It was a list that never shrank, just grew. For instance, a recent training exercise, Eager Beaver, aimed at hastily paving runways in the Arctic, led not to conclusions but to a cascading series of new problems. The panel ultimately did not learn whether runways had even been successfully built; instead, members heard about the need for better ways to deal with foggy windshields, inadequate drilling techniques, and insufficient training for technicians. Finally, there was the rather bleak question about whether there existed a simple and fast way to determine the thickness of ice. Too many vehicles had crashed through and been lost while the engineers tried to build runways.[3]

If every excursion into the Arctic raised operational questions, few critical questions followed. No one asked whether a program to land planes on ice runways was actually needed or worth it; no one wondered whether all the ice and snow samples being compiled might be used for nonmilitary studies. No one raised questions about simply dumping materials—including those vehicles that crashed through the ice—out in the wild. Nor did they discuss the potential impact of military activities on local populations. The assumption was that the answers to these questions did not matter. What mattered was the list of "unsolved problems" that continued to grow, a logic that pushed more and more resources to the Arctic. Any concerns voiced were about access to resources, not hubris and waste. No one in the room was willing to point out that perhaps a massive base on the edge of the ice was not necessary or might cause damage beyond what they could then perceive.

Samuel W. Boggs was a great fit for a committee concerned with environmental ambiguity. Boggs was not in the military, nor was he a man particularly interested in the extremes, but he had a penchant for mapping what was unknown, for creating visualizations that highlighted the degree of knowledge about a place, not the place itself. Boggs was not some crank whose ideas would be dismissed as idiosyncratic intellectualism. Since the 1920s, Boggs had been a mapmaker for the State Department, and by the time he showed up as an affiliated member of the Panel on Arctic Environments, he was a cartographer of some renown. A contemporary explained that Boggs was recognized "as one of the world's leading experts on international boundaries and also as an authority on the subject of map compilation, editing, and cataloguing."[4]

In 1949, Boggs made a sweeping claim that the world needed an "atlas of ignorance." He explained that "for the first time in history there are world problems," yet the world lacked adequate expertise for dealing with them. Boggs argued that the best starting point for mastering environments everywhere, all at once, was to have a clear sense of which places and aspects were already known. By creating maps of what was not known, researchers would be motivated to find out more. Ralph Bagnold, after all, had pressed out into the Sahara because his imperial maps ended with phrases like "edges of dune field unknown." Filling in maps—showing knowledge—that was a clear imperial project. In the 1950s, knowledge itself had become the new terra incognita, the new terrain to be conquered. The engineers might have a simple topographical map of Alaska's Seward Peninsula, but that did not mean they knew about the flora and fauna, or the layers of permafrost down below, or other environmental hazards. As Boggs pointed out, to know what you did not know so you could make sure to know it, that was power.

Boggs understood that people believed in the authority of maps. But as someone who had devoted his professional life to the trade, he was uncomfortable with how well maps could account for the scope of knowledge needed. Most maps, Boggs explained, perpetuated a "delusion of adequacy and completeness," preventing the reader from understanding that a great deal of data remained obscure.[5] They tended to smooth over uncertainty, completing lines and filling in color blocks even if the exact contours were not resolved, cultivating an illusion of authority and thus obfuscating what remained unknown. Had he known of the Military Evaluation of Geographic Areas maps, Boggs would likely have been horrified—color blocks based on abstracted data and comparison were not what he had in mind. *Cartohypnosis* is the term that Boggs came up with for the mistaken sense of accuracy one had from looking at a map made with incomplete or imprecise information.[6] What if authoritative maps convinced people of a world that did not actually exist? There had never been a complete map of the world, Boggs declared.

The ice that covered the Arctic was exactly the type of ambiguous and mysterious subject that Boggs wanted better charted. In fact, it was "problem 1," for the RDB's Panel on Arctic Environments. Knowing when seas were passable was important to mobility. Local data were all that had been necessary for small communities living in places such as Thule. The Inughuit needed to know when it was safe to move across the ice and along the edges of the sea. But the United States wanted global information not only to dominate certain places but also to connect the world. When and where ships could pass was strategically and commercially important. The work of the RDB was tied up into those intertwined concerns, and civilian and military members made it so.

For Boggs and his peers, they knew it might be possible to create a map of pack ice from the best available data and to hand it out to naval captains to help them navigate the seas. But if that data were collected from only a few locations over a short period of time, it might be worse than having no data at all. An incomplete map might suggest that in early June there would be no ice on the water in a place such as Baffin Bay, for example, even though the ice was often so thick that navigation was impossible. Or perhaps incomplete data about when an inlet refroze in the fall would lead to ships being iced in, frozen in place where they were ill prepared to stay. The ghost of wartime aviation engineers like Brigadier General Donald Davison lurked in the background; Boggs and his peers knew what incomplete information could do.

Boggs's atlas of ignorance captured the conundrum of global environmental knowledge at midcentury. Not only had new technologies made the world more interconnected than ever before; they had also opened up new scales of thinking about Earth. There was an emerging consensus on the need for more global knowledge and awareness in the 1950s and 1960s. Boggs acknowledged that postwar international agencies needed more refined data to deal with globalized issues around agriculture, health, poverty, and more. In the 1960s, soil scientists working through the United Nations began a twenty-year project to classify and map the world's soils, creating new ways of thinking about the global environment in the process. US development programs, too, were deeply implicated in trying to "master" various environments so they could be manipulated.[7] But in the end, only strategic programs had the truly world-spanning capacity to see it through.

For military planners, the Cold War demanded that the whole Earth be immediately known, while air-age globalism required knowing Earth in all directions. Even the "fourth dimension, time" was vital. Boggs explained that to know the whole Earth, they also had to go back millions of years in geological time and forward to predicting climate and weather. This was a lot of information to accumulate and process. Early Greenland missions had been interested in something similar, in digging down into the ice to think about past climates and sending balloons into the sky to predict the weather. But Boggs was suggesting something on a far grander scale, and in the 1950s, the US military was the only organization that could even approach the magnitude of what Boggs was gesturing toward. Rather than thinking about specific places in relation to local environmental issues—for instance, how to grow rice or stop flooding in a particular region—the US defense apparatus had the capacity to grapple with planetary cartography.

In an article, Boggs listed thirty different topics to be addressed and charted, all vital to global environmental knowledge. Some dealt with political and social issues—such as population statistics, history, folklore, and language families—but the majority centered on the physical sciences, with the biological sciences a close second. Boggs's topics, in fact, tracked closely with the RDB's own list of subjects that required greater attention. For example, basic topographical data had to be gathered; so, too, did data on global geology, mineral resources, soils, and water. Boggs called for maps that showed where there was not enough information about ocean currents, tides, the ionosphere, and magnetism. Meteorology was a big one. Boggs highlighted the large patches of the world where weather data were insufficient. This included the polar regions, much of the world's oceans, and large parts of the Southern Hemisphere. Without the data, Boggs explained, weather could not be truly understood. They needed local data to make global processes visible, and they needed global processes to understand what could happen in a particular place at any given time.

Whether or not driven by Boggs's call, the Panel on Arctic Environments took on the problem of ignorance with fervor. There was so much to map and to know, starting with the most basic question of all: where was the Arctic? Surely it was not simply a circle inscribed across the far north. As was the case for the hot desert areas of the world, defining relevant terms was a crucial undertaking for the Panel on Arctic Environments. Just like engineers were defining and classifying soils and sands, everyone needed a shared language and set of transferable tools for use around the world. The stickiness of local terms or professional discourses would only hinder globe-spanning knowledge.[8]

The answer to the question, What is the Arctic? was important in on-the-ground, practical terms and for organizational clarity. Was *Arctic* the right geographic designation for the RDB's panel? At one point, the idea of using *polar* arose; at another, *cold climate*.[9] Would *northern* be better? The encyclopedia that the board had already commissioned was going to be called *Arctica*, a possibility that linked the polar North and South.

The Air Force's recently reconstituted Arctic, Desert, and Tropic Information Center (ADTIC) lamented that the traditional meaning of the Arctic Circle—as inscribed by lines of longitude—had become irrelevant, because it "does not coincide with any of the natural boundary lines occurring" that reflect "variables such as climate, vegetation, permafrost, and sea ice."[10] Definitional problems were compounded on the ground. What was the difference between firn and névé? Taiga and tundra? Fast ice and sea ice?

The question of definitions got to the heart of the concerns Boggs had highlighted. On the one hand, without the ability to dole out money, the RDB panels could claim a sort of intellectual and lexicographical authority. At the same time, an agreed-on language was necessary to be able to move and operate across the world's different terrains. Upon the firm recommendation of the panel, passed through the general committee, the Department of Defense allocated funds to create the Roster of Arctic Specialists, an *Encyclopedia Arctica* that might run as high as eight million words, and a detailed set of bibliographies that would be pitched to particular military needs, such as Arctic construction, mobility through ice and snow, survival skills and training, Arctic culture, and more. In bringing this material together, the Panel on Arctic Environments would become "a true center of polar information."[11] It would also help foster conversations across regions and areas that shared these environmental features.

Through the ADTIC, the Air Force took the lead in crafting the glossary of Arctic terms. In 1955, the ADTIC published a compendium of over four thousand terms that included not just geographic and geological features, but also cultural and botanical ones, and more besides. The glossary cross-referenced the terms that different individuals and groups used for the same thing. People did not always need to share the same language—either literally or figuratively—but they had to be able to translate what everyone else was talking about.[12]

Crucially, the US defense apparatus was defining terms and sifting data. It was through a strategic lens, even if not always direct strategic programs, that it would frame the Arctic.

The case of permafrost highlighted how the military might deal with a largely alien concept and accentuated the considerable pitfalls associated with wrestling a nascent and scattered scholarship into a coherent vocabulary. *Permafrost* was first used in a 1943 Corps of Engineers' report to replace the expression "permanently frozen ground." The entire topic remained classified until after the war when, in 1946, the term was declared too simplistic and misleading.

A brief look at some of the conversations about terminology reveals the stakes. The Harvard geologist Kirk Bryan proposed *cryopedology* and a new vocabulary for what he termed the new "subscience" of frost action and frozen ground. In Bryan's lexicon, *congeliturbation* would replace *frost-heaving*; *congelifraction* replaced *frost action*. A 1948 symposium on the subject in the journal the *Military Engineer* confirmed that Bryan's terminology was "strange, and seemingly awkward," but it otherwise drew few concrete conclusions. In 1949, the RDB discovered no fewer than twenty-five different programs looking into permafrost, with no coordination between them. The results, the board noted, were "not encouraging." There were two things everyone did seem to agree on: the Soviets knew more about permafrost than anyone else and the importance of getting to know permafrost and its capricious nature.[13]

For Colonel Morton Solomon and other engineers sent to work on Arctic projects, they knew that Thule was going to be built on permafrost, but a multitude of questions were still unanswered. Was the ground in Greenland like frozen ground in parts of Alaska?[14] How much of the ground thawed in the summer? How deep was the permafrost? In some places, the ground was frozen for thousands of feet.

Permafrost is a layer of ground that has been continuously frozen for at least two years, although most of the world's permafrost is far older than that, having been frozen for hundreds of thousands of years. It has been described as one of the "weirder concoctions of the Earth's Ice Ages."[15] The frozen layer can be a mixture of soil and frozen water; ice that forms in wedges, needles, veins, or any number of arrangements depending on the ratio of water to other materials in the ground. Permafrost can extend down into the earth a few feet or thousands of feet. It covers nearly a quarter of the Northern Hemisphere.

The "active layer" is the layer of soil that rests on top of permafrost. This layer seasonally thaws and freezes, sometime disguising the deeper permafrost below. It is the dynamics of the top layer that leads to some of the odd geographical feature of permafrost lands: trees listing this way and that ("drunken forests") and polygon depressions across terrain that open in the summer and fill with water. This is all because as the top layer melts, there is no place for the water to flow because the surrounding ground remains frozen tight.[16]

Permafrost is a shape-shifter, which can make it a befuddling construction medium. It is also why highly local information was important to knowing it. Aerial photos and mapping might have been useful for identifying certain features of permafrost—location, continuity, depth—but if the engineers wanted to work in the stuff, they needed to know what it was. This is what most seriously concerned Solomon and other engineers building in cold places. Not all permafrost is the same. When frozen, permafrost is harder than concrete, because the rock, soil, and debris frozen below has solidified into a solid mass. But if disturbed, permafrost does not always stay frozen. What once seemed solid ground can thaw. These sorts of changes underground wreak havoc on things built above. It might take a season or a few years, but things built on permafrost often list and slump, and sometimes cleave in two. Human activities can alter Earth's thermal equilibrium. By midsummer a building might list to one side or begin to tilt in the corner. Often, the effects of permafrost cannot be known until it is too late.[17]

Postwar Alaska was full of examples of wartime structures literally going sideways, signs of what the RDB rather benignly called "special engineering problems." The post office in Nome, Alaska, for example, was settling unevenly due to its permafrost foundations, which had not been properly considered when it was built. "Standby crews" of men were on call day and night along the Alaska railroad because the tracks could sink or slide at the rate of a foot an hour."[18] The Alcan Highway, built to ferry supplies from the lower forty-eight up to Alaska, was buckling and rolling in certain places.

To make matters more complicated, military engineers working in the Arctic and sub-Arctic were coming to realize that not all permafrost was the same. The type of soil, rock, and sediment mixed in with the ice mattered. The size and shape of soils and granular materials mattered, too.[19] These data were important for understanding the rate and depth of thaw, which, in turn, were vital to getting foundations right for building there. No wonder the RDB and the Snow, Ice, and Permafrost Research Establishment had called for maps of global permafrost conditions; not just maps that read "edge of permafrost," but maps that showed things like depth, age, temperature, consistency, underlying material, and more. Boggs, of course, would have advocated for maps that showed precisely where all of these data were unavailable so that engineering teams could devote their time to collecting data and training there—not to training in places they knew, but mapping out places still unknown.

The expanding scope of US military plans for the far north demonstrated how important these data were going to be. In addition to facilities in Greenland and across Alaska, the Pentagon was eyeing partnerships with Canada to build more weather stations, test sites, and a string of radar installations, called the Distant Early Warning Line. Rising abruptly from the icy lands, these giant radar installations and a smattering of support facilities required great care in construction precisely so they would not list and buckle. Wonky radar would provide bad intelligence. The stakes were all the higher given the expansive reach of the US Armed Forces and the tendency of military installations, once constructed, to become gateways to even more construction and experimentation. If building a military base required a lot of data, then maintaining and expanding the infrastructure demanded even more detailed environment information and know-how.

The defense apparatus's interest in permafrost was obviously angled toward strategic needs: how to operate effectively in and through the stuff. Permafrost could even be a strategic asset, a secure place to dig down into and bury things. As a result, Corps of Engineers' reports were pitched to that audience most directly, but commercial interests were attended to as well. Corps files were meant to be shared, a public good for companies and private interests that could benefit from research funded by the US government. This was not only so US companies could be hired for military construction but also so they could press out around the world with broader commercial interests: building pipelines, roadways, even new factories in places where none had been before. Corporate influence would track along the edges of strategic power, a symbiotic relationship meant to amplify US security needs.

It was with these joint interests in mind that the Corps of Engineers excavated a permafrost tunnel outside of Fox, Alaska, in the 1960s. The engineers had experimented with a permafrost tunnel at Thule: they excavated a shaft back into the frozen ground where they fashioned furniture out of "perma-crete." But the facility in Alaska was far more accessible and convenient for military and research needs. The first tunnel was excavated straight back 120 feet into the permafrost and used to study Cold War concerns: how to excavate and work with the odd material. Samples were taken from various places and depths and examined for tensile strength and creep. What were the geotechnical qualities of this odd stuff? Could permafrost be used as a shelter to insulate people from conventional explosives or nuclear bombs? What were the best ways to extract resources from down below?

Decades later, in the early 2000s, the tunnel was expanded and revamped. The type of science and research shifted as well. Early interest in engineering processes has been supplanted by "studies focused on climate change and paleoclimatology."[20] Today, that tunnel is used for important permafrost studies because it is now understood that, in addition to being an odd material, permafrost is vital to climate regulation. For millennia it has captured and stored carbon and methane, both greenhouse gases. No one in the 1950s and 1960s would have noticed these connections because the men reading permafrost samples and layers were caught up in the strategic apparatus. But as Earth's temperatures rise and vast fields of permafrost begin to melt, those materials are released back into the atmosphere, leading to dangerous feedback mechanisms of more warming, more melt, and so on. Scientists are also interested in understanding how the melt of underground permafrost will affect aboveground structures and landscapes, how life will change in places like Alaska as the ground liquefies. Important work, to be sure, but why did it take so long? Until well into the twenty-first century, strategic funding animated permafrost research, limiting the sorts of questions asked and the information found.

"Reliability maps" seem to have been the Corps of Engineers' response to the tension that existed between the need for global capabilities and the reality of limited intelligence (fig. 9.2). Throughout the 1950s, the Corps of Engineers created hundreds of terrain assessment documents for areas of strategic interest. These included deserts, to be sure, but also increasingly areas in the far north. Alaska, for example, received increasingly more attention in engineering documents. Engineer Intelligence Guides were prepared to provide the "systematic presentation of data on terrain and construction material" in areas "of military significance." The goal was to have handbooks that engineers and others could use when and if they had to deploy to a specific place. In addition to area-specific studies, the

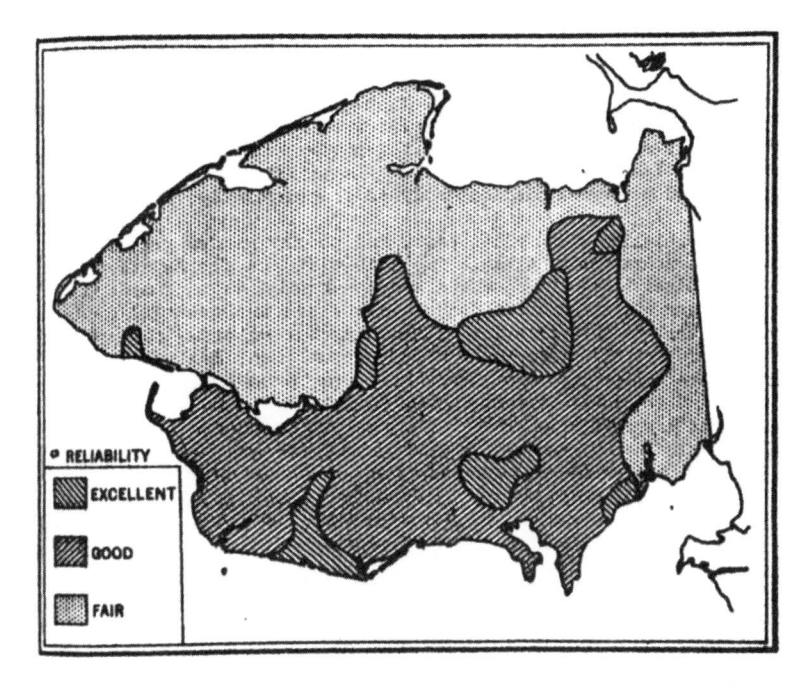

FIGURE 9.2 Engineer reliability map, for Seward Peninsula, Alaska. The Corps of Engineers created maps such as this to show the extent of what was known and how reliable information was at the time of printing. This was part of an effort to better understand far-flung places where military engineers might, some day, have to build. Records of the Office of the Chief of Engineers, Records Group 77, Engineer Intelligence Studies Reference Library Files, National Archives, College Park.

Corps also created general interest documents, such as a glossary of natural terrain features and the *Collection of Information on Coasts and Beaches*, and studies on specific types of construction such as "underground installations" and "precast concrete construction in the Soviet Union."[21]

Given their work manipulating materials, engineers would likely have appreciated Boggs's atlas of ignorance. It was far better to show up in a place knowing that one knew very little than to show up assuming one knew a lot. The reliability maps that began appearing in terrain documents were aimed at that sort of transparency. For example, a 1959 report on the Seward Peninsula of Alaska printed a reliability map. The efficacy of the study, the report explained, "depends largely on the adequacy of the basic data and the validity of interpretations and extrapolations from them." Although part of the peninsula was well examined (see "excellent" on the map), others were barely known at all. For the "fair" areas, the only intelligence had come from aerial photos; useful, to be sure, but not always reliable. The reliability map for the Denali area was slightly more refined, with four categories instead of three. Such maps were not to preclude people from using the data, but they were meant as a caveat to how certain the content was. Perhaps, as Boggs would have intended, the "poor" or "fair" categories would be an invitation for ever-greater study and refinement.[22]

It was going to be hard, time-consuming, and expensive to master the world. Conquering the extremes was iterative, with lessons incrementally accrued and piled on as more and more absurd things were proposed. As engineers and experts grappled with the conundrum of how to map every place without knowing everything, the limits of what could be done began to become clear. Rather than hide from them entirely, engineering centers and reports gradually began to accept uncertainty, incorporating it into how they assessed and presented the world. Highlighting what was not yet known was not a sign of defeat but rather an invitation.

★ 10 ★

PITUFFIK

There was no advance warning. One day the sky opened up and the Americans came down (fig. 10.1). "Thousands and thousands of Americans," reported Odaaq, an Inughuit hunter who lived at Uummannaq, the village near Thule station. And that was just the beginning. They kept coming, Odaaq marveled, "from the sky every day."[1]

It was mid-June. The ice was finally melting from North Star Bay. Cracks in the ice zigzagged out from the shore, hints of the first pathways that would lead to open water. Kayaks could soon move along the channels to reach the seals sunning themselves farther out on the edge of the ice. Many of the Inughuit had just returned to Uummannaq from their winter camps and hunting trips, joining their families who had stayed there for the lean season.[2]

Odaaq would not have been surprised to encounter foreigners. After all, since 1910, a rotating crew of Europeans had operated a small trading post there. Indeed, Odaaq had worked with many of the explorers and traders who came to the northwest. He was something of a legend, even to the Americans who came to scout the area for their giant military base. Although they seemed to know little about the coastal environment, they noted the presence of "Oodaq" or "Ootaw" who lived at the Thule village. They knew him as the man who had helped Robert Peary and Knud Rasmussen move across the ice and snow.[3]

FIGURE 10.1 A Greenlander with dog sledge looks at Air Force radar installation, Thule, Greenland (1966). NF/AFP via Getty Images.

But this time, Odaaq understood, things were different: the numbers were staggering. It was an invasion. "You can't count them," he told Jean Malaurie, the French anthropologist who had been living and traveling in northwestern Greenland all year. It is through Malaurie that we have some of Odaaq's comments and local Inughuit reactions to the base.

The French visitor was stunned at how quickly the Americans could rearrange the natural world. Malaurie noticed the smell; a hint of something sour and harsh, the odor of leather and oil. A "great yellow cloud" rose over the valley, he wrote, as engineers excavated and relocated materials that had been deposited millions of years before.[4]

Colonel Morton Solomon also returned to Thule in June, and, like Malaurie, understood that the Americans were to forever transform the area, but their agreement would have ended there. Malaurie lamented the destruction of the Inughuit way of life; Solomon marveled at the arrival of modern security needs. With blueprints in hand, Solomon could squint and see a modern facility rising from the seemingly empty land.

Malaurie reported that rumors were swirling. The locals whispered that the Americans had brought the atomic bomb (not true, at least not yet), that their canned food was full of salt (partially true), that the Americans planned to melt the sea ice so there would be no more winter (not true, although they did try to bomb it into breaking and melting), and that dozens of their ships had sunk out at sea (not true, although they were stranded and waiting for the ice to melt). The noise was incessant. Malaurie wrote of churning and grinding, of tires spinning and engines roaring. The Americans were going to bulldoze the area into a new and unrecognizable place. What was clear to Malaurie was that given the size and scale of US plans, nothing would be the same again. "They don't want to explore the Pole anymore," the French alpinist Robert Pommier told Malaurie. "They exploit it."[5]

Watching the construction of Thule, Malaurie was more precise still. The Americans, he wrote, were each "engaged in a definite job aiming at efficiency and result. They worked, ate and slept, and nothing else." The place they had entered mattered less than the task at hand.[6] Most of the men did not bother to wander over the ice of the tundra to peer to the east over the ice cap that created odd weather patterns. They worked at moving the land, they ate in hastily assembled mess halls, and they slept on boats anchored in the bay. They came with the spring thaw and left just as quickly before polar night set in and the bay froze over. But they always left their gear behind, which meant they would be back to do more rearranging.

The US military project was a curious thing. Malaurie reported that the residents of the village found it especially odd that there were no women. Nor did the Americans seem interested in the resources that the ocean provided. For the first time anyone could imagine, groups of men were coming to the far, far north who seemed to have no interest in the far north at all.

There was truth to Malaurie's interpretation. The Americans invaded Greenland despite its resources, not because of them. To avoid the very Arctic battles that kept men like Siple up at night, they carried out a battle of their own. Their weapons were strategic plans, bulldozers, and a construction team of five thousand—all driven by a sense of urgency written into exceedingly tight deadlines. The engineers were going to overcome the small piece of northwestern Greenland they had been told was called "Thule." They were going to press and pull that bit of land into an air base where a man could get a hamburger. There was no effort or interest in learning from local practices or attending to the oddities of the Arctic environment. Those were irrelevant to the job unless they interfered directly with it.

It would be easy to imagine this as a story that fits the mold we know well: an outside power arrives to rearrange a place—to modernize it—only to discover that it needs local expertise to get the job done. That is not the story here. The US defense apparatus was overwhelmingly successful in turning a glacial valley into a modern base that has been in operation since 1952. In 2023, in fact, the facility was renamed Pituffik and given a new mission as a US Space Force base. This was a success anchored in a logistic network that seven decades early was brought, improbably, that far north.

Success is a relative term, of course. The US strategic project has been violent from the start. Even the research programs that have spun out in its service must be seen as such. Appropriating land for military bases, labeling terrain as "empty" and "useless" so that it could be trampled and desecrated in the name of strategic interests, has had tremendous consequence, borne most immediately by local people and places. It is a story that was repeated around the world in the 1950s and 1960s as the US defense apparatus spread out.[7]

Thus, there is no surprise ending to the story of Thule. The Inughuit were dispossessed of their settlement at Uummannaq. Odaaq and his community were forcibly removed to a site sixty miles to the north. The US military's view of the area prevailed. US military activities in the Arctic, from training exercises to research missions, have left irreparable scars on the region from the literal dumping of materials to the longer-term impact on the world's climate. This, too, is the story of Thule.

In early 1951, the ads started appearing in newspapers across the upper Midwest: "Construction Engineers, Construction Superintendents" needed for "work outside the Continental Limits of the United States in a Very Cold Climate." In the largest type of all: "TOP SALARIES." Another ad explained that if a man between the ages of twenty-seven and fifty could answer three simple questions with a yes, a job was his. "Can you take cold weather?" "Ready for a long trip?" "Do you want quick or big money?" More than twenty-five thousand men replied in the affirmative. Harold "Oakie" Priebe was one of them. He admitted that all he knew of the job was that it had a high "wage scale and that it would be cold." He figured the job would be in Canada. Others guessed Korea.[8]

It appears that Malaurie was not wrong in assuming that the men coming to Greenland had no interest in the island at all. The five thousand men who signed up to work on Operation Blue Jay, the code name given to Thule's construction, committed to the project before knowing where they were going. Many would not know the location until they loaded onto ships or planes and were handed cold-climate clothing.

Of course, not knowing did not make the actions of the Americans any less significant to the locals. Malaurie reported that the pastor at Thule, a native Greenlander who was "usually very calm," floundered when he realized the magnitude of American plans. Scratching figures on a chalkboard, he tried to calculate the value of all the stuff arriving. Hundreds of thousands of dollars, surely—no, millions. In the end, it was more than $200 million, far more than Denmark had spent on Greenland during centuries of colonial rule.[9]

Had they known how much the workers were making, they may have been even more stunned. People in the United States certainly were, which is why applications poured into Blue Jay hiring offices. Workers were being promised nearly $2,000 a month (roughly $18,000 today). In Minnesota, where much of the recruitment and training for Blue Jay was centralized, carpenters made $2.12 per hour. On Blue Jay, they would make $3.70. That topped even the pay scale for Alaska, which was around $3.14, or for North Africa, which averaged $2.45. Malaurie ran into one man who was making $5 per hour counting cars.[10]

Seasonal work was not unusual. Back in the United States, there was nothing odd about construction workers following projects from state to state as the weather changed. In the 1930s, for example, Priebe had fled Oklahoma and the Dust Bowl (thus his nickname) and "hoboed" to California seeking work. He became a water boy on a Kaiser project and moved up from there, following the next major dam or construction project anywhere the work took him.

Most of the five-thousand-strong workforce would come via ship later that summer. As a construction supervisor, Priebe arrived as part of an advance crew in April, when the ground was still hard and winds particularly fierce. His crews were to help set up additional tents and facilities that would serve as mess halls and medical facilities for the thousands soon to follow.

Priebe arrived just about the time that the number of men at Thule began to outpace the supplies at the weather station. He reported that construction crews had begun to run out of food. Regularly scheduled flights continued to land on the airstrip, but they were all packed with construction equipment. "One plane," he explained, "met eagerly by hungry men, was filled from one end to the other with empty Dixie Cups." There was no food anywhere in its hold. One of the construction workers "actually cried with frustration and anger."[11]

Early logistics were wonky, and it took time to iron out the who, what, where, when, and how of moving things from point A to B to C through lots of intermediaries. Once in place, it became difficult to imagine dislodging what had been built. Still, even when the US military apparatus knew where they were, work was difficult in the capricious extremes. Interestingly, despite their hunger, no one that winter seemed to contemplate doing what the Inughuit had long done: hunt and forage for food.

The place that the US engineers called Thule Valley was known to the Inughuit as Pituffik. The valley was broad and flat, about three miles wide, sloping gently up toward the east and to the ice. It had been carved out of rock by the grinding, scouring movement of the ice as it pulled back, leaving a small glacial trough nestled between two bedrock ridges that the Americans unimaginatively called North and South Mountains.[12] Solomon was told that it was one of the few good sites in northern Greenland to place an air base. The rest of the coast was jagged cliffs and glacial ridges.

The Inughuit tended to cross the valley in the winter, when the ground was firm. It was when they set and checked their fox traps. Fox meat was often consumed in the winter when food caches ran low and little else was to be found. But the fox was most important as a commodity to trade at Thule station. Furs could be exchanged for luxuries like tobacco and coffee.[13] Uummannaq, the town that Malaurie called the "capital" of the Inughuit, was just a few miles up the coast, past the weather station, along the curve of an inlet, at the foot of Mt. Dundas, the flat-topped mountain that was the most recognizable and most photographed feature of the area (fig. 8.2).

The Inughuit avoided the valley in the summer, for reasons the Americans would soon discover. Spring turned the ground into a squishy morass as the active layer of soil thawed and left soggy puddles. The telltale polygons of permafrost dimpled the valley like a honeycomb lattice created by ice wedges that angled down to unknown depths. The Corps of Engineers described the ground between the wedges as coming to a "boil"; it was hardly the ideal place for a major construction program.[14] It would be hard to move across such ground, and not even light sledges could manage. Already US military trucks were getting stuck in the mud, tires spinning. The "muffled grinding of ceaselessly turning motors" punctuated the air.[15]

Avoiding difficult terrain was not part of the American program that year. Indeed, even if Odaaq had told Solomon that the valley turned to mush precisely as the Americans wanted to build, Solomon would have barely heard him. They were going to dig in regardless. The emphasis was not careful attention to the place but the best application of tools to navigate and manage it.

When Solomon returned that spring, joining Priebe and others at the makeshift huts that had been set up, he had a "scope of work" list that included dozens of buildings: a ten-thousand-square-foot "Wing HQ," an equally large post exchange and sales commissary, a gymnasium seven times as big as that, and a 364-seat theater. So fast was everything constructed that, in June, Malaurie reported they screened *Kon-Tiki*.[16] In April and May, the advance teams had mapped out the areas, driving stakes into the hard ground to mark off roads, hangars, and the base's dump. The buildings and facilities would be connected by more than 2.8 million square feet of roadways and parking lots. Utilities had to be established as well: water distillation plants, deep-well pumping stations, and means of disposing the waste of a small city, most likely by carrying it out into the ocean.

By July 1951, like the July before it and the July before that, the ice on North Star Bay had cleared enough for the Inughuit to launch their kayaks from the beaches and move out into the waters to hunt. In 1951, however, they went to look at the Americans who had scared all the game away. The boats had not all sunk at sea, as rumor suggested, but had been "stranded" in the ice for an extra week—the Americans had been amply warned that the ice would likely not clear until July 1, but they had pressed their luck and hoped for an earlier melt. The big American ships had finally made it through Baffin Bay and were ready to offload men, machines, and food. But first the Inughuit kayaks moved in, "circling like eels round the high steel hulls." A telegram sent to Blue Jay headquarters on July 11, from the USS *Shadwell*, reported that "several natives in kayaks speaking Eskimo words" had come looking for cigarettes.[17]

For a time, the Inughuit enjoyed excess American rations. Malaurie reported that a group of construction workers came to the village to trade a crate of oranges for local trinkets. Still more brought tins of corned beef and ham, jam and magazines.[18] It was not long before the Inughuit began finding use for Blue Jay refuse; discarded boards and crates were appropriated and used to shore up sod houses.

The Inughuit had a long history of adapting imported items into their lives. The sledges that carried Malaurie's expedition, for example, were made of wooden planks, not the bones and ivory of a century before. With the arrival of the Euro-Americans that far north in the late 1800s, the Inughuit swapped traditional sledge technologies for lighter, faster, and bigger devices. The foreigners brought the only supply of timber and wood, which was, as the US engineers quickly realized in the 1950s, completely unavailable in northwestern Greenland where the tallest plant was the Arctic willow, rising only three inches from the ground.[19]

New sledges had meant the expansion of hunting and settlement patterns. The archaeologist Bjarne Grønnow has re-created the routes and settlement patterns that Odaaq and his son Kutsikitsoq moved across from the 1920s through the 1950s. Together, they used at least a dozen winter sites as they relocated to try out new hunting grounds or returned to known areas. In traveling with Kutsikitsoq in the winter of 1950–1951, Malaurie recorded going north to Smith Sound and then across the ice bridge to Canada.[20]

Whatever maps that Solomon had with him when he arrived in 1951 would not have had those sledge routes drawn in. They had not been mapped at all, and the Americans did not know them.

"Memorandum #1" arrived via telegram on July 7, 1951, just days after the first big ships began to unload and crates of oranges began appearing at the village. Drafted out of Operation Blue Jay headquarters in New York, the memo was addressed to all units involved in the Greenland construction program. The subject: "Relations with Native Population." Just five instructions followed. Point 1 dispensed with any nuance: "It is prohibited . . . to trade or have any contact with the native population." Point 2: "All personnel will remain clear of the native settlement." Point 4: "Every effort will be made to avoid *any* contact." The final point indicated that the directive would be "rigidly and continuously enforced."[21]

The memo and enforcement mechanisms were a clear sign that existing policies had been breached. All foreigners who arrived in Greenland at midcentury were given clear instructions on when and whether they could interact with local people. In the northwest, those interactions were seriously proscribed and regulated through the office of the Danish colonial administrator, Torben Krogh. Krogh lived in a small white house with his wife and young son on the edge of the inlet that curved away from the valley toward Dundas Mountain. Krogh appears to have been only thinly informed about the arrival of the Americans; certainly he was not told about the vast numbers and quantities of equipment and resources.

One of the early American survey crews had visited Krogh and received a sheet of strict instructions about interacting with local people. The rules, according to the Danes, were meant to keep the northernmost Inughuit isolated from too much outside contact so they could preserve their traditional ways of life, which included hunting, frequent migration, and seasonal gatherings.[22] The rules were clear: any contact with local people was forbidden unless it had been cleared directly with Krogh. Under no circumstance were alcohol or firearms to exchange hands; trade in polar bear parts and products was also prohibited. The purchase of small trinkets was permissible, but only with permission.

The rules made sense when there were a dozen foreigners in the area—a few weathermen, the occasional anthropologist—but they had not been updated for hundreds and then thousands. Indeed, no official word had come from Copenhagen because an official agreement on the base had not yet been made: not when Solomon came in February, not when Priebe arrived with a crew in April, and not even in May, when the equipment and gear were being shuttled from around the United States to ports on the East Coast to ready for shipment. The US military prepared for a base it did not have permission to build because it assumed that Copenhagen was in no place to refuse.[23] As had been the case in Libya and many other sites around the world, the United States took advantage of colonial and postcolonial relationships to create its empire of bases.

"Memorandum #1" was an amplification of Krogh's rules and a recognition that he could hardly be expected to manage five thousand Americans or more. Malaurie described a combination of "college boys from Harvard and Berkeley" come to "'see' the Arctic," along with "sailors and working-class fellows," all of whom came to barter for souvenirs and photos. In the end, it took military intervention to keep US personnel away from Uummannaq. While official US documentation on the episode is scant, Malaurie explained that "rifle-bearing soldiers on patrol" began to police an "iron curtain" between the base and the village. He reported that two Americans who had breached the policy had been sentenced to a lengthy detention on base.[24]

Official rules might have kept groups from interacting, but they could hardly keep the sounds, smells, and sights of construction from encroaching on Inughuit lives. The women told Malaurie that hunting would be more difficult; the water was tainted with oil, the seals were troubled by the incessant noise. Right away, the Inughuit Council of Hunters complained that their hunting grounds were being damaged, that the animals were being scared off, and that traditional routes had been bisected. "With all the noise," Malaurie recorded one Inughuit saying, "who can sleep anymore, or hunt!"[25]

While Malaurie worried and Krogh sought tighter controls and US Air Force officials contemplated how big their Greenland footprint might be, the engineers continued to dig and to move things. They ordered construction crews to hollow out huge pits in one place and move dirt and soil and rocks and gravel to another. They looked for a magic mix of aggregates that would adhere with cement to make concrete. In the coming years, they would operate their own cement plant. They blew up bits of hillside and built ramps and foundations on the snow and ice. A total of 1.5 million cubic yards of earth was excavated in June alone (150 million bowling balls' worth). Perhaps more would have been moved, but the construction crews ran out of hard-faced drill bits and had to wait for resupply.[26]

The crews slept and ate on the boats that bobbed in the bay, and then they arrived via landing crafts and rafts to work twelve-hour shifts. Then they returned to their ships, passing their replacements each way. Back and forth, in and out, on and off the shifts. There was no natural rhythm to follow because the sun never set; it just circled the horizon in permanent Arctic summer. The work was feverish and constant. One worker recalled that on each shift, it took a few minutes to get one's bearings, because so much had changed in the intervening twelve hours.[27]

On land, the engineers went down the list of facilities and utilities that were needed. They slashed a ten-thousand-foot runway across the valley. They dug into the permafrost and laid a six-foot insulating layer of gravel before pouring over the pavement, smooth and careful. They dredged out a harbor and set long piers out into the water.

They filled in some places and drained others. They brought bulldozers and backhoes, huge cement trucks and machines to crush rocks.

Nothing seemed to disrupt their work, although their work interrupted plenty about the natural world around them. Their concrete foundations and roadside culverts disrupted the flow of glacial water in the summer and the formation of ice in the winter. Roads snaked from the beach back to the base and to the trash dumps. They laid pipes and communication lines and drew fresh water from the ice cap. The roads began to snake east from the base. The US military, once on location, tends to grow.

Although engineering documents are rich in detail about dirt moved and roads paved, they are surprisingly quiet about contact with local people. Solomon and the military engineers in Greenland, unlike their colleagues down in North Africa, did not have to deal with property acquisition or displacement. No one was making a claim to the valley where Solomon was digging pits and drilling out cores of ice and snow and permafrost to sample and test. Where Troxler had to negotiate the worth of wells and date palms, Solomon was working against time. How long did they have before the cold closed in on them?

Solomon may have thought his work had nothing to do with the Inughuit. But that was not so. The engineering work destabilized a careful balance that had been managed for centuries in northwestern Greenland. Traditional Inughuit hunting paths and trails were bisected with roads and runways, piers and shipping lanes; hunting areas were fenced off; and access to the Pituffik Valley was completely proscribed. When Malaurie had predicted that the Americans would change Thule, he had no idea how far they would go or how careless they would be.

Henri Bader and Carl Benson, both of whom worked with the Snow, Ice, and Permafrost Research Establishment (SIPRE), arrived in early 1952 not to build things, not to meet with the area Inughuit, and not to conduct SIPRE-related basic research, but to help the defense apparatus sprawl out. As soon as work on the Thule runway began, men back at Air Force headquarters realized that such a valuable facility needed protection. What if the Soviets launched a surprise attack against a base that was totally unprepared? Shouldn't such an attack be assumed in the absence of protection? Radar sites were needed, and Bader gave his recommendations. The first site was about a hundred miles north of the base; the second was where he had been stranded in July. Construction commenced the following summer, and by the fall of 1953, twenty men were stationed at each remote radar site. They lived in metal tubes that were expected to soon be overcome with snow and ice.

This was the absurdist logic of the Cold War. The men at the center of this story—the Siples, Baders, and Wilkinses—all wrote and spoke about burying men in tunnels under the ice as if it were the most obvious plan. But it was completely ridiculous—unnecessary, costly, and incredibly disruptive to the lives being led on top of the world.

It was the perpetual movement and continual expansion—an American restlessness—that led to the forced and abrupt removal of the Inughuit from the area. In May 1953, as construction began on the radar sites, the Danish colonial authorities ordered the Inughuit living in Uummannaq to pack their things and move north to Qaanaaq, where they would spend that summer in hastily constructed tents. Over the next few years, they would have to establish more permanent homes, never again allowed to return to Uummannaq. In all, 116 people were moved, their belongings packed on sledges.[28] The Danish carried out the expulsion. Unlike in Libya, where Troxler helped negotiate prices for land, wells, and homes, in Greenland, the Inughuit were offered no compensation, even though the hunters' council lobbied for payment of some kind.[29]

On May 31, a telegram from the Danish station at Thule to Copenhagen explained that the "village population of total of thirty families now all removed to mouth of Inglefield Bay." At the time, there were conflicting reasons given for the relocation. Many of the Americans at Thule seemed to have believed it was to protect the natives from disease.[30] The *Washington Post* ran a comforting story that a few Greenlanders had traveled to Copenhagen to "request" that their village be moved away from the roar of the planes. The fish were scared of the noise. Even if this were true, of course, having to ask to move is different from choosing to. An American stationed at Thule later recalled that he heard the "problem was the noise from aircraft." But he added that "it really didn't seem right to move them."[31] The engineers were silent on all of it.

The reality was an all-too-familiar tale that was repeated around the network of US bases. The US defense apparatus needed room to grow, and the Inughuit were in the way. The Inughuit Council of Hunters sued for compensation in 1960 and again in 1985, but their claims were repeatedly refused. They wanted recognition of both their lost hunting grounds and interference with their rights. Not until 1999 did the Danish government issue an apology and open a conversation about compensation, although the issue of land rights and return to their lands was denied.[32]

Of all the animals in the Thule area, the only one that seemed to like the Americans was the Arctic fox. As Thule construction continued in 1951 and then again in the summer of 1952, the Inughuit noticed a steep decline in the local walrus, seal, and fish populations, just as the women had predicted, but a marked uptick in the foxes that scavenged in the American trash heaps.[33]

In addition to rearranging the land, the Americans brought and therefore left an abundance of stuff: boxes, cans, wood, plastics, food scraps, and containers. They produced a lot of trash. Eventually, they would also leave piles of metal car frames, airplane parts, and machines, some buried in the winter snows, and others thrown into the nearby waters. Less obvious—but no less significant—was the hazardous waste from the daily operations of the base: used oils and solvents, paint sludge, plating residue, asbestos, jet fuel, PCBs, battery acid, and more.[34] Most of the items the Americans shipped to Greenland were on one-way tickets. Even paperwork and electrical components were left behind.

The 1968 crash of a B-52 bomber trying to land at Thule is indicative of how insignificant the Greenland environment was to US interests. The aircraft was carrying four thermonuclear bombs. Although the devices did not detonate, the crash into sea ice caused the conventional explosive aboard to ignite, scattering radioactive waste around the area. The US military hired Inughuit sledge drivers to help them get to the affected area. A major cleanup operation commenced, but winter conditions made it exceedingly difficult. Contaminated ice was shipped to the United States for disposal. One of the four thermonuclear devices was never found.[35]

★ 11 ★
FIELDWORK

It is hardly surprising that from the moment the US military arrived at Thule to build an air base, some of the men began looking east to the ice cap. The Greenland ice cap rises like a dome, a giant mountain of ice seven thousand feet thick in places, spreading across 684,000 miles. It contains nearly 10 percent of the world's fresh water; only Antarctica has more. The ice covers the island like a great white cloak, and the US military was not entirely sure what it concealed.

Varyingly described as "eternal," "timeless," and "stable," the ice has come and gone. Midcentury scientists established the cyclical nature of historic ice ages, evidence of which was written across Greenland's landscape in deep fjords and flat valleys like Thule edged with debris that had been bulldozed by the grinding and crushing of moving ice.

In the 1950s, the engineers and research teams that came to Greenland wondered not about the past or deep geological time but about what was happening to the ice in that moment, whether it was growing, shrinking, or in some sort of equilibrium, just staying the same. These were interesting scientific questions, to be sure, but they had strategic significance as well (fig. 11.1). If the ice was melting—as many people thought—there would be implications for navigation and control of the sea lanes at the top of the world. Just as the air age had opened up a new vista, an ice-free Arctic would be a seismic shift. The Americans who scouted the site for the base in 1951 were

FIGURE 11.1 Landing on the ice. The most efficient way to get from the base to the top of the ice for research was via airlift. Across Greenland, research programs depended on the US military for transport and supply. Image courtesy of the Minnesota Historical Society.

told that the previous generation of Inughuit had remembered a time when the ice came all the way down to the water's edge. In 1951 it was fourteen miles back; it has since receded even farther.

Today it is clear that the ice is melting and that the melt rate has precipitously increased over the past few decades. Twenty-first-century scientists are engaged in urgent and ongoing studies to calculate how fragile the Greenland ice sheet might be, which includes identifying whether there might be a tipping point that could lead to catastrophic collapse. Rather than gradual melt, for example, entire glaciers could crash down all at once. Even without the worst-case scenarios, the ice melt will cause sea levels to rise by many feet. No one knows how fast or by how much—hence, the urgent need for scientific investigation.[1]

These questions and problems were far from the minds of the military engineers and the smattering of scientists interested in ice and snow and glaciers in the 1950s. Their job was to create the tools and practices to facilitate operations into extreme environments. But being able to build that expertise, as Henri Bader and the glaciologists knew, meant reading it across time, using the land and the layers to create more detailed stories about a place so few had ever visited.

From the moment that Bader and the teams of researchers arrived in Greenland with the Air Force, science and strategic needs about Greenland and the ice became intertwined. Greenland had the potential to provide answers to the strategic and scientific questions swirling about the Arctic and polar regions. It was where the Research and Development Board's (RDB) wish lists about what needed to be known might get a hearing. As one 1954 report from the Snow, Ice, and Permafrost Research Establishment (SIPRE) explained, even if the "work done was mainly of a basic research nature . . . problems were chosen that sooner or later could be expected to have some military implication."[2]

Of all the glaciologists and environmental scientists working under the auspices of the US military, Bader carried the greatest reputation. If someone had a question about operating in icy and snowy terrain, he was the one to call. Trained in Europe during the interwar years, the Swiss glaciologist emigrated to the United States after the war and took a position as a research professor of minerology at Rutgers University. He arrived in the United States precisely as the defense establishment was becoming aware of how little expertise it had in environmental sciences. In 1947, the Pentagon hired Bader to report on the state of the field in ice and snow studies. His first step was to return to Europe, where most of the research on ice and snow was happening. The Swiss had long been concerned with glaciers, avalanches, and mobility in the Alps, and up in Scandinavia, the cold was a perpetual reality. The Soviets and countries behind the Iron Curtain, of course, also had a robust interest in the same questions the Americans were beginning to ask, although Cold War tensions kept Bader from traveling there. Everyone else, the Americans worried, was already better poised to master the cold. The United States reportedly had just a few dozen individuals who identified as experts on ice and snow; the pipeline wasn't exactly flowing.[3]

Bader's assessment of the field was bleak and reminiscent of how strategic planners spoke about desert knowledge on the eve of World War II. "There are so many things we do not know about ice that a fruitful general discussion is very difficult," Bader wrote. Because the field lacked sufficient "observational and experimental data," too many people were prone to "hypothesizing" about things they didn't understand.[4] His report was passed along to the RDB's Panel on Arctic Environments, which recommended that the Corps of Engineers create a center devoted to the study of ice, snow, and permafrost, much as World War II–era engineers had wanted a desert testing site. SIPRE was the result, and Bader was put in charge.

Bader threw himself and SIPRE into the Greenland programs. Despite his outsize role, detailed information about Bader is spotty. Pictures show a man with thinning and combed-back hair, dark-rimmed glasses, and a well-manicured mustache and goatee. He reportedly spoke with a thick accent and smoked heavily. The journalist Jon Gertner described him as "sometimes irritable" and often "intimidating." Yet he was also prone to florid philosophizing. He had once remarked that "snowflakes fall to Earth and leave a message."[5] He devoted his life to determining what that message was.

In his professional ambition, Bader was not unlike David Bruce Dill, the wartime physiologist who had so successfully pitched the work of the Harvard Fatigue Lab to the US Army. Both men understood that defense funding was key to being able to carry out extreme environmental research. The US military was the only organization with the logistic and financial capabilities to support wide-ranging and long-term studies in such unfamiliar environments. After all, who or what else could drop a man 220 miles from base, on top of the ice sheet, and keep him well fed and sheltered for two weeks?

Bader was brilliant at linking his interests to military questions. If the Army asked how to best move over ice or build on snow, Bader explained that he could help find answers as long as his teams could focus on the basics. In SIPRE's first official report in 1950, Bader made this point directly. "Judging from past experience," the report explained, "basic research in snow, ice and permafrost is essential for the satisfactory developments of many phases of arctic engineering practice."[6] Extreme environments created barriers to efficient operations. In service of efficiency, SIPRE set itself a vast agenda, including compiling maps of snow, ice, and permafrost around the world and creating description and classification systems for types of snow, ice, and ground.

Some of SIPRE's proposals admittedly had "limited practical value." But what were apparently tangential projects would prove vital to solving practical problems down the road. For example, an Army commander was unlikely to find the military value in studying snow crystals or their behavior, or to understand why a program to map the structural patterns of ice and snow needed lab space. But SIPRE insisted that it was precisely those studies that would allow engineers to classify and describe particular types of snow and ice, which in turn could lead to successful runways and structures.[7] SIPRE even created charts outlining the intersection, albeit somewhat tenuous in many cases, of military utility and research science (fig. 11.2). Without the basics, you could not expect to manufacture the unexpected. By way of example, General Arthur G. Trudeau of the US Army explained that the ability to launch planes and, ultimately, missiles from ice or snow first required knowing how much pressure the ice could take before shattering.[8] It might seem hard as concrete, but it could also melt and fracture. The same could be said of compacting ice into roads and runways.

A child building a snow fort or preparing an arsenal of snowballs intuitively understands the properties of snow. Fluffy, dry snow is terrible for compacting. Like working with sand, molding and forming snow into structures and shapes requires just the right level of moisture, a kind of Goldilocks malleability. Too much and the snow is heavy and caves in or turns into a slurry that makes one's fingers soggy and cold. Too little and the flakes of snow will not adhere. Now imagine planning to land a plane on the stuff with a wheel load of seventy thousand pounds, or trying to pack and ship the right equipment for digging down into it.[9] Or, God forbid, burying a nuclear power plant in a tunnel of ice. What happened when all this took

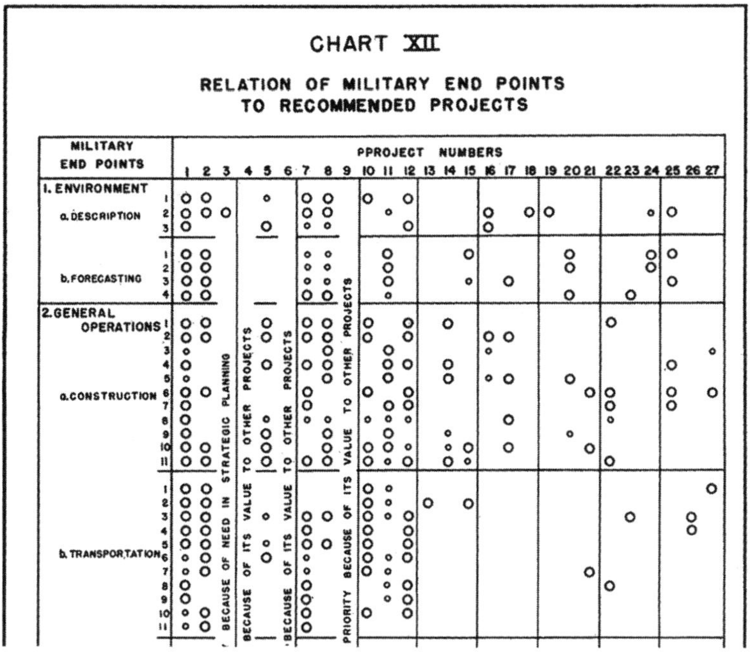

FIGURE 11.2 Science and strategy meet on the ice: excerpt from a chart in SIPRE's first report. The size of the dot indicates "relative importance" of the project to strategic needs. Of SIPRE's initial twenty-seven projects, many were pitched at operational needs, including mobility and construction. Not all research had obvious strategic end points, but the chart shows that research would be additive, and one small project might lead to larger, more strategically vital work. "Interim Report to Snow, Ice and Permafrost Establishment," SIPRE Report No. 1 (1950), US Army Engineer Research and Development Center.

place when the average temperature was negative twenty degrees Fahrenheit?

Bader and his fellow engineers and researchers wanted the same type of information about snow as Lieutenant General Donald Davison and others had wanted for the world's soils during wartime: clear definitions and classifications. Standards were needed to make sure the same things could be done at different points around the Arctic. This required equipment and devices for measuring established characteristics, and then scales and tables to define agreed-on qualities. The work was not confined to Greenland but rather spanned all

SOLID PRECIPITATION
(falling snow, hail, etc.)

Type of Particle		Code symbol F	
Code 1	Graphic Symbol	Term	Remarks
1	⬡	Plates	and combinations of plates with or without very short connecting columns.
2	✳	Stellar Crystals Cr1s	and parallel stars with very short connecting columns. conng
3	▭	Columns	and combinations of columns. ccon
4	↕	Needles	and combinations of needles.
5	⊛	Spacial Dendrites	
6	⊢⊣	Capped Columns	
7	∿	Irregular Crystals	
8	♀	Graupel	
9	▲	Sleet	
0	△	Hail	

FIGURE 11.3 Snow classification, excerpt from Bader's snow classification system. Early on Bader wanted to create classification systems for snow and ice. The image here was accompanied by a simple one-pager that researchers would take into the field for data collection to take measure and evaluate what they were seeing in Greenland and elsewhere. "First SIPRE Snow compaction Conference," SIPRE Report No. 2 (1950), US Army Engineer Research and Development Center.

of the Arctic and icy areas. It is why the RDB and various agencies wanted a glossary of Arctic terms. But Bader's interest was more specific: a terminology for types of snow and ice.

In one of the SIPRE's early reports from 1951, Bader set out his case: "The groundwork has yet to be laid," he wrote, for understanding the "principles of snow mechanics." SIPRE was just beginning to test snow classification systems, including evaluating standard methods and instruments that could be used to determine things like tensile strength and moisture content. What was the difference between old snow of a glacier, or firn, and fresh snowy powder?[10] Creating a clear and consistent classification system was essential to enhancing engineering and military programs (fig. 11.3).

In an age of big data, it is hard to grasp the idiosyncratic nature of how these classification systems were constructed, which, of course, makes doing so all the more important. In retrospect, the accumulation of information looks surprisingly simple: Men with pencils (ink would freeze) headed out with data sheets—created by Bader—to mark off certain characteristics and qualities of snow and ice. The sheets were not unlike the data sheets being sent out to the world's deserts by Wilkins and the Corps of Engineers. Bader later added a "hardness data sheet" and "meteorology data sheet" to the list of items to bring and record. Basic data points were to be scratched into tables and descriptions scribed onto forms. Each researcher was to assess materials, using his eyes, ears, and hands to get a feel for the place. Bader outlined more than a dozen specific qualities that needed to be measured, including density, permeability to air, tensile strength, hardness, grain size, structure, liquid content, and more.[11] Fieldwork was considered vital because, although crystals can be examined under a microscope, how a particular type of snow or ice or permafrost behaves depends very much on where it is from in the world and how it compares to its neighbors.

Like desert studies, then, men working in the Arctic placed a strong emphasis on human experience and expertise. Only boots on the ground could allow for the bits and pieces of environmental data to come together. Greenland certainly upped the ante. It was a place where individual survival and comfort were even more difficult to attend to. Logistics were crucial. One might be able to survive a few days in the desert, but the Arctic increased the stakes.

SIPRE teams created a number of instruments, guidelines, and techniques for studying the materials of the Arctic. Bader issued a simple field classification system for observing snow that built on the draft international snow classification system, akin to the modern Casagrande soil system. In his writings, the expertise and ability of the individual remained central. To measure hardness, for example, researchers were to use a hand (gloves recommended) to push down into the snow. "Soft" snow was defined by "four fingers" being able to easily be inserted into the snow; "very hard" snow was so hard a knife could barely break in. Moisture, too, was to be examined with a gloved hand. "Dry snow" was snow that could not be made into a snowball; "slushy snow" was measured by the fact that water could be pressed out. As the historian Janet Martin-Nielsen has explained, thirteen years of work in Greenland allowed Bader and his colleague Daisuke Kuroiwa to publish *The Physics and Mechanics of Snow as a Material*, the primary reference for military and civilian glaciologists.[12]

But in all cases, the point was the same: to begin creating tables and charts of global ice and snow crystals. The defense apparatus would foot the bill because the underlying logic was that this knowledge of the world's various environmental materials would make it easier to travel and to build everywhere and anywhere.

In January 1954, the chief engineer held a three-day conference at the Corps of Engineers' headquarters in Virginia on "proposed Greenland operations" for the year ahead. In the same year, engineers in Libya were planning to excavate roads and runways at the Air Force's new Sahara training range at El Uotia, a land of sand and more sand. It promised to be a challenging year.

In Greenland, the air base was complete, the two radar sites on top of the ice were occupied, and it was time to determine how to make the operation sustainable and perhaps even scalable. If twenty men each were able to live on the ice cap at each of the radar sites, how would two hundred fare at a possible new installation? Or, as a SIPRE engineering report suggested, could they put an entire air base under the ice? What about towns and research centers? How could they keep supplying the stations with goods and new technicians given the challenges of mobility?

The problems of how to move across ice and snow were paramount. It was the dominant topic of conversation at the conference in Virginia, which featured a who's-who of polar operations. In addition to SIPRE personnel, representatives of the various engineering and government units involved in cold-climate construction and operations were on hand. This included engineers from the Engineer Research and Development Lab, the Waterways Experimental Station, the Frost Effects Lab, and the Arctic Transportation Group. Academic consultants and private contractors came and went as their expertise was required. Siple was there, and so was Bader. The agenda highlighted an eighteen-point program of high-priority projects that needed to be initiated or accelerated in order to best exploit icy terrain.[13] Mobility topped the list, but construction came in a close second.

The meeting notes read like a road map for how basic research would be meshed with strategic needs. The focus on mobility dovetailed with Bader's argument about the need for basic science. To move across permafrost and ice and snow, he said, they first needed an intimate understanding of the nature of local materials. The snow of Greenland was different from the snow of Alaska. Similarly, building a road or a pipeline across the marginal zone required detailed long-term studies of the movement of the ice sheet. The question of how to compact snow and ice into runways and roadways led backward to the basic mechanics of the materials in question. For example, even snow that appeared to be hard could be loose and fragile, unable to bear the weight of a truck or plane.

The assembled men cast about for comparisons and examples. The usual suspects appeared: research being done in Alaska, Canada, other parts of Greenland. The work of foreign experts, too, was considered. The specter of Russian capabilities was always looming just over the horizon.

But woven throughout the meeting notes is also a surprising thread of information on deserts. "Blowing snow is like desert sand," the men assembled to discuss "project 3" agreed. The theme was "compaction of snow roads," and so they were also interested in drainage—"no different than western desert country"—and drifting. How could they stymie blowing snow and sand? Both were granular materials, so were there similar ways to keep the grains from heaving and covering new roads? Was building foundations in both materials similarly complicated? How much would the engineers have to improve or change the material conditions of a particular location to allow for the construction of their bases? Already at the Yuma desert station, engineering teams were testing out a new sled meant for the snow and ice.[14]

The *Military Engineer* was never far from these conversations. It frequently printed new techniques and best practices for constructing in extreme environments. A story published in late summer, titled "Army Engineer Climatic tests," took on the desert and the Arctic, including a two-page photo spread of various types of tests (fig. 11.4). Pictures on the right showed materials "in the desert"; pictures on the left were "in the Arctic." Categories included experimental housing (a domed wooden tent structure in the sand; an igloo-like structure in snow), "typical test roads" that showed mounds of both snow and sand and paths snaking through them, and road construction equipment that featured rollers of different sizes and types designed to meet specific conditions.[15] Equipment and practices were being adapted to meet the global needs of the US military. In trying to work through best practices for the arid extremes, engineers were looking for continuities and comparisons that would help them grapple with environments defined more by their material foundations (snow or sand) than anything else. In each place it was the granular processes and formations that had to be accounted for when thinking about paving roads and runways, managing blowing materials, quickly erecting prefabricated structures, and more.[16]

Sir Hubert Wilkins, who visited Greenland and Camp Tuto in the late 1950s, made similar connections. In his examination of how environments might affect operations, he explained that "the temperature relationship between ambient conditions and comfortable body temperature in deserts and polar regions is comparable." In both places—deserts and the far north—they also had to watch for the effects of radiation, as intense exposure to sun that could lead to unexpected warping of materials, deterioration of supplies, and damage to the human body. Sun glare and sun blindness were another menace common to dry and icy arid environments. Both required careful attention to water supplies; men in both places were prone to dehydration, and a reliable source of water needed to be found far in advance. Most of all, blowing desert "dust and driving ice and snow" were hazards to operations.[17]

In these extreme environments, the basic materials—sand or snow—mattered. In each place, the world was stripped bare, laid out like bones. Machines and men had to be rejiggered to meet the uncommon challenges. Wilkins understood that how men would function in such barren and unfamiliar places depended a great

IN THE ARCTIC **IN THE DESERT**

Experimental Housing
View of an arctic hut showing entrance of tunnel through the snow / Wooden tents modified to a double shell making the hemispherical shelter

Typical Test Road Sites
An early experiment in road building in snow. Four different construction methods are now being tested. / Rocky terrain through which test road was constructed with grader-mounted V drag

D7 Tractor Tests
Almost hidden is the winterized tractor dozing snow / Testing with hydraulic starter, winter front, and special greases

FIGURE 11.4 Equipment for the desert and Arctic extremes. Equipment for both were often compared and developed in tandem. This example of that synergy appeared in a two-page spread in the *Military Engineer*. Engineering centers sought to make connections across materials so that they could more efficiently and effectively design equipment. Image originally appeared in the *Military Engineer* 45, no. 306 (July–August 1953). Reprinted with permission of the Society of American Military Engineers.

deal on how the strange could be made somehow familiar, or at least manageable. Men had to be prepared to deploy to such landscapes where distance was hard to decipher, where dehydration could happen quickly, and where isolation could be maddening. They had to be well trained in caring for their equipment, and the equipment itself had to be designed to weather the extremes. Again, duplication was important. No man could be sent into the extremes without backup.

Of course, the similarities could be overstated. The Arctic was far more grueling than even the most extreme desert conditions. As military survival training manuals made clear, a man could survive a few days in the hottest desert with enough water on hand; but if he was poorly equipped in extreme cold, the chances of survival were slim. Frostbite set in quickly, and hypothermia was a problem. It also became clear that, just as men at the Desert Training Center had learned they could drown in the desert, people could die of dehydration in the High Arctic. With water frozen tight, one might be stranded atop a giant dome of ice and have no way to drink. That, at least, is what happened to some of the explorers who first tried to cross Greenland. But in sandy and snowy places, the unknowns and their implications meant that logistics were crucial. The improbability of being able to build anything in the extremes first linked the places in military thinking, but material similarities and environmental effects would keep them tied.

As in Libya, where military and research programs snaked out from Wheelus, they did so at Thule. The meeting in Virginia that first month of 1954 was a precursor of what was ahead. Figure 11.5 shows the first effort to identify all of the programs the US defense apparatus was involved in during the 1950s and the early 1960s. The information is pulled from a variety of sources, including SIPRE reports, RDB meeting notes, the recollections of men once stationed at Thule, newspaper stories, and maps. Although others have written about specific programs on the ice, with particular attention given to the most sensational, the totality of the programs has never been collected.[18] But it's this totality that is vital to understanding the stunning audacity and breadth of the US infrastructure, and the lengths that

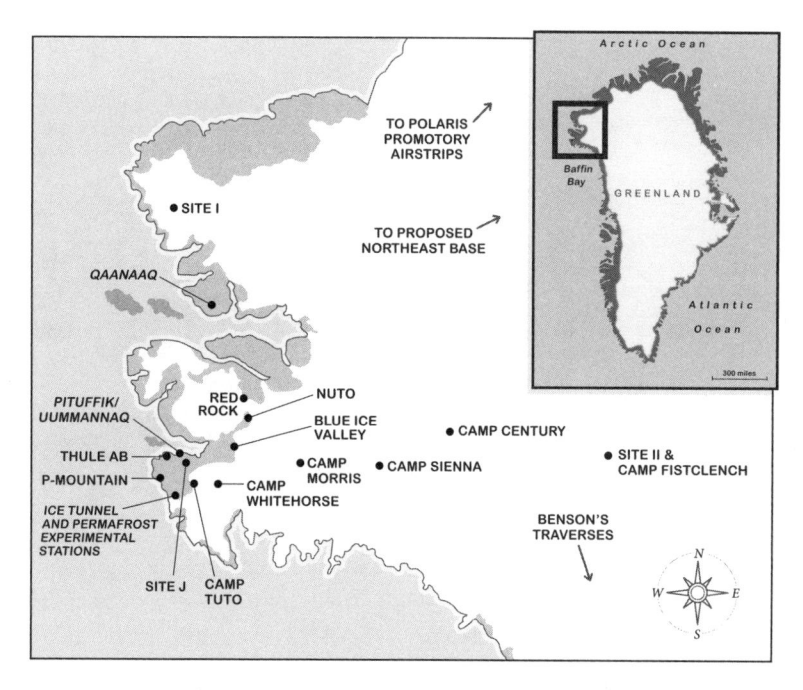

FIGURE 11.5 US defense-related installations in northern Greenland. From Thule Air Base, strategic facilities spread out across Greenland. This map attempts to show the variety of locations where research teams and military units tried to dig in. Some locations are best guesses based on existing materials. This is the first time these facilities have been mapped together. Map by Alejandro Gonzalez.

engineers would go to master the extremes. Many of their studies were not technically necessary; they justified them after the fact. Yet each one sought to answer specific questions not only about how to operate in the ice but also about what the ice was up to all on its own.

Program creep was somewhat predictable. The construction of a major modern air base necessitated radar stations that could protect it, which is what Bader was scouting. In January 1952, the Air Force also asked the chief engineer to locate a research station somewhere on top of the ice cap. This "semipermanent" site would provide weather observations, air rescue training, equipment testing, and "scientific investigations as appropriate."[19] More engineers came to build roads and ice runways. It was a pattern repeated around the world as facilities led to new roads and pathways that supplies could

FIGURE 11.6 Aerial photo of Camp Tuto, June 4, 1961. This was the research camp created about seventeen miles to the southeast of the air base. It was active from 1954 to 1966. Hundreds of research programs were launched from the camp. SC-592972, Box 1359, no. 12, Records of the Chief Signal Officer, RG 111, National Archives, College Park.

be carried along, logistic networks that could bring not just emergency rations, but also crates of steak and potatoes.

From the air, it looked like a road to nowhere: a raised gravel ramp that moved up and up for a mile, linking the bare ground near a place called Camp Tuto—short for "Thule takeoff"—to the ice that loomed above (fig. 11.6). Seventeen miles to the southeast of the base, this was where the Corps of Engineers decided to mount an assault on the ice. The thin road began on the gravelly, brown earth—brown, at least, in the short summer months—then rose steadily up onto the ice, and ended rather abruptly, not in a parking lot, but on the glacier. Construction began in the summer of 1954 and continued for years after as SIPRE and engineering teams worked out the best strategies for creating a road on top of an unstable surface that could handle repetitive loads.

It was not an easy road to build. The Corps conducted numerous experiments in the area: digging pits and trenches, backfilling with different types of materials, grains, and gravels. It borrowed heavy machinery from Thule, drove it out to Tuto, and then drove it back and forth for days over practice roads before taking samples with test kits. How much stress and strain could the experimental roadways take? What was the best way to build a road on top of ice that could and would melt or decompress, slump or shatter?

A road was built because other ways of mounting the ice had failed. In a mash-up of local practices (sledges) with US-mechanical force (imported machines), the Corps' Arctic Transportation Group tried to drive directly onto the ice. They turned back. They then called in the French expedition leader and wartime US consultant Paul-Émile Victor, who had previously managed to haul enough equipment and resources up the middle of the ice to let some of his team spend the winter there. The engineers brought in new equipment, like the M-29 cargo carrier—known as "The Weasel"—designed and developed during World War II to get across difficult terrain.[20] That, too, failed. By late spring, the ice was "rotten," a term adopted from explorers, meaning it was too soft and unstable to adequately support the movement of anything.

Camp Tuto became the base of US research operations in Greenland. What started as a few temporary huts turned quickly into more permanent structures. The Tuto orientation pamphlet noted that, while Thule had many more amenities, Tuto had its own laundry and bottle shop, basically the last bar before heading out into the wild. The men at Tuto were also warned that, should they want to head back to Thule—just seventeen miles but over an hour away—they had to shave. The Wild West–style beards and sideburns allowed at the research station were frowned upon back on base. To the men who spent time at Camp Tuto, or the installations even farther afield, Thule was the place they went to get fresh oranges and a ready supply of beer, bread, and mail. Thule had a bowling alley and a movie theater, a well-stocked hobby shop. Thule, the wild and improbable base perched precariously on top of the world, was a refuge from the ice. The men took to calling it "downtown."[21]

From Camp Tuto, facilities—and the refuse that was carried with them—spread, creating the paths along which more and more research would follow. Men wanted to study caves and permafrost tunnels, test pits, and best practices for building on top of or under the ice. Arctic experts like Vilhjálmur Stefánsson and Hubert Wilkins made appearances, lending authority to the work. A flurry of oddly named projects and places emerged: the projects SnoComp and Dogsled, Crystal Party and Snowman; Camp Fistclench, Camp Redrocks, and Camp Century. Wilkins reported that the military was also interested in a base on the eastern edge of Greenland, some seven hundred miles from Thule. Given the difficulties of supply, that base was never built, although over-ice traverses to the location took place as early reconnaissance.[22] Camp Tuto grew in step with these interests, and by 1958, it had over two hundred residents. A temporary camp had turned semipermanent as the US defense apparatus dug in.

Within months of the construction of Thule Air Base, Greenland had become a vital node in the US government's quest to know the world. Bader's 1952 trip to the top of the ice was the first step up and out; from then on, more building blocks were laid, moving out from Thule in various directions as the engineers sought the best paths for learning. Working in the polar extremes encouraged expansive thinking. The farther the military and scientific programs spread across the ice, the more the engineers began thinking about other extremes where they were working, such as the desert. Like a giant Venn diagram, circles labeled "Arctic" and "desert" overlapped across the map of the world. This was an important shift in how engineers were thinking about Earth. Not only were they linking specific environments together in new maps of the world; they were relating material conditions in new configurations. How much could exceedingly cold and dry snow be treated like sand? As the environmental categories and systems that the engineers were producing expanded, they spilled over into and intertwined with those of seemingly vastly different locations.

★ 12 ★

LIMITS

As soon as they figured out how to get onto the ice, the engineers began to dig (fig. 12.1). They dug test pits for snow samples. They dug tunnels that led to rooms where a man could stand and test and work, all the while examining the layers of snow and ice. They dug trenches, some of which they covered and some they left open. They dug straight back into the ice and straight down through it. Men held religious services at an altar they carved in the ice tunnel. They drilled a tunnel into the permafrost. They used their hands and pickaxes, simple shovels, and eventually machines. Back in the United States, engineers designed new tools to dig and test, poke and prod. As long as the engineers were up on the ice, they dug. Even when they thought they were finished, they dug out what had already excavated because the snow never stopped blowing and filling the holes and pits and caverns they had created just months before.

The more they dug, the more they learned, not only about the ice and about Greenland but also about their own limitations. No amount of logistic control or mechanical might could conquer the inexorable environment. There was no victory to be had against the snow, the ice, the darkness, the relentless cold.

These were the unyielding limits that the US military engineers met on top of the Greenland ice sheet. The work they did was audacious, to be sure. It was also incredibly wasteful. They put structures

FIGURE 12.1 Entering Camp Century. US Army Colonel Walter H. Parsons (center), chief of the Snow, Ice and Permafrost Research Establishment (SIPRE), and visitors climb up to an escape hatch to enter Camp Century. Photo by US Army/Pictorial Parade/Archive Photos/Getty Images.

on stilts that could be manipulated every year to rise above the snow. They put men and materials under the ice in shelters that could house hundreds, complete with hobby shops and hot showers. They imagined that someday entire communities could live hidden high above the Arctic Circle. One of the researchers with the Snow, Ice, and Permafrost Research Establishment (SIPRE) said that he thought an entire air base could be buried below the surface, with only a smooth ice runway maintained above.[1]

But in the end, they abandoned it all. They built big and then left. Not simply because the far north became less strategically vital as time went on, but also because what they built could not be maintained.[2] The hubris of believing they could do it all collided with the reality of resource limitations. It was simply too costly and time-consuming to keep shuttling goods to men living in a place where no one thought they should be.

When the Americans pulled back from the ice cap, they did little to cover their tracks. They assumed that the ice and snow and environment would do that for them. The road to the top of the world was simply abandoned, barrels and materials left where they had been put as beacons. Camp Tuto was closed; ice tunnels and test pits, deserted. Nearly everything left behind—metal bars, wooden benches, piping and waste—left to be crushed by the ice. By the late 1960s, of the facilities constructed in Greenland, only Thule, the true gem, remained open. It remains open today—albeit with a new name. Despite local efforts to have it closed, it is a center for ongoing scientific research in the Arctic and an important terrestrial node in the burgeoning space force.

At the 1954 Greenland research conference, failure and abandonment were certainly not the objectives. Nor were environmental costs on the engineers' agenda. Scientists and researchers lent their expertise to military projects, providing numbers and theories for what was going to happen. At every construction site and testing area, the engineers carried with them strings and ties, hooks and nails, sensors and ways to measure what happened to the material world around them. How fast did the glacier move? How fast did ice and snow deform? If they built a tunnel, how long might it last? Knowing it would not last forever, how could they make it last long enough?

Time was one of the biggest problems. Odd in a place that was so often was described as timeless; where light ran into dark and then wandered back into light, where the line between earth and sky, often featureless, sometimes stark, felt as "forever" as any place could be. Jean Malaurie wrote that, for the Inughuit, "hours and the days flew by according to an elusive continuance."[3] Village time was different from hunting time, which was certainly different from military time. So how would the American project function? They were measuring things with wooden yardsticks and hours. How long did they have?

Three years—that was how long the first structures lasted. When they were built in 1953, the facilities at radar Site I and Site II—where Henri Bader and Carl Benson had been digging in 1952—were "the most sophisticated installation to be built on a polar ice cap," but they were not built to last. The engineers gave them an expected life of five years.[4] They underperformed.

Building on the ice cap required addressing two different sets of extreme problems. First was the climate. How to insulate men and equipment from cold and wind was a problem being worked out across the Arctic and, as we have seen, in lab spaces closer to home. The second problem was the ice itself. Permafrost was difficult enough, but at least there was gravel and dirt and rock to work with. Ice was odd. Especially an entire land made up of it. As a viscoelastic material, it acts part liquid, part solid. It slips and moves; it is hard to contain or mold. As with permafrost and sand, it was one of those materials that was hard to work with because it was so ill defined.

In thinking about large-scale construction projects in an environment dominated by ice and snow, engineers had a thin archive to draw from. The Inughuit tended to avoid the ice cap because there was nothing useful to them to be found there. The two European scientific expeditions that had overwintered on the ice had each done so by digging down into it. In 1930, a German team at Eismitte (meaning "mid-ice") excavated a cavern where they could escape the incessant winter wind and blustering snow. The glaciologist on the team, Ernst Sorge, described his winter hole as an icy white "crypt." The platforms they carved out to serve as beds made them feel like they were like sleeping in a "sarcophagus." He spent the dark months worrying the ceiling would cave in.[5]

Nineteen years later, Paul-Émile Victor's French Greenland expedition dug out its own version of a snow fortress. The Frenchman and his team carved out a long trench into the ice that they filled with prefabricated structures. A series of carved tunnels and caverns served as additional storage, but daily life was carried out in the relative comfort of insulated rooms, which could be heated to sixty-five degrees Fahrenheit. Sorge's cave from nearly twenty years earlier had never reached above freezing. Victor's "Station Centrale" was no doubt more comfortable than the German cavern, but men still complained about the sense of isolation and anxiety that resulted from such remoteness.[6]

These expeditions taught US engineers how not to establish and maintain large-scale subglacial living conditions. The US military had no intention of sending men out onto the ice to live in deprivation. The military's goal, consistent with the work being done in laboratories and simulated spaces closer to home, was to insulate the men from the environment as much as possible by replicating more familiar structures and amenities. Most of the US personnel sent to the radar sites were not trained in Arctic research; they had not been selected because of some predisposition for extreme exploration. Radio technicians and airmen had not signed up to conduct extreme research in an extreme place. The Corps of Engineers had some glaciologists and specialists, to be sure, but those men did not make up the rank and file. Like the soldiers of World War II, the men heading to the most extreme environments in the world in the name of Cold War security had no experience to draw on nor much desire to be there.

Therefore, like facilities in the desert, those in the Arctic were to be as regular as possible. Complete with mess halls and rec rooms, hot showers and record players. A man might be able to imagine he was anywhere but where he was.

How could an ice cap facility be made livable for a very average man? A submarine under the ice was the basic model, or more accurately, a series of submarines connected with tubes. At Sites I and II, the engineers nestled giant, eight-foot-diameter corrugated metal tubes into the snow (these were the "submarines") and then filled them with prefabricated rooms—think trailers—that held bunks, a kitchen, toilet, sturdy tables, wooden chairs, and a few games. The rooms were suspended in the tubes and surrounded by empty space, a pocket of insulation meant to protect man from snow and cold as much as snow and cold from man. As the snow eventually buried the tubes, each tube was equipped with an escape hatch, jutting up and out from the icy white like a periscope (fig. 12.1). The goal in the early designs was not to adapt to the local environment but to find ways to separate living quarters and equipment from it. To keep man and materials separate long enough that a job could be done.

Which meant it was just a matter of time. As they built Site I and Site II, the engineers lined the tubes, passageways, and pipes with devices that would alert them to how much the ice and snow were pressing against the metal and wood. Copper thermometers were installed to monitor subsurface temperature changes. They knew the heat of human life would influence the snow, but how much and how evenly? Would some parts of the buildings settle faster than others? Would the movement of the glacier eventually compress or move the tubes? What about the weight of the snow falling and drifting? The engineers returned over and over again to take readings and check the rulers and guideposts.[7]

Site I was abandoned first. By 1956, the structure was listing—yes, some places could heat more and settle faster than others, especially when human waste was involved. Measurements also showed that the tube walls were beginning to press in, stressed by the amount of snow accumulating outside. Original plans predicted that "lateral movement will be negligible." Those plans were wrong. Site I had been built too close to the edge of the glacier, where the ice was moving as much as 8.5 meters (nearly twenty-eight feet) per year.[8]

Site II (where Bader had been stranded during his first summer in Greenland) was abandoned the following February, not the best time to move men on and off the ice, but it was necessary because so much snow had accumulated on top of the site that it was deemed "uncomfortable to live in." The decision to abandon the sites was also strategic: the radar installations were no longer deemed essential. In fact, new structures were already being designed and occupied nearby.[9]

The submarine-like tubes were also unnecessary. That was one of the lessons that the engineers took from the radar sites. The key to building in the snow and ice was going to be using the snow and ice, not just as cover from the elements—a veneer of protection—but as building blocks of modern military construction. This was not about simply digging and cutting, but also about managing and processing the materials. If they could manipulate snow to make runways, then they could manipulate snow to make walls and ceilings. In 1950, a SIPRE conference presenter noted that the Canadians were working on creating "snowcrete" by mixing snow with sawdust.[10] The task for the military engineers in the 1950s was to figure out how to build stronger, longer-lasting structures out of ice and snow that were, as Bader said, "consistent with modern military standards of comfort and utility."[11]

In 1954, Bader implemented a five-year plan to "design criteria for military installations on high polar ice." His use of *polar* rather than *Arctic* was hardly accidental. As for many others in the military and scientific community, Bader's interests extended across the Arctic and down to Antarctica as well: work done on the Greenland ice cap was seen as having application to the larger ice cap at the South Pole. Nearly everything that Bader and SIPRE teams did on Greenland was applied to projects on the other side of the world.

Bader's plan involved three different types of construction: a snow house, cut-and-cover trenches, and tunnels (fig. 12.2). The above-surface igloo was created out of huge snow blocks. The house was to combine "the best elements of both the native and American art" with modern military techniques.[12] The resulting structure was more modern than artisan. Yet despite the modern upgrades—metal beams for the roof, insulated liners on the walls—it was not particularly useful for military purposes. Only one snow house was ever built.

The other modes of ice-sheet construction involved going down below the surface. The engineers excavated tunnels into the firn, the upper layer of snow on top of a glacier, down to thirty feet, where the firn turned harder, and then to one hundred feet. From the bottom of that tunnel, they drilled another fifty-eight feet. This type of building

FIGURE 12.2 Life in buildings made of ice and snow at Thule Air Base (1953). Across the ice sheet engineers tried various ways to shelter troops. Ice houses, such as this one, were ultimately considered insufficient for long-term deployment, although they could serve as emergency shelters. Subglacial facilities could better protect men and machines from the severity of environmental conditions in Greenland. Photo by George Silk for *Life* magazine.

turned out to be scientifically interesting, but not ideal for cultivating comfortable living conditions.[13]

In an adaptation of practices at the radar sites, the third construction technique had the most staying power: a cut-and-cover trench program. Out at Site II, in 1954, the engineers started with hand shovels and saws, hauling chunks of snow out of the trenches with sleds. The next year they returned with Peter snow millers, which were used in the Alps to clear avalanches. Instead of leaving the tops of the trenches open, the engineers covered them with a variety of premade arches—metal, wood, plastic—and then covered those with processed snow. As the snow hardened into firm, stable slabs, the prefab arches could be removed, leaving white tunnels with clean lines and crisp edges.

By 1957, this experimental program had turned into the large under-ice research facility called Camp Fistclench (figure 11.5 shows the location of all of these facilities). In the cut-and-cover trenches, the engineers put up prefab buildings that had been shipped across the ice from Thule, stopping for a while at Camp Tuto. The buildings were used as barracks, labs, mess halls, and machine rooms and to house diesel generators. Over one hundred men could stay at Fistclench during each summer research season; it was never adapted for winter use. It was from this camp that many of the SIPRE ice-cap programs were carried out, including studies of whiteouts, dune formation, drilling techniques, and water purification programs.[14]

Fistclench, in many ways, was proof of concept for what the engineers had been working on since the early 1950s. It showed that it was possible to move men and materials up onto the ice and back down from it, to have people live and work hundreds of miles from any place anyone could identify on a map. By 1957, they had worked out means, albeit imperfect ones, for moving things, and they had started putting into place the necessary components for seasonal subglacial living. The next step was going to be to figure out how to keep men there all year long, not in the metal tubes that buckled under the snow, but in snow pits and tunnels.

"At first glance," *Popular Science* reported in February 1960, "the unnatural life within the confines of a buried snow town four blocks long, and three blocks wide looks forbidding, depressing, killing in its monotony."[15] Time spread out for the men who were stuck on top of the ice. The magazine was describing Camp Century, the under-the-ice city that the Corps of Engineers had started to build in Greenland in the summer of 1959. The article went on to explain that the men deployed in the icy wastes had luxurious subglacial living conditions, with "spacious dormitories," "hot and cold showers," and "a laundry and a barber shop." They also had nightly movies, radios and record players, and plenty of steak.

To fresh eyes the camp may have seemed bold, but to the engineers who had been working on questions of subglacial digging and living for years already, it was a culmination of their work. Camp Century was to be a subglacial base capable of housing 250 men year-round. Power would be provided by a portable nuclear reactor to keep the lights on and the water hot. Century was closer to Thule than Site II and Fistclench were, about 120 miles onto the ice—just far enough to avoid the fast-moving ice at Site I and a bit closer to the coast than Site II so logistics were easier.

The base was anchored by a 1,100-foot trench that was twenty-eight feet tall and twenty-six feet wide. Twenty-one tunnels linked to the main artery, each containing prefabricated rooms that served a variety of purposes. A large mess hall took up one trench; in others there were bunk rooms and latrines. Trench 12 was used for research, and Trench 16 held a theater and chapel. The library, enlisted men's latrines, and the club were in Trench 9. At the end of the camp was the power station, where the portable nuclear reactor was installed. Pipes and communications lines were strung up along the sides of the trenches, snaking down into prefabricated buildings that made up living areas and workspaces. The rooms were kept comfortably warm, and men were often pictured in T-shirts and sweaters; the tunnels, however, remained freezing (fig. 12.3). The temperature was well below zero, and tunnels were vented with shafts to keep them cold. A room made of ice and snow stayed a room only if the material stayed frozen.

Army public relations worked hard to promote the camp. In October 1960, two Eagle Scouts, both eighteen years old, were brought to overwinter at Camp Century. According to a later report, their experience seemed to move from excitement at something new to boredom at days so monotonous. There was little for the scouts to do. Life under the ice out in the middle of nowhere was about as interesting as one might imagine. They might spend a few days helping research teams, observing work with the nuclear reactor, or perhaps helping

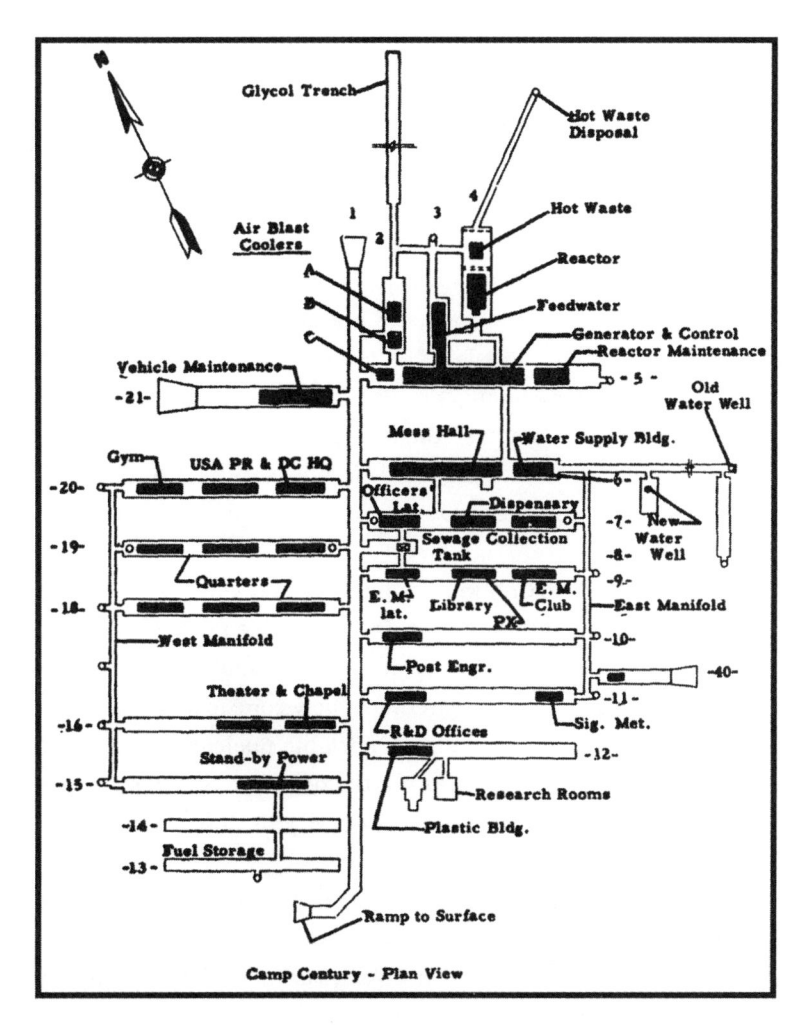

FIGURE 12.3 Blueprint for Camp Century. Trenches were carved out of the ice, roofing placed over the top, processed snow poured over the top, and then the roofing materials removed, leaving tunnels made of ice and snow. Portable rooms were put inside the tunnels for men to live in. Plumbing and electricity ran through the tunnels as well. The portable nuclear reactor was in trench 4 (top right); research and drilling took places in trench 12 (bottom right). "Main Street" (the horizontal tunnel in the middle of the image, labeled 1), was more than 1,100 feet long, 26 wide, and 30 feet tall. US Army Corps of Engineers.

out in the kitchen or arranging books into a small library. They were also given the task of measuring the rate at which the walls of their tunnels were caving in.[16]

"What is Camp Century's role in this polar development program?" an amiable Walter Cronkite asked Captain Thomas Evans, the head of the Camp Century construction program. Tucked into his Army-issued parka with fur-lined hood, the journalist had scored a trip out to the ice cap in 1960 to preview the "miracle" taking shape on top of the world. It was there on the "deadly," "desolate," and always "shifting mass" of ice that not even the native Greenlanders dared cross that man would "continue" the "battle against nature," Cronkite said as the camera panned across a plain of windswept ice that seemed to go on forever.

Evans responded to Cronkite's dramatic, if softball, questions with half-truths. Camp Century was a research facility, he declared, with three goals, none of them defense related. The engineers wanted to create a new base for ice research, test new concepts in polar construction, and practice operating a portable nuclear power plant. That all happened, but this was not the dream of SIPRE scientists and engineers. In fact, the engineers already had a research facility at Fistclench.

Camp Century was a military base (fig. 12.4). It was designed to test the possibility of a program the Army called Ice Worm, which was a fantastic plan to shuttle nuclear missiles around on railcars under the ice. The plan is comprehensible only in the context of two things: the nuclear arms race and interservice rivalries within the US defense bureaucracy. The advent of intercontinental ballistic missiles capable of delivering nuclear warheads to targets halfway around the world in thirty minutes transformed strategy and weapons procurement in the United States. This, combined with nuclear deterrence, meant that the military had to find ways to deploy weapons so as to deter the Soviet Union from launching a surprise attack. To that end, the Air Force had come up with its Minuteman intercontinental ballistic missile system, one thousand missiles planted in the America heartland, and the Navy had Polaris, submarines shuttling around

FIGURE 12.4 Construction of Camp Century. This trench was carved out of the ice, a metal roof was placed overhead, and snow was thrown back over it. Eventually, when the ice and snow hardened, the roofing would be removed. Camp Century was built 120 miles out onto the ice cap; it took three to five days to get there from Camp Tuto. August 1960. SC-586833, Box 1337, no. 54, Records of the Chief Signal Officer, RG 111, National Archives, College Park.

under the ocean carrying intermediate-range missiles. The Army's own solution was Ice Worm.[17] If they could bury missiles under ice and move them through hundreds of miles of invisible tunnels, they would have an unmatched deterrent force. It was utterly absurd but also completely in keeping with Cold War thinking. The project also meant that the Army—if successful—would get a large slice of the pie of nuclear financing.

As was the case with other military bases, of course, SIPRE and other engineers and research teams figured out how to best pitch their research interests into defense funding priorities. Meanwhile, as Evans seemed to suggest, the Army was all too happy to have the engineers and research teams as cover for the work they really wanted done.

What if they ran out of stuff? It was a very real possibility and a challenge for any facility out in the middle of nowhere without local natural resources to fall back on. The problem of supply preoccupied nearly everyone who tried to travel onto the ice. Even if they figured out how to turn snow into building blocks or hollow out a glacier into caverns, every piece of equipment, every conceivable provision, had to be brought in. Fuel, food, toilet paper, drill bits, playing cards.

The polar explorers spent months planning and packing. They knew that every item had to be tucked into another, saving space and weight, like modern box furniture. Even Bader's colleague Carl Benson was cognizant of weight when traversing in the 1950s with the support of the US military. They thought about how to decrease the weight of their rations, too, because every pound was significant when it was to be loaded, transported hundreds of miles over treacherous terrain, unloaded, and then unpacked.

Worse than bringing too much, though, was bringing too little or misjudging what was needed. The American explorer Robert Peary had calculated wrong in the 1890s. He had to shoot some of his sledge dogs to keep the others alive. Knud Rasmussen lost a man to starvation on his 1913 expedition and had almost died himself. At the German station in the 1930s, the meteorologist Johannes Georgi, who spent months alone, had the habit of looking to the west whenever he emerged from his tent, looking for signs of relief and resupply parties. He constantly worried about running out of stuff.[18]

Timing was the key to proper provisioning. When Bader put in place his plans for construction out at Camp Fistclench (the precursor to Camp Century), his teams would order supplies months in advance. Drill bits, for instance, had to be ordered sometime the winter before Greenland's summer construction season so that they could be made and shipped from wherever they were sourced. The drill bits would then head to one of the disembarkation points—Hampton Roads, Virginia, if going by boat, most likely, or Westover Air Force Base, Massachusetts, if taking to the air, which was unlikely outside of urgent need—and then packed into crates and set onto pallets before being loaded. Two weeks at sea led to the Thule pier, where crates were unloaded, stacked, processed, stored, and hopefully left untouched, lest another team take what was sent.

What of the logistic bridge from Thule to the ice? The trip to the ice-cap stations took days, sometimes weeks. The first stop was Camp Tuto. Only seventeen miles from Thule, the road was rough and jagged, etched into the rocky terrain left behind by melting glaciers. On a good day, a truck could reach fifteen miles per hour; on bad days, transport stopped when the winds picked up and the weather turned to a "phase 4." A "phase 4" generally meant that men should stay put and take whatever shelter they could, to hunker down and wait until one of the sudden storms passed. Men and vehicles could be stranded in hundred-mile-per-hour winds and plummeting temperatures. For those who were stuck en route, the road was dotted with rescue and emergency stations.

Tuto is where most of the research teams joined up and collected their gear, checking and rechecking it before the arduous journey up onto the ice. Had all the packages made it from Thule, packed and labeled correctly? If the crates passed inspection, they would be driven up the Tuto ramp, the raised gravel road built by the engineers that led up onto the ice, and then transferred to large tracked vehicles that would make the ice traverse.

The construction of Camp Century required hundreds of these tractor-drawn sled trains. In 1959, they transported 1,000 tons; the next year, 4,600 tons. In the summer of 1960, the portable nuclear power plant was carefully hauled across the ice before being lowered into its trench. Most of the supplies were on a one-way ticket, making it out to the ice and never returning, although the valuable nuclear plant did make it back.[19]

This all took tremendous effort, which of course made failures all the more devastating. At best, the route from Tuto to Century (128 miles) took four days. Often it was much longer. In addition to crevasses, weather was a constant wild card. Whiteouts and thick fogs, something like "being in a bottle of milk," could arise out of nowhere. There was no way to predict them (though the engineers did have projects to try and eliminate them). Storms were fierce, fiercer on top of the ice than near Thule and Tuto. Men were told never to leave their vehicles if stranded. You could freeze in minutes. In his broadcast, Cronkite called it the "most dangerous road in the world." But it wasn't a road; it was an icy path marked out with red flags and oil drums.

Always there was the lingering "What if?" It was a matter of time before things failed. The winter of 1962 hinted at what that failure might look like. It was a miserable winter, worse than normal. No swings—lines of tractors and cargo carriers that slowly made their way across the ice together—or cargo planes could get onto the ice. At first this did not matter, because Camp Century was equipped for sixty days. But in January, the steam generator broke and there were no extra parts. To conserve power, the camp's laundry was shuttered. By early February, the showers had to be closed. By mid-February, the men were low on food, although the cooks were making do. When an airdrop finally made it, the men at the camp had to scramble to collect items that had been tossed off crates and pallets. The mailbags often burst open, sending letters hurtling across the ice.[20]

Every summer through 1960, SIPRE teams drilled into the edges of the ice. They did this seemingly without regard for any potential implications of the work; they dug. Up north at a place called Red Rocks they drilled down and back; closer to Thule they excavated a tunnel in the permafrost where they practiced mixing "perma-crete." They even shaped tables and chairs out of the material. Nearer to Camp Tuto they dug their ice tunnel back five hundred feet; the next year they came back and added caverns and rooms. Two years later they added a longer, more complex tunnel just above the first. Perhaps they would try to live there. They filled the space with carpets and bunk beds, a mess hall, waste pits and latrine facilities—all etched into ice that was tens of thousands of years old. They sent pipes and wires through the tunnels. They strung lights up along the walls, and the heat of the bulbs led to little light caves. The ice glowed blue, seemingly from the inside. Perhaps it was this feature that led some contingent of men at Camp Tuto to create a chapel in the tunnels, a glowing, bluish altar.[21]

The documentary record on these programs is thin.[22] Piecing bits together shows that the engineers tried out a number of techniques that suggested progress toward train tunnels. They stored fuel in large pits and rigged up electrical lines. They worked on ventilation and studied the effects of rail tracks on the ice below. How much insulation would be needed? Could under-ice tunnels be made to curve along with rail tracks?

One of the largest rooms excavated—sixty by sixty feet with an arched ceiling twenty-five feet high—was designed to house large pieces of military hardware.[23] Other rooms contained lined pits where the engineers left diesel oil over the winter to see whether the ice could be a storage container. Fresh food was left out for months to see what would happen. Could the ice be a huge refrigerator for an emergency food stash?

All the while the engineers rigged the walls and ceilings with their grids of string and pegs. If one edge of the grid began to slump, they could tell the ice was pushing in from that side, or if one side of the grid seemed to be listing down, they could tell pressure was coming from above. They could thus tell which part of the ice was moving fastest. It was science meeting military construction.[24] At the edge of the glacier the ice was moving some five feet per year, but the engineers were interested in whether it was all moving at the same speed. What did tunneling do to movement? While the openings of holes in the ice seemed to close in faster than tunnels, would the tunnels last? In all cases, it was clear that a large gelatinous mass of ice was moving and shifting, squeezing in. So much for the subway.

The Scouts were asked to mark time. With tape measures and charts, they went from trench to trench measuring how much space remained between the floor and the ceiling and from wall to wall. They noticed the effect of "squeezing in" right away. Their measurements would serve as a baseline from which engineers could continue to chart how fast their walls were caving in. Everyone knew that the ice would move and compress eventually; the question was how long "eventually" might be. The engineers had thought a decade, but time did not look to be on their side. Søren Gregersen—a Danish scout—wrote in his diary that "the tunnel sides are coming in, the floors are

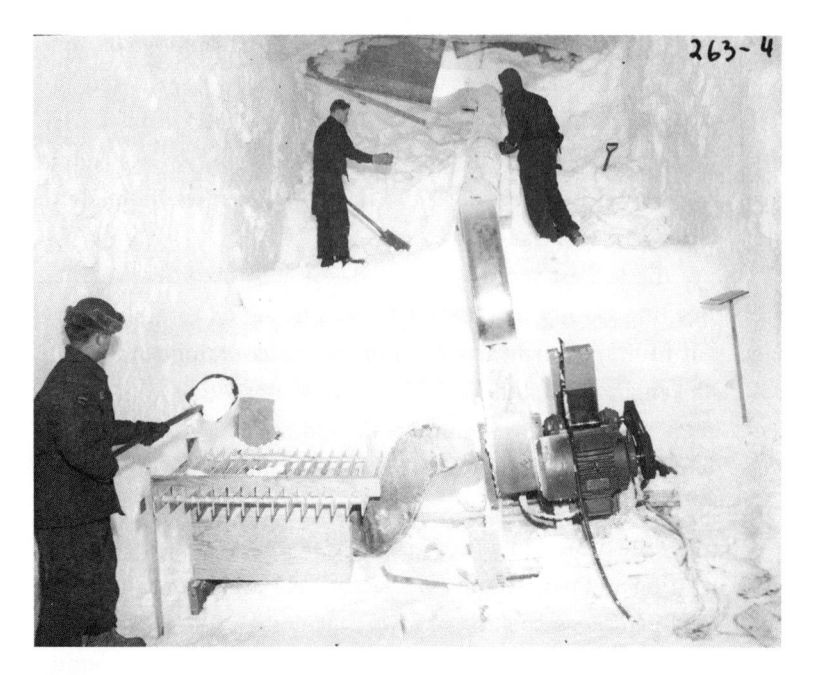

FIGURE 12.5 Around-the-clock excavation (1961). Just months after moving in, many of the men deployed to Camp Century spent most of their time clearing the snow and ice caving in around them. The military abandoned the camp just a few years after building it, leaving nearly everything behind and assuming that materials there would be entombed forever. SC-592995, Box 1359, no. 38, Records of the Chief Signal Officer, RG 111, National Archives, College Park.

coming up, and the ceiling coming down."[25] The mess hall alarmingly began to quickly slump, probably from the heat of bodies and cooking. And this was all in the first year.

By the second winter at Camp Century, cleaning out, shaving off, and ensuring the integrity of the tunnels was a full-time job that employed many of the men sent to live there (fig. 12.5). By 1963, the roof of the nuclear reactor trench was particularly worrisome, so much so that the reactor was deactivated and removed. Many of the trenches were closing in by thirty inches a year.[26] After just three winters of use, Camp Century closed and would open only in the summer.[27]

The United States abandoned the ice cap in 1966. All the most sensitive materials were removed: the nuclear reactor, the cores, the radar equipment, the men. Everything else—PCBs, lead paint, diesel fuel, and radiological waste—stayed behind, entombed, it was thought, in snow and ice. Three years later, a reconnaissance mission to the camp photographed what remained: crumpled buildings and frames, tunnels crushing in on the chairs and tables, metal shelves tossed to the side. More recently, a 2016 study warned that climate change might mean that Camp Century's leftovers will emerge far earlier than anyone predicted releasing the toxins long thought to be locked below.[28]

When he had taken his trip out to Camp Century in 1960, Walter Cronkite had wondered why the Army would build a base in the middle of an ice cap. In doing so, he also hinted at what was likely a secondary reason for Camp Century, one not talked about openly: building in even more extreme places. Just as the subglacial fortress was being entrenched and inhabited, the Corps of Engineers was consulting the Army and Air Force—and later, NASA—about lunar construction. The plan was to bury tubes in the lunar surface and provide power in them with a portable nuclear generator. The questions that remained were logistic ones, such as, How could they get enough stuff to and from a place so far away? For a time, ice-cap living and building seemed a sign that more was possible. Publicly, at least, it was a success.

★ 13 ★
SECRETS OF THE ICE

There was a point about a million years ago when the snow stopped melting. At least, that is the most recent estimate of when the Greenland ice cap began to form. More snow fell each year and accumulated until the snowpack was so heavy and firm that it pushed and compressed everything below it into hard ice. Each century the ice grew thicker and thicker, and it became so thick that, by 1952—when Henri Bader was stranded on top of it—it was over four thousand feet deep. Like a tree trunk, its layers told a particular story of the past.

Things besides snow got stuck down there as well. Tiny airborne particles like dust, ash, pollen, and trace elements that, in large enough concentrations, left dark lines and smudges in the crystal ice. Even cosmic dust was found there. In 1952, 160 feet below the surface, a thin line of sulfate particles represented the eruption of Krakatoa in 1883. Below that, there was a thin residual reminder of the explosion of Tambora in 1815 and, below that, Kuwae in 1453. In some places, the ice filled with bubbles and speckles, appearing shadowy and milky to the naked eye. The bubbles are like tiny fossils that hold remnants of carbon dioxide, methane, and more. The deeper the ice, the clearer it gets, because the intense pressure of all that sits on top compresses the bubbles so they become indistinguishable specks. Once the necessary methods and tools were developed, scientists realized that they could reconstruct past climates by examining the ancient atmospheric gases and particles trapped in the core's ice layers.

Henri Bader knew about this icy archive even before he arrived in Greenland in the summer of 1952. He wanted to know the story left by all the snow over a hundred thousand years. That was as far back as one could expect to go, because the ice flows and moves, grinding out toward the sea over time. For decades already, scientists had been trying to find the best ways to access and read the ice. But never had anyone had the logistic power and potential that Bader had to do it.

We have seen that science and research spun out across Greenland alongside US defense interests, and that engineers set down pathways through which the work could travel, granting access to places so remote that scientists had deemed them impossible to reach. We have seen how this global enterprise wrought environmental expertise that could be translated into space, and we have seen Bader's role within it. What we have left to explore is the program that is likely the most important research effort to come out of US military interest on the ice: drilling ice cores (fig. 13.1). Cores are long, continuous rods of ice that would eventually help scientists decipher hundreds of thousands of years of climate history. If there was a single project in the Arctic that transformed how the world was seen and understood, this was it.[1]

Transformation was not necessarily the intention of the men working on ice cores in the 1950s and 1960s. They wanted to gather as much data as possible from the ice, but they did not yet know exactly what the data might mean or where it might take them. That they were able to gather data was not the same thing as knowing how to read it or use it. After the SIPRE team had pulled the first longish core—a few hundred feet of ice—from Camp Fistclench (in 1956), Chester C. Langway Jr., a researcher hired by Bader, had to try to figure out what to do with it. How, exactly, could they read those messages contained within the ice archive? In 1958, Langway asked for help. "Interested persons," he wrote in the *Journal of Glaciology*, "are cordially invited to submit proposals for further studies."[2]

Over the following few decades, proposals flowed into SIPRE's

FIGURE 13.1 Examining ice cores. Beginning in 1956, researchers attached to SIPRE went to Greenland to drill down and retrieve long ice cores. Researchers hoped to use the cores as an archive of Earth's climatic history. US Geological Survey, National Science Foundation.

headquarters and circulated around the world. Research teams submitted ideas, samples of cores were sawed off and sent, new cores were drilled, new techniques established to assess them. By the early 2000s, building global climate models was possible because of the information pulled from the ice. Every year, it seems, new data is layered on top of the ice core information and the models of what might happen are refined. The ice at the top of the world continues to reveal suggestions and surprises.

And so it was that a US military engineering program that began by drilling down into the Greenland ice sheet set the foundation of what became an international effort to map out the relationship between polar ice and global climate. It started with Bader's particular interests; his background in the mechanics of ice and snow crystals then fostered a community of researchers able to use the sinews of US military power to travel to the edges of where was reasonable, to dig and drill down, and then to return again.

Research on Greenland tended to happen in a vertical plane. Weathermen with balloons and radiosondes looking up in the sky to read atmospheric phenomena while engineers drilling down below to read the ice.[3] The result was a new way of thinking about the fourth dimension: time. Researchers wanted to connect the climate above with the size and shape of the ice. They were particularly interested in the question of ice-cap equilibrium: Was the ice shrinking, growing, or staying the same? Which factors and bits of data would help them figure that out? Measuring just one area was not useful in getting a sense of the whole.[4]

Although much of this story of research on the ice cap has been told elsewhere, it is important here for four reasons: First, it reminds us how intimately tied were scientific research and military programs. Second, it highlights the way that science continues to operate in the space carved out for it by the defense apparatus. Third, it shows us the importance of knowing and seeing the extremes to grasping the basics of climate change science. Finally, it shows how engineering practices and knowledge created broader ways of experiencing and seeing the world and Earth systems.[5]

The science of ice cores in Greenland, of climate change, and of the ice sheet itself continues to evolve. Since about 2010, in fact, basic assumptions about the ice cap have been reformulated as historical data is layered with modern ways of measuring and seeing to create new information. What we know is that the ice is melting and that the rate of melt has been accelerating. Scientists now look to various feedback mechanisms within the ice itself that might accelerate the melt. The cores have also shown something that paleoclimatologists initially found shocking: rapid climate change. About 14,700 years ago, the temperature increased by twenty degrees Fahrenheit in just fifty years. As the geologist Richard Alley exclaimed after evaluating the core data, instead of happening slowly over centuries, as had been long expected, "the ice age came and went in a drunken stagger."[6]

Carl Benson liked to read the snow. He read it for rate of accumulation and melt, he read it for signs of odd weather events—like the unusually warm summer of 1954—and he read it, ultimately, to see whether he could determine the balance of the glacier itself, whether the ice cap was growing or shrinking. Armed with Bader's tools, instructions, and data sheets, Benson knew how to read the snow. And for the four summers he traveled to Greenland, he did little else.

When Benson arrived in Greenland, he was just twenty-four years old. His first interest had been minerology, but fieldwork in the Sierra Nevada led him down an unexpected path, first to Alaska, then to Greenland, and ultimately to a life of ice and snow. It was not an unusual path to the cold extremes. Many stumbled into it through a job opportunity, an influential adviser, a last-minute spot on a research team doing fieldwork far, far away. Benson recalled getting hooked when he did fieldwork with the US Geological Survey in 1950 in the Colorado mountains. He was intrigued by the same questions that had lured many explorers to extreme environments in an earlier era: How to understand a place that is so vastly different from what we think we know? What can we learn about ourselves when we are there?[7]

Such people had always been a relatively small cohort. In part, Benson later acknowledged, this was because when most people contemplated spending so much time in such remote places, it seemed "a terrible thing to do." But for those who caught the Arctic bug, an entire world of interesting possibilities had opened, particularly during the 1950s, when the US government was so keen to learn about operating and surviving in such environments.[8]

Benson spent much of his first summer in Greenland in 1952 trying to get Bader—and later Bader's equipment—off the ice. When they had been stranded in June, Bader and the pilot stripped their plane to nothing—pulling out all the gear and materials inessential to flying. They were trying to make the plane light enough to take off once again. When the helicopter came to finally rescue them, they left it all behind to be collected later by men such as Benson. Benson returned the next summer and then the following summer, using his work with the Snow, Ice, and Permafrost Research Establishment (SIPRE) and the US military to begin to map the first snow accumulation record at the top of the world.[9] The military sanctioned the missions because of ongoing interest in ice runways and polar construction.

Years later, Benson recalled that "the program was overwhelmingly big. I just can't get over that." The military allowed him to do things he would not be able to do even today. "There was never a nickel exchanged," he remembered. "There was never talk of money. If we needed Weasels, if we needed an aircraft, we just put in the request and we got these things." Chester Langway noticed the same support. His teams at Site II were fed, sheltered, and supported by the Army. "That was the route to success," he admitted.[10]

To dig a research pit on top of a glacier, one needs a surprisingly simple set of tools: a hand shovel and saw are sufficient. Bader's preference was that the saw be firm rather than flexible; the teeth spaced widely, only about nine per inch. Because ice had a way of wearing down even the sharpest, most robust tools, a sharpening file was recommended.[11] Before he joined SIPRE in 1950, Bader had worked out a system for carving ice samples during glacier studies in Alaska and the Alps. It took about five hours of relatively heavy work to dig out a nine-foot pit in which a man could stand and read the layers in the snow. The top layers of snow were relatively light. Cutting into them was like cutting into Styrofoam. But the snow got heavier and harder the deeper they dug, and it gradually turned to ice.[12] An ice pick and hammer might become useful, though not just any ice pick. Bader had specifications for that as well.

Bader and Benson continued to expand and refine their guidelines for examining ice and snow. The test wall should be on the southern side of the pit to avoid direct sunlight. This was particularly true in the summer months, when twenty-four-hour sunlight meant that solar radiation could easily deform a test wall, which would fare better in constant shadow. Rope ladders were useful, or perhaps even stairsteps carved into the ice; later Benson would recommend a twenty-foot ladder. When not in use, the pit should be covered, because unpredictable wind threatened to fill in the hole within just a few hours. Benson added more supplies to the glaciologists tool kit: pencils (#4H and #2H) and grid paper, a ruler, a brush for clearing excess snow off the test wall. For the collection of more detailed data, the researchers carried instruments for taking density and temperature readings. They lined the test wall with tubes and string and wires and crouched at the bottom to jot down what they saw.[13]

What were they looking for? Because Greenland had been undisturbed for so long, and because snow falls and compresses year after year, they could read the layers of firn like a tree's rings. Benson could see layers in the snow that corresponded to seasons and then to years. Benson was not the first person to do this sort of work, and he was well aware of the debt he owed to those who had come before. Just twenty years earlier, the German expedition led by Alfred Wegener had traveled to Greenland to demonstrate the utility of ice pits for scientific research. Wegener's plan was to operate the first year-round research station at a place he called Eismitte, literally, "Middle Ice." There a small team would collect weather data from the surface of the ice cap and evaluate annual rates of snow accumulation and temperature down below. Wegener's mission ended in tragedy when he and his Inuit guide, Rasmus Villumsen, died trying to make it from Eismitte back to the shore before the winter. But the men who stayed all winter—including the glaciologist Ernst Sorge—were able to take their readings, not to discover Wegener's fate until the summer.[14]

While they waited for resupply and contact, Sorge dug a shaft fifty feet down and started measuring. Summer snow was different from winter snow, he recognized. It meant he could track yearly snow accumulation by counting the layers. This sort of data was an important start. But twenty years later, in the 1950s, SIPRE research teams were determined to refine that data and to dig more. The SIPRE teams explained that summer layers are coarser grained and have generally lower density and hardness values than winter layers; they can also show evidence of surface melt. The more specific and refined the data, the more opportunities there were for scientists to start measuring what they found below. Importantly, that meant that, with the right tools, they might be able to read the layers as they dug to depths where the snow had compressed into ice, with bands no longer visible to the naked eye. That happened at about seventy feet below the surface.

Benson's contribution to advancing knowledge about Greenland was additive. Digging a pit in one place was useful for a particular end, but digging pits across the ice cap and comparing them could provide something of a time-lapse map for how the ice behaved. Instead of seeing how much ice accumulated and melted in one site, the SIPRE researchers could create maps of the entire ice sheet. It was like collecting meteorological data: small data points were interesting, but they became revelatory in combination.

For months, Benson and his team would zigzag across the ice on tractors and sledges. They covered a distance akin to walking from Maine to Minnesota. Nothing about the ice sheet was smooth and easy. The topography was an incredibly varied assortment of dune fields, plains of crusty ice that cracked, soft and slushy snow, even areas of dramatic hoar formations. At planned stops, they dug holes—146 in all—where Benson took temperature and density readings, and evaluated grain size and hardness data. Benson used the results to craft his 1960 dissertation, which was an important step toward better understanding of the structure and balance of the ice. His conclusion was that, in the mid-1950s, the Greenland ice sheet was stable, meaning that it gained as much snow each year as it lost.[15]

Sixty years later, a team of researchers from the Cold Regions Research and Engineering Laboratory—SIPRE's name after 1956—retraced Benson's steps, comparing their own readings to his charts and tables. Like Benson, they came and left from Thule, their mission supported by the supplies and materials of the air base. Air Force cargo planes moved them about, dropped supplies to them, and picked them up off the ice when it was time to be done. As they went, they found no trace of Benson's tracks or pits. The ice and snow and wind had erased any evidence of his work. Besides, the ice had shifted "a few hundred yards." The 2013 team used GPS—rather than the sun compass Benson used—to track his path and pinpoint the

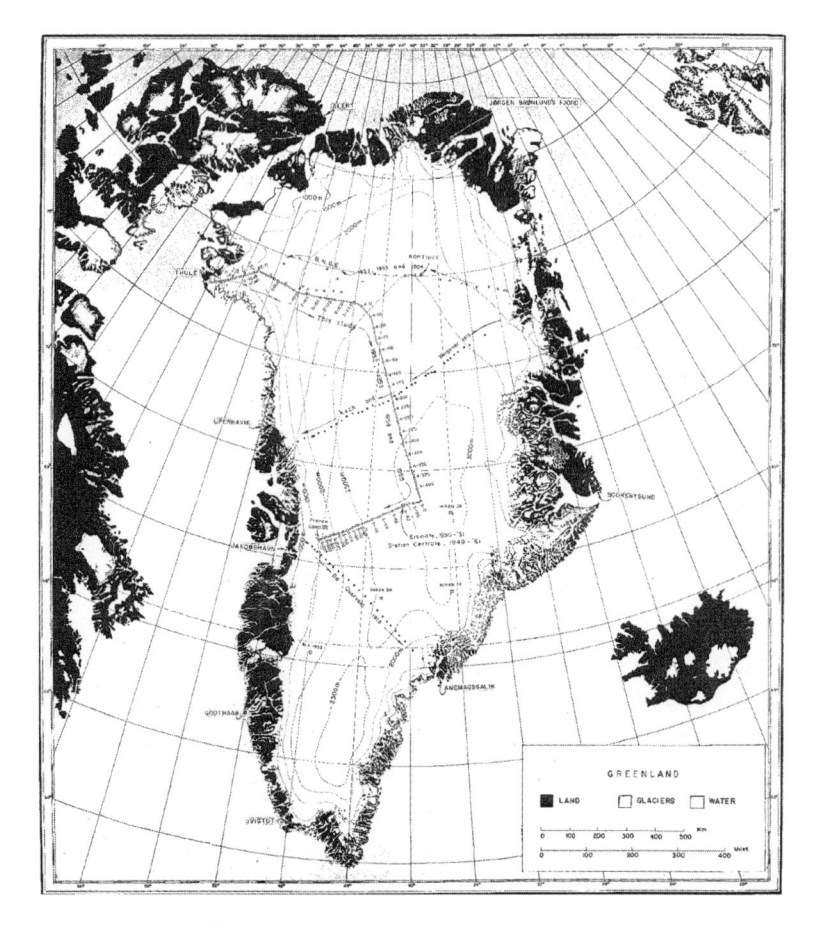

FIGURE 13.2 Benson's Greenland transverse routes. Each summer he traveled a distance equivalent to that from Maine to Minnesota. To complete scientific work, he relied on US Air Force support and logistics. US Army Corps of Engineers, from Benson, "Stratigraphic Studies."

plots that he marked on his own map (fig. 13.2). They struggled with the vagaries of the ice surface and spent days stuck inside because of wind and snow. While Benson had communicated with the outside world through Morse code, the twenty-first-century researchers kept a blog. Day in, day out, they did pretty much the same things Benson had done. Technology could help them see and measure from above to get a broad view of the world and its systems, but to really know the ice, they still had to dig down into it. They dug test pits and took measurements and test samples. They might not have found Benson's tracks, but they found traces of human activity: The ice was warmer than it had been just sixty years before.[16] The warming climate was warming the ice. In 2013, the ice sheet was no longer in equilibrium, and it clearly had not been for some time.

FIGURE 13.3 The drilling room at Camp Century. For six summers, men from the Corps of Engineers' SIPRE came to Greenland to pull ice cores from the ice sheet. In 1966, they hit bedrock. Photograph by David Atwood, US Army-ERDC-CRREL, courtesy of AIP Emilio Segrè Visual Archives.

Chester C. Langway arrived at Site II in the summer of 1956. It was still a long trip from the United States, though each season it got a little easier. He would have landed first at Thule, and then taken the road to Camp Tuto to check supplies and to make sure he had the right gear. Then he would have headed out on a heavy swing, a line of tractors and cargo carriers that slowly made their way across the ice together, for the research camp at Fistclench, two hundred miles out on the ice.

Like Benson, Langway had been sent to Greenland to read the snow, but he was to go deeper. He was there to test how legible the ice layers were as the mass of ice got heavier and the layers of annual ice thinner and thinner. In 1956, dispensing with the hand shovels and saws, the SIPRE teams brought an oil drill adapted for use in ice. "Much knowledge has been gained about the surface characteristics and stratigraphic layers of the upper few meters of firn," Langway wrote. "Below the thin surface shell of the ice sheets lies an almost completely unexplored region about which little is directly known."[17] He would be the one to explore it (fig. 13.3).

The year 1956 was when Bader earnestly tied military construction support to scientific research, not least because it was at Site II (Fistclench)—the site of his five-year construction program—that the first drilling would be done. While his teams relied on US military logistic support, Bader also wanted scientific backing and resourcing. He found it through the work of the 1957–1958 International Geophysical Year (IGY).[18]

The IGY was an international effort to conduct research and share findings widely to better understand Earth and its planetary environment. For the Americans, the work was firmly anchored in strategic thinking and planning, in trying to continue gathering intelligence about the atmosphere and Earth systems. Why else was the Arctic so important to the program?[19] For his part, Bader linked drilling ice cores to the country's research plan. While the bulk of the IGY's marquee work would be in Antarctica, Bader argued that Greenland was the best place to practice techniques and technologies. Because the US military was already in place there, the strategic community agreed. After all, his teams had already built test facilities and there were proven ways of getting men and equipment to locations that, just a few years earlier, would have been considered out of reach. The IGY affiliation was important to bringing scientific attention to the program, as well as more money.[20]

If a simple, undisturbed snow pit could proffer data about the previous few decades, a core hundreds of feet long promised an even richer trove. Good climatological records had been kept for only a short time, a decade or more on the Arctic ice, perhaps a century in other places. But scientists interested in understanding a longer history of climate required data of greater longevity and consistency rather than sporadic accounts culled from journals, ship logs, and elsewhere. Ice cores were only one possibility—others included tree rings and ocean and lake sediments—but they provided data going back the farthest, and that was probably the most informative. The ice on Greenland had been accumulating over a hundred thousand years' worth of trapped particles and chemical traces. "Ice cores are one of the best archives that the Earth provides," explained the geologist Bess Koffman. They are like "a gigantic Library of Congress filled with important information. The key is learning how to read it all."[21]

That first summer in 1956, the drilling team recovered a 305-meter core. Langway recalled that the team would do nothing but "work, eat, and sleep." At the end of the season, they returned to Camp Tuto, where Bader and the SIPRE team conducted a training camp for seventeen international researchers who would be involved in IGY programming in Antarctica. Langway was invited to go along but decided to stick with reading the Greenland cores that winter. Bader wanted all the participants to know how to use SIPRE's snow-study kit and become proficient in the drilling techniques that Langway's crew had used. The crew that went south took with them the rotary well-drilling rig.[22]

The following summer, Langway and the Greenland team returned and drew up a longer, more consistent core of 411 meters (1,348 feet). The core came up in fragments, a few inches to a few feet at a time. The ice cylinders could be clear or milky or dirty, or other times full of bubbles. Sometimes the cores hissed and popped as they came up into the air.

The nightmare for Langway, and for anyone pulling up rods of ice, would be to accidentally jumble the order of the cores, messing up the entire timeline being assembled. A system emerged to avoid such a mix-up (fig. 13.1). Immediately as a core was pulled up, it was placed in a wooden trough with the "top" marked with a small hole drilled into the upper part. The trough was then placed on a light table so the core could be examined. Subtle stratigraphic differences were noted and logged, and eventually marked on long strips of paper. The scientists cut the core into logs that could be stored and sawed them in half lengthwise so there were copies to ensure continuity should a shipment go missing. The cores were then bagged, identified by cards inserted into each bag, and placed in specially constructed aluminum-lined shipping tubes. In 1957, the cores were sent to Wilmette, Illinois, where SIPRE had its headquarters and where Bader could look at ice samples under his microscopes. Langway would spend the winter examining the archive. He counted annual layers reaching back 174 years. Using new tools, he would go back further still. Other US researchers were slow to appreciate these findings, but for Langway, "the value of investigating deep ice cores is almost limitless."[23]

The upper layers of the cores could be read with the naked eye, much like ice pits.[24] But the deeper layers and lines were harder to discern, which is why Langway's 1958 article announcing the cores was accompanied by a pitch for assistance. Fortunately, scientists around the world took notice. An international cadre of research teams began crafting strategies for mining the ice for information. This included Willi Dansgaard, a Danish paleoclimatologist who would work with Langway over the coming decades.[25]

For Bader, it was just the start, a proof of concept more than a done deal. He wanted to drill all the way down through an ice sheet to hit rock bottom, and Camp Century provided him the opportunity. After all, the under-ice base, powered with a nuclear reactor, provided a more consistent and comfortable place to continue the drilling program. SIPRE had access to two trenches—Trench 11 and 12—where the drill could be set up and lab space commandeered.

The drilling room was a rectangular white chamber with walls that narrowed as they angled toward the surface (fig. 13.3). The corrugated metal roof remained intact. Wires and lights were anchored into the icy walls, drooping and dangling, the jerry-rigged setup a reflection of the temporary nature of the work. The men all wore winter parkas and hats with earflaps. They needed gloves. The tunnels had to be kept cold—ideally well below freezing—so that the walls and floor and ceiling would not melt. Some days, men had to make room for the engineers who would scrape out the ice and snow that was creeping down, closing in.

Drilling all the way down took six years and a lot of mistakes. The team usually headed out to Camp Century in April or May and left in August before the weather turned too grim. From down below, they worked long shifts. Drills broke or were mechanically insufficient for the task; cores shattered or got stuck. New parts had to be ordered constantly. The effort was painstaking, and men were anxious at the prospect. What would they find? No one actually knew what would happen when they hit bedrock, or if there was bedrock. There were rumors that perhaps it would lead to a massive explosion as air or gas escaped. Bader pointed out that there were three options: solid ground; a mix of solid and gaseous material, such as dry natural gas; or a mix of the above with some liquid thrown in.[26] Bader thought it more likely they would hit solid ground or rocks the glacier had ground out of the bedrock.

Bader was proved right on July 4, 1966, when the team at Camp Century hit the bottom of the ice. The team had snaked through nearly a mile of it, 4,550 feet, and they continued down into 12 feet of subsurface material.[27] The men in the drilling room took a chip of ice from the core they estimated was the year of Christ's birth, poured some Drambuie onto it, and toasted to their success.[28] They packed up their equipment and left as the walls slumped in.

The Camp Century core was not the last core pulled from Greenland, nor was it the most useful one. On the contrary, international teams would continue to draw cores from Greenland and Antarctica well into the twenty-first century. At the same time, the old cores, archived, could be reinspected as new ways of examining them were discovered. But the collapse of Camp Century and the relatively rapid US military withdrawal from much of the ice cap meant that ice core collection efforts became limited. Langway, for example, wanted to continue drilling, and in 1970, he met with a group of scientists and engineers from the United States, Denmark, and Sweden to plan a new program, eventually called the Greenland Ice Sheet Program.[29] The program was ambitious and would last over a decade, hitting bedrock in multiple locations so cores could be compared across the ice sheet, providing foundational data for climate research. But it was slow in getting started and there was difficulty in accessing the ideal spots for coring.

Camp Century had been an amazing place to perfect the techniques and tools, Langway understood, but it was too close to the edge to be the perfect location for drilling. That site would be flat and smooth, untouched, and, importantly, as close to the middle of an ice sheet as possible. Eismitte, where the earlier French and German expeditions had gone, was an example to be emulated. Such a site would limit the amount of movement of the ice down and out, as it pulled toward the edges. But the US defense apparatus was not willing to foot the bill. And the National Science Foundation, which had agreed to help fund the Greenland Ice Sheet Program's team, required that it operate mostly out of existing military facilities for cost purposes. One of those was a still-operating radar station called DYE 3 that had been constructed in the late 1950s. Only later, in the 1990s, would teams be able to access the better locations, once new, post–Cold War funding streams had been established.[30]

Even as direct US defense agency involvement in ice-core programs receded, the infrastructure of US military power remained central to the scientific undertakings that could be attempted. Even today, survey flights to assess the health of the ice sheet are conducted out of Thule Air Base. For over a decade, NASA's Operation IceBridge, to monitor changes to polar ice, used Thule as a home base. Josh Willis, a researcher monitoring the ice, explained: "There's a lot of [Greenland] that you just can't reach from anywhere else. Thule has been a great hub for us to bring in equipment and swap out crews."[31]

There were no immediate revelations from the cores pulled up in Greenland in 1960, nor from those in Antarctica. It took time not only to read the cores but also to develop the best tools for doing so. It was a truly international scientific undertaking, involving small labs and research centers around the world. Samples of the cores were cut and sent, data shared, and findings discussed at conferences. The frozen material that was pulled from deep below spread across the world, linking and connecting scientific teams who wanted to know more about how the Earth system worked.[32]

What did they learn? Temperature data going back thousands of years was the most notable thing. But perhaps far more useful has been the ability to read the gases and particles trapped into the ice—aerosols such as dust, ash, carbon dioxide, methane, pollen, trace elements, and sea salts—which provided a more nuanced view of what was going on in the world at any given time. According to the climate scientist Allegra LeGrande of the NASA Goddard Institute for Space Studies, "Scientists can directly measure the amount of greenhouse gases that were in the atmosphere at that time by sampling these bubbles."[33]

In putting the pieces together, in layering the data and comparing one set of cores to another, scientists realized that the ice archive was far more surprising than the monotonous cold of the Arctic might suggest. Not only could they find residues from "terrestrial and extraterrestrial" remnants down there—big events like the clash of meteors or the eruption of volcanoes—but they also could see changes to the ice itself. Most fruitful has been the use of this data to create climate models; what LeGrande describes as "laboratories inside of computers."[34] All known data is fed into these virtual labs, where it becomes a closer and closer approximation of what was and has been, and therefore of what might become, leading to a new level of simulation that extends through time.

Ice cores "opened a new portal to the past," Langway explained. For the first time, they provided scientists a "continuous long-time prehistorical and geological age records of precipitation, climate, rapid climate change, and natural, artificial and baseline atmospheric chemistry conditions, and more." Scientists could compare preindustrial atmospheric gases to postindustrial conditions, highlighting humanity's production of the greenhouse gases that have led to rising temperatures. "By studying these archives," suggests the geologist Koffman, "we see that the rate of change that humans are enacting on the environment is faster than anything that we see in the geological record."[35] The cores still are providing vital data to models of our climate future. Indeed, as more refined tools are developed to read them, more information is gleaned.

Combined with other ways of seeing and sensing the ice, studying ice cores led to new ways of understanding Greenland and Earth more generally. Ice is no longer seen as "eternal" and "changeless." It is clear that the ice has come and gone, and it may do so again. Various satellite programs have been launched to measure the mass of the ice, including NASA's Gravity Recovery and Climate Experiment, and others have been used to evaluate the internal plumbing, or to measure sloughing off along the edges due to interaction with ocean currents and temperatures. The ice cap is now understood to be a world unto itself, a vibrant and dynamic ecosystem still little understood but also clearly vital to larger Earth systems. Scientists study the feedback mechanisms whereby the darkening ice, for example, leads to warmer temperatures because the glaring white long reflected the sun, creating an albedo effect. But without the albedo, temperatures rise faster. There remains much to be studied, but there is clear consensus on this: Greenland is melting faster now, and human activities are to blame.[36]

Although the US military largely abandoned the ice cap in the mid-1960s, the networks it set down, the reports it generated, and the tools it crafted to master the unknown have all continued to animate research into and on the ice. Research teams continue to depend on US military infrastructure. By the 1970s, engineering activities around the world had so completely integrated the Earth into a system that it was never even a question if they could or should or would be able to deploy to its extremes. The US military now had the capacity to deploy anywhere; it simply lacked the patience to stay.

III. STARDUST

PRELUDE

The *Selene* was lost. The "dust-cruiser" had started off normally enough, taking a group of visitors from Earth across the Sea of Thirst. Despite the fact that the sea was "completely flat and featureless," best known for its "monotonous uniformity," it was a tourist draw. Where else could one cruise over a vast expanse of a not-quite-liquid, not-quite-solid material that had no known analog on Earth? The "sea" consisted of a dust "fine as talcum powder" and as dry as "the parched sands of the Sahara," but it flowed as smooth as a liquid.[1]

The Sea of Thirst was always placid. Nothing caused ripples or waves. Until this day, when a moonquake threw the sea off balance and the *Selene* sank below. There were no ships to rescue the crew because the *Selene*, not quite boat, not quite bus, not quite spaceship, was the only vessel of its kind. In fact, Arthur C. Clarke, who invented the *Selene*, described it as "not unlike the Sno-cats that had opened up the Antarctic a lifetime ago." Drawing on extreme environments closer to home to envision worlds in outer space was hardly unusual. Clarke's dust-cruiser was something like the wanigans used on the Greenland and Antarctic ice caps, part "sledge," and part "bus."[2]

A Fall of Moondust was Clarke's 1961 contribution to lunar science fiction. By then he was already a famous science-fiction writer and futurist. He was named the "Prophet of the Space Age," a moniker that certainly stuck after he completed the screenplay for *2001: A Space Odyssey* (1968). *A Fall of Moondust* was not a stellar example of

FIGURE P3.1 Chesley Bonestell's *A Lunar Landscape* (1957) highlights how the surface of the Moon was imagined before humans had eyes in the sky. The full mural hung in the Boston Museum of Science from 1957 through the 1970s. It was recently restored and is on display in the National Air and Space Museum's exhibit *Destination Moon*. Image courtesy of the Smithsonian's National Air and Space Museum.

his work, or widely read, but Clarke's lunar landscape is a reminder of all that remained uncertain about the Moon in 1961, even as the United States began planning to send people there. If the cartographer S. W. Boggs had drawn a map of the lunar surface, it would have been colored in brown blocks marked "unknown."

Conversely, in 1961, there were few people able to authoritatively refute Clarke's description of the lunar surface. To be sure, geologists and astronomers thought that the Moon was a solid surface, but because nothing had actually yet landed there, they could not be certain. Nothing about Clarke's world was particularly unbelievable. There were no strange monsters or ghouls, no Sea of Thirst creatures that rose up to suck the humans down. That the Moon might be made of unstable materials—like quicksand—was plausible, and it seemed to make the idea of space travel even more precarious.

The proposed conquest of space threw into high relief the boring but crucial lesson that rocks and dirt matter. They might not be shiny things that make the news or topics that generate works of art, but sand, snow, and stardust (in this case, moondust) were all increasingly central to the modern world, and not just in military spheres. Modern technologies and construction practices meant that more and more people would encounter environments and conditions that were only partially understood. The Moon was an extreme and distant version of that, but like all extremes, it tended to reflect what people had been thinking about closer to home.

The people on the *Selene* survived, their rescue not dissimilar from the real-life effort to retrieve the stranded glaciologist Henri Bader in 1952. In both cases, there was limited capacity to stage a rescue, yet it was done by a small cadre of men who defied the odds.

Could the US military engineers shift those odds further still on the actual Moon? That was the question in the 1960s. As a new space engineering team coalesced in the Corps of Engineers' Virginia headquarters, they understood that a few key things stood in their way. Until someone could assure them that the Moon was more solid than liquid, more rock than Clarke's Sea of Thirst, they could not fathom how it could all be done.

★ 14 ★
DESERTS IN SPACE

In the fall of 1962, the office of the chief engineer of the US Army Corps of Engineers received an urgent memo from one of the country's newest agencies: the National Aeronautics and Space Agency (NASA). There was nothing particularly odd in receiving requests from federal bureaucracies. The Corps remained the nation's main construction agent and was accustomed to fielding calls to do things like build an air base in Greenland, pave runways in Morocco, sink missile silos into the Great Plains, or help build a dam in California. In fact, the Corps was already working with NASA on the space agency's administrative offices in Houston and launch facilities at Cape Canaveral, Florida. But the new request was most intriguing: NASA wanted a how-to guide for building on the Moon (fig. 14.1).

No one at Corps headquarters seems to have found this exceptional. On the contrary, a team was quickly assembled and tasked with creating a lunar engineering model. To be sure, the engineers understood that the obstacles to "space age" construction would be significant—no air, no water, no known resources—but just as they had built things under the Greenland ice cap and in the middle of barren deserts, they could find a path to keeping men on the Moon. "Since shortly after our first satellite went into orbit," the chief of engineers explained in 1962, "the Corps has been conducting preliminary studies to determine what will be needed in the way of oxygen, shelter, power, water, sanitation and other facilities before

FIGURE 14.1 The Corps of Engineers' *Lunar Construction* model and plan (1963; cover shown here) was created at the behest of NASA and crafted over many months by engineering teams. US Army Corps of Engineers Archives, Alexandria, VA.

lunar explorers can live and work in the inhospitable environment of the moon."[1]

Much has been written about the US space programs—about the machines and technologies that hurled men into the heavens, and the individuals who made the journey. Less well known are the plans that emerged in the late 1950s and early 1960s to build bases and facilities on the lunar surface.[2] Corps teams worked hard to make space construction possible.

In thinking about lunar construction, Corps of Engineers teams drew from two intertwining lineages. One was their ongoing work in extreme environments. Many of the tools and designs for outer space were remarkably consistent with work done in the Arctic and deserts. But in imagining even more extreme places—where no human had ever set foot, let alone collected a soil sample—the engineers also drew on widely popular conceptions about space that circulated at midcentury.

Indeed, a full decade before NASA's quiet memo, many Americans already had a pretty good idea of what colonization of the Moon and Mars was supposed to look like. In the pages of magazines like *Collier's* and *Popular Mechanics*, a group of men called "space boosters" or "space enthusiasts" spun tales and painted dramatic pictures of exploration and colonization. These boosters played a role similar to the explorers in World War II who consulted with the Quartermaster Corps. Each group always pushed for more. Like the explorers, space boosters were an eclectic bunch. Their ranks included science-fiction writers, amateur rocketeers, aerospace engineers, journalists, and a small cadre of German scientists, including some former Nazis, who had been brought into the US military at the end of World War II as part of the scientific and research intelligence program called Operation Paperclip. What they shared was the conviction that space exploration was happening, and the US government should be the one to make it happen.

The connections between the terrestrial and cosmic extremes were not entirely new at midcentury. In the late nineteenth century, Percival Lowell believed that one had to travel to the desert in order to understand space. The heir to a Massachusetts textile fortune, Lowell had spent the first few decades of his life traveling the Far East and writing about his experiences. A chance encounter changed the trajectory of his work and life: In 1893 someone handed him a copy of Camille Flammarion's *La planète Mars*. From then on, Lowell devoted his considerable fortune and energies to peering into the sky.[3] In 1894, he traveled to Flagstaff, in Arizona Territory, where he built an observatory so he could look at Mars. The Lowell Observatory is still in operation today.

There were two reasons Lowell chose the Arizona desert. The first had to do with the art of seeing. "Water vapor is a great upsetter of atmospheric equilibrium," Lowell explained. Humid places were characterized by "commotion in the air" that was "the spoiler

FIGURE 14.2 Article on Percival Lowell's canal thesis. Lowell's thesis seemed to "prove" that there had been intelligent life on Mars. Astronomers were suspicious of Lowell's ideas, but they set the tone for Martian studies for decades. From the *New York Times*, December 9, 1906.

of definition." The high, clear skies of the desert made it good for stargazing.[4]

At the same time, Lowell believed that Mars itself was a "vast Sahara." The Bostonian found strange serendipity in basing his telescope in an environment that was a precursor to the world he watched above. Lowell (like many others) believed that Mars had once been lush and green but that it was dying through a process he called "desertism," which he saw as "the state into which every planetary body must eventually come." Mars was telescoping Earth's future. The world's deserts, Lowell suggested, were a harbinger of things to come, something like "the first gray hairs in man."[5]

Lowell saw proof of desertism all around—in the petrified forests of the Sonora, to be sure, but even more so in the Sahara. On the shores of the Mediterranean, he wrote, "at the edges of the great Sahara, are to be seen today the ruins of vast aqueducts." (This is precisely where the US military built its massive air bases in the 1950s.) Earth was drying out.[6]

Central to Lowell's fascination with Mars was his certainty that intelligent life had—at some point—been there (fig. 14.2). The proof was in dark lines that angled across early images of the planet, lines that Lowell and others misread as purposefully constructed canals.

In Lowell's writings, an ancient Martian civilization had constructed the canals as the planet dried to shunt water from the polar regions down to Martian settlements.

Although astronomers were suspicious, Lowell's ideas set the tone for Martian studies for decades, and even today. The idea that the shadows and markings suggested life was compelling lasted in popular culture well into the midcentury space age. The 1951 film *Flight to Mars* featured Earthlings visiting the Red Planet only to find a dying civilization. More plausible seemed to be the theory that Mars housed some sort of low vegetation that had adapted to the extreme environment, much like desert plants had done. The Harvard astronomer Fred Whipple said as much in the *Collier's* series on space. "Through telescopes," he wrote in 1954, "astronomers can clearly see Mars's reddish deserts, blue-tinted cloud formations and–especially conspicuous—its distinctive polar caps." As they melt each spring, he continued with great certainty, "strange blue-green areas develop," only to turn brown "some months later." Whipple did not come out and say it was plant life, although his descriptions strongly implied that.[7]

The power that Mars held over stargazers had something to do with its place in the cosmos. It is close enough that it can be seen with the naked eye, a red light in the dark sky. Yet it is still too far away to be seen with any real clarity. Even as telescopes grew in sophistication, Mars remained fuzzy, which kept it all the more intriguing, especially compared to the Moon. Mars was vague enough to be suggestive, a place where anything was possible—even life. No wonder all sorts of fantasies were read into the blobs on its surface.[8] The most accurate, detailed image at midcentury was like that in figure 14.3. When the conditions were just right, astronomers had far better views. But the photographic evidence—seemingly objective and authoritative—remained hazy.

The relentless certainty that life would be found somewhere was a reprise of how early explorers approached the terrestrial extremes. Ralph Bagnold and his band of desert explorers created the Zerzura Club in a nod to the mythical city of the sands. Europeans hoping to cross the great Greenland ice cap long imagined that there might be

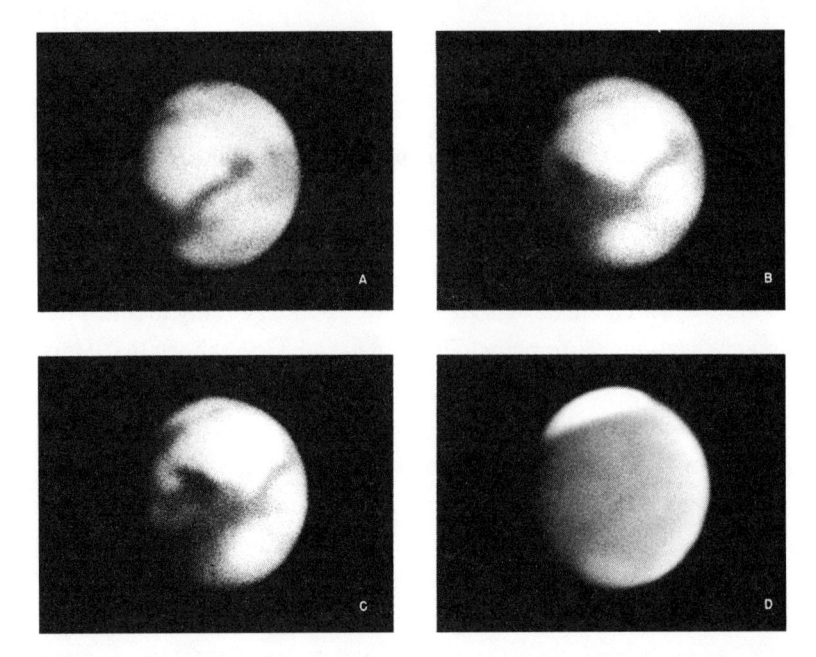

FIGURE 14.3 Midcentury Mars, September 11, 1956. These images were taken from the hundred-inch telescope at the Mount Wilson Observatory, California. These were the best pictures of Mars until NASA obtained images from the Mariner missions in 1965. Space boosters, scientists, and the popular press all promoted the idea that the shadows could be clouds or vegetation. NASA.

a temperate oasis in the middle of the island. It was not just the quest for something unknown and different that animated these dreams but also the desire for something familiar and knowable. They wanted to cross a dangerous and vast frontier to get somewhere safe, not to simply find another edge of not knowing. The quest is never for less. Ideas about Mars proved that point over and over again. A dead Mars was also not consistent with Western ideas of expansion and empire that had so long pushed people to pursue the far flung.

Lowell did not spend much time thinking about the Moon, although the people who followed him in dreaming of Mars certainly did. By the late 1940s and early 1950s, the community of space boosters looked to the Moon as the first—if not the most interesting—stepping stone to everywhere else. Lunar exploration was framed as training wheels for interplanetary travel. If they could master the Moon, they could master anywhere.

The problems of lunar exploration were not going to be that different from the problems at the terrestrial extremes. At least that was the argument the space boosters pitched in the eight-part series "Man Will Conquer Space Soon," which appeared in *Collier's* between 1952 and 1954. As Wernher von Braun and the astronomer Fred Whipple explained, the problem with the Moon was the "danger of the unknown." Much like the extremes at home had once been feared, more knowledge and data about outer space would help people embrace the next frontier. Readers were told that science and technology had finally caught up with science fiction. Within their lifetimes, the authors declared, humans would have a base on the Moon. The urgent question was who would build it and why. Would such a base be "the greatest force for peace" or one of "the most terrible weapons of war"?[9]

Collier's was clear on the answer: the United States must build "space superiority" or "somebody else" would. And that somebody else was the same foe motivating US military construction across Earth's most extreme environments: the Soviet Union. Whoever controlled the Moon, just like whoever controlled the terrestrial extremes, "will control the earth."[10] If Soviet capabilities in the Arctic and the Sahara were seen as threatening, the idea of cosmonauts looking down from above from was downright terrifying. The Cold War strategic thinking that sent US teams to every nook and cranny on Earth traveled easily into space.

The popular weekly magazine was hardly alone in promoting space programs and presenting them as urgent security problems. Writing in *Popular Science*, Arthur C. Clarke, author of *A Fall of Moondust* and the head of the British Interplanetary Society, explained that advances in rocketry and nuclear power meant that "experts no longer argue whether men will someday ride rockets into outer space, they've moved on to planning how." For Clarke the question of where man would first go was clear: "Obviously," he wrote, "to the Moon."[11]

A rich literature, both popular and academic, exists on the connections between space boosters and government policy, between science fiction and plans for the cosmos. Missing, however, has been the overlap between construction programs for building on Earth and in the heavens. Precisely as the US military was extending its reach into the extremes on Earth, it was also contemplating extraterrestrial basing. The year of *Collier's* first Moon article, 1952, was the same year that Henri Bader was stranded on top of the Greenland ice cap, which was also about when Paul Troxler and his staff of military engineers struggled to keep Saharan sand from burying their new desert runways. It was thus the case that even as military engineers and research teams labored to master the most extreme environments on Earth, there were some already making plans to take things to the next level. From Earth to the Moon, *Collier's* and the space boosters who wrote for it promised, and ultimately beyond.[12]

But for men like von Braun and Whipple, it was clear that the United States was not doing enough. The government had not yet committed itself to gathering the scientific expertise, technological abilities, and basic information that would be needed to soar into the heavens. In this they echoed concerns of the Research and Development Board's various panels about the lack of scientific tradition regarding desert and Arctic knowledge.[13] Despite the Moon's proximity, maps and charts of it at midcentury were not that much clearer than those of Mars. The maps that did exist had been cobbled together from Earth-based observations, photographs, and telescopic views, some one hundred years old. These were fine for armchair space enthusiasts, but landing men up there and keeping them there required precision.

Collier's tried to fill in the particulars. Spread across two issues in October 1952, pieces about the Moon were penned by a who's-who of the space community. Von Braun was the Nazi scientist who had developed the V-2 rocket. He had been plucked out of Germany through Operation Paperclip and sent to work for the US Army's rocket program. By this time, he was one of the most visible advocates for space exploration. He would later work for the Walt Disney

Company and NASA. Whipple's Harvard connection brought gravitas to the enterprise.[14]

Finally, there was Willy Ley, renowned for his outer space visions. Born in Germany, where he developed a keen interest in rocketry, Ley fled the Nazi regime in 1935 and ended up in the United States. During the war he wrote about the intersection of society and technology. Ley's books on space travel were internationally renowned, including *Rockets, Missiles, and Space Travel* (1944) and *The Conquest of Space* (1949). *Collier's* described him as "the best-known magazine science writer in the US today." His biographer declared he was "the man behind the curtain" whose ideas and works were foundational to the space age. In Ley's oeuvre, space travel was filled with great opportunity; it was a realm where people could harness terrifying technological progress to subdue new worlds.[15]

These emerging giants of the space age came together in the popular press to pitch the future to the American people. The *Collier's* articles were all written in the first-person plural, and decidedly terrestrial descriptions were used to help the reader settle in. "We have arrived," Whipple and von Braun wrote, imagining a rocket ship touching down in "Dewy Bay, not far from the lunar north pole." As "we" stepped onto the lunar surface, a plain extended like an "expanse of broken ice." Elsewhere the horizon was edged with towering, jagged mountains, cliffs that were "blinding white against the pitch blackness of the sky," almost like icebergs. The first few steps out of the rocket would be done carefully, of course, because no one wanted to "break through the crust" and fall into a "great crevasse." That is why, von Braun and Whipple acknowledged, the "keen eyes of the experienced geologists" would be needed. Even in space you needed a man who could read the rocks.[16]

Von Braun expected that as many as fifty men would be part of the first landing team of scientists and military officials. He estimated that might happen within twenty-five years. As in unknown environments on Earth, early work would be focused on general reconnaissance to better establish construction and survival practices. Fieldwork included drilling cores, collecting samples ("dust . . .

mineral specimens, rock and lava fragments"), and mapping terrain. Such an expedition also had to be carefully planned to coincide with the sun. Von Braun wanted the crew to arrive at the beginning of a lunar day, which, given the rotation of the Moon, meant two weeks of constant sunlight. Like construction crews sent to Greenland to build Thule Air Base, lunar teams would take advantage of the sun's light and warmth; crews would work in twelve-hour rotating shifts. The authors explained that daytime temperatures would thus be "livable" at forty degrees, whereas the two weeks of lunar night would see temperatures plunge to negative forty degrees. (Eventually they would discover their calculations were wrong—the Moon can range from 250 degrees in daylight, to negative 208 when night falls.)[17]

The space boosters were presenting the lunar environment as a mash-up of the Earth's deserts and the Arctic. The Moon was "dust-covered, drab, silent," with "the frozen stillness of a faded backdrop."[18] Men would "overwinter" on the Moon, just as they overwintered in the Arctic. The surface was both strange and familiar, a more extreme version of the extremes. The accompanying images in *Collier's* reflected these motifs. The lunar landscape looked alternately like a desert sand sea and an Arctic vista. The Moon was spare and rocky; the horizon, ringed by jaunty peaks.

In 1957, visitors to Boston's Museum of Science would have gotten a true feel for what the space boosters envisioned. In March, the museum unveiled Chesley Bonestell's *A Lunar Landscape*, a mural forty feet wide by ten feet tall that took up the entire wall near the planetarium (see fig. P3.1). It was intended to be immersive: when viewers stood at the right place, they would feel as if they were on the surface of the Moon, looking out across a classic moonscape. "No spaceship reservations are needed for a startlingly realistic visit to the Moon," the museum declared.[19]

Bonestell's space art became iconic. By the early 1950s, he had created the visual lexicon of the space age. Born and raised in California, Bonestell had initially trained as an architect. He worked on some of the most recognizable structures of the early twentieth century, including the Golden Gate Bridge and the Chrysler Building. In the late

1930s, he felt the pull of Hollywood and turned to making film sets for movies such as *Citizen Kane* (1941). In 1950, he combined his amateur interest in astronomy with cinematography and designed sets for the film *Destination Moon*. He was a frequent collaborator of Ley and illustrated some of Ley's popular books, including *Conquest of Space* (1949).[20] Many of Bonestell's images were in the *Collier's* series as well.

Yet Bonestell's images were firmly rooted in earthly conventions. He worked in the style of landscape paintings, a genre closely tied in America to nineteenth-century westward expansion and ideas of the majestic wild. *A Lunar Landscape* was familiar but strange, resonant of a desert landscape with rocky ridges and jaunty rock formations, yet painted in the colors of the Arctic, pale gray and white. What gave the location away, however, was the glowing Earth, hanging just over the edge of the horizon. Bonestell painted *Earthrise* before such an image could actually be captured.

The connection between extreme environments at home and those in the cosmos was not confined to Bonestell's palette or the pages of *Collier's*. Space enthusiasts and planners alike grasped that moving humans into outer space was going to require the same sort of logistic infrastructures that had been created to move militaries to the far-flung extremes. The first space pioneers were not going to be lone explorers heading out into the unknown; they were rather going to resemble well-organized and well-supplied military teams heading to the extremes. Space required the sort of methodical planning and redundancies that existed for the conquest of the extremes closer to home. Highly refined logistic networks were needed—the *Military Engineer*'s bridge to Mars comes to mind—carefully stacked building blocks to outer space. Space travel was simply not possible without the backing of organizations capable of moving things across vast distances with speed and consistency.

Of course, the prospect of building on the Moon amplified the stakes and dangers. Military engineers understood this immediately when they were asked explicitly to weigh in on such plans. In 1958 and 1959, the Air Force and Army, respectively, initiated top-secret studies of building military facilities in space. In many ways, these were knee-jerk responses to highly visible Soviet space programs, which seemed far ahead of US abilities. But they also demanded close attention not to rockets and aerospace technologies, but to the particulars of place. To the ability not just to get somewhere but to stay for a while.

The October 1957 launch of Sputnik, the first Earth-orbiting object, turned space from the subject of glossy magazines into a strategic problem. The Soviet satellite was little more than a bleeping metal ball, but it led to fierce recriminations and finger pointing in Washington, DC. What was next? If the Russians could weaponize orbiting platforms or other planets, there would be no way to stop them. "Who Controls the Moon Controls the Earth," the next day's headline read, echoing the *Collier's* issues from earlier that decade. Indeed, von Braun and his peers had been sounding this alarm for some time already. The American military needed to mount a Manhattan Project level of interest and funding to beat the Russians to the Moon.[21]

Sputnik made a rapid response possible. "To be second to the Soviet Union in establishing an outpost on the moon," the Army proclaimed in Project Horizon, its plans for lunar bases, "would be disastrous to our nation's prestige and in turn to our democratic philosophy." The Air Force silently agreed. In its own top-secret plans for a lunar base, Lunex, it projected being on the Moon by 1968.[22] The fact that each service undertook its own Moon-base planning was indicative of the interservice rivalries and budgetary battles that plagued the defense establishment during the Cold War. That it was inefficient and wasteful is not in doubt; but this approach also highlights the ways that strategic and military leaders channeled space programs to their own ends, driving contracts and conceptual plans about outer space well before NASA was able to engage in the same questions.

Regardless of who was doing the planning or which agency tried to control the narrative, it all came back to the same question: How could it be done? How could the United States build in a place no one had ever been? For engineers and government planners, the simple reality was that in 1958, 1959, 1960, and beyond, no consistent data had been collected about the Moon's surface. Astronomers and geologists could provide highly sophisticated theories of what was up there, but the engineers wanted physical proof. At the same time, administrative turf wars kept information from flowing freely. The most up-to-date information was thus not always shared quickly. To the engineers, it seemed that the closest images of the Moon remained far away, hardly the up-close terrain analysis they had become accustomed to on Earth. Engineering teams would have to extrapolate from the reports, images, and hypotheses of the men who had been writing about and examining the extraterrestrial already and, in some cases, from the pictures and stories that splashed across the popular press.

Of course, a few myths could be quickly dispensed with. The Moon was not made of cheese. There was no man on the Moon (yet), nor was there a rabbit, or Nazis, or the Selenites that H. G. Wells had written of in *War of the Worlds*. Yet rumors and groundless suggestions abounded. There was a hypothesis that rubber boots would react violently with the interstellar dust that coated the lunar surface, leading to miniexplosions. Others believed—as Arthur C. Clarke had written in *A Fall of Moondust*—that the Moon was covered in an ocean of fine-grained particles that would swallow anything that landed. In some accounts, the Moon was made of a fluffy, ash-like substance; others suspected it was rockier. Newspapers mused that the astronauts might "need snowshoes" or that a lander would need pontoons to hover on sloshy, unstable ground, not quite liquid, not quite solid.[23] NASA and the engineers had to consider them all. Everyone understood that an early lunar disaster would spell a very early end to space exploration.

Not even the 1961 *Handbook on Environmental Engineering* could provide definitive data. Designed to be used by Air Force teams and industry partners involved in construction and operations, the handbook outlined all known environmental features on Earth and in space. Sections on Mars, Saturn, Venus, asteroids, and the Moon were included. These were defined as "exotic environments" that were "strange or foreign to the Earth" and required special attention.[24]

Of them all, the Moon remained something of a riddle. More was known about Mars. Basic statistics about the Moon were available—distance from Earth, diameter, volume, density, gravity, albedo, and orbital speed—but the details of what men would find once they landed were scarce. "It is believed," the handbook said, "that the Moon's surface is covered with pulverized rock and dust" (a belief that turned out to be accurate). Yet, beyond that, "no definite statements can be made about the mineral composition of thickness of the upper layer."[25] Hardly inspiring stuff.

To fill in the gaps, the Army personnel involved in Project Horizon, including von Braun, looked to expertise closer to home. Men working in the deserts and Arctic were quickly invited to consult space programs on things such as materials, equipment, mobility, and navigation. Paul Siple, for example, the Arctic expert, was called in to help evaluate the psychological effects of a crew being isolated and remote for so long.[26] Climate chambers were repurposed for "hyperenvironments" or "exotic" environments. The sort of analog mapping and analysis that Hubert Wilkins had worked on for the world's deserts was pulled up into the cosmos.

It was hoped that extrapolating to space would be a matter of degree, not a wholesale rethinking of what was needed. For example, a drill that could bore down into ice and snow might work on the Moon, even if the strength required to maneuver it would be different. Engineering centers were enlisted to test new types of tires and treads, new drills for working in odd materials. What about Earth could be abstracted and known and then applied to military planning? Could the environmental categories the military had constructed be extended into space?

Data collection was not going to be left to best guesses, of course. Once the United States had made the decision to send men to the Moon through the Apollo program, earnest efforts to gather intelligence about the lunar surface commenced. President John F. Kennedy wanted boots on the ground before the decade was out, and everyone involved wanted to make sure those boots landed on a spot that could support the man wearing them. No one wanted a *Fall of Moondust* sort of scenario. Losing a man in space because he fell into lunar fluff would be a catastrophe not just for the man, but for the nation. For a space program built in prestige—getting to the Moon first, besting the Soviets—these were not insignificant considerations.

Throughout the 1960s, increasingly sophisticated probes and orbiters were thus sent toward the Moon, sometimes ramming into it, other times circling around. The images that were returned and the data collected would help engineers and scientists come to a better understanding of its surface and terrain. But as far as the military engineers were concerned, the information was streaming in slowly, far too slowly for the type of programs they were being asked to build.

Indeed, the gap between what was expected and what was known became especially acute on November 27, 1962, when the chief of engineers received NASA's invitation to build a manned lunar base. Overlooked in the documentary record until now, NASA picked up where the Army and Air Force had left off: when and how could the United States build some sort of base on the Moon? Presumably a NASA-base would be for research and testing, more like Camp Tuto than Thule Air Force Base, but who really knew? In its 1962 query to the Corps, NASA indicated that the space agency hoped to have a temporary base in just seven years, with the capability of hosting twelve to eighteen men by 1975. "Time," the Corps explained with some understatement, "is critical."[27] Yet it was not on their side.

★ 15 ★
MOONDUST

At 10:03 a.m. on May 10, 1960, the House Committee on Science and Astronautics convened in Washington, DC. The chief of engineers, Lieutenant General Emerson C. Itschner, had been invited to provide insight into the Corps' role in the country's space programs. Itschner traveled with a small entourage to help extol the work of the engineers. His team included a military construction officer, the Corps' chief scientist, and a man from the Engineer Research and Development Lab.[1]

There were four areas in which the Corps had capacity to support extraterrestrial programs: mapping Earth; mapping the Moon; terrestrial construction in support of space activities; and construction on the Moon. As the government's leading construction agent, Itschner explained, the Corps had abundant practice with sprawling, technically sophisticated facilities.[2] In fact, as Itschner spoke, ships were being packed for the trip north to Thule, filled with materials needed to build Camp Century.

It was not just construction that the Corps and the Army were good at, Itschner insisted. Ongoing efforts to understand the entire planet were also part of the Corps' purview. Itschner gestured toward his agency's support for early satellite programs. So much remained unknown about Earth as a planet. New Corps programs would help the government finally measure the actual size and shape of the Earth, as well as pinpoint the location of remote Pacific islands.[3]

"Satellite-born cameras," he explained, were vital to "mapping on a global scale." It was that very ability that he explained should be sent to the Moon.

Itschner's remarks punctuated the tight links between military and civilian space efforts even as authorities were trying to publicly disentangle the two. NASA had been created specifically to avoid the whiff of militarization; space was supposed to be for "all mankind," not just for men with guns. But the reality remained that the US military—because of its global expertise and Earth-spanning infrastructure—was the organization best poised to move into the planetary realm. Itschner spoke to this tension directly: "I wish to make it clear," he explained, "that these proposals are not predicated on a military requirement for bases on the Moon."[4] Rather, he continued, the interest of the Corps was in making the most of existing competencies.

Indeed, Itschner declared that the Corps could and should be called on to do much more. The real point of his remarks was to make sure that in the flurry of space spending, the engineers were not left behind or overshadowed by a louder, brasher agency (i.e., the Air Force). This was doubly important because the rise of funding for space programs coincided with a decline in money for research programs in extreme environments. New programs had a way of siphoning resources off from older ones. To an outsider, defense budgets might look endless, but any limits were usually fought on precisely those sorts of margins.

Leaving aside the question of money, Itschner pointed to the global infrastructure of bases and facilities that the Corps had already put in place as proof of their abilities. "The Corps of Engineers can perform, design, construction, and [provide] real estate services for NASA," anywhere in the world, Itschner explained, just as engineers had recently done in Libya and Morocco. An additional example often thrown around was the construction of Air Force ICBM sites, thousands of missile silos being constructed around the country from which rockets could blast off into the heavens, carrying nuclear warheads. The knowledge gained through terrestrial military construction and the "worldwide district organization"

FIGURE 15.1 The US Army's proposed Moon base. In the late 1950s, the Army's Project Horizon crafted blueprints for a lunar base that bore striking resemblance to Bonestell's space art and to Camp Century plans. General Itschner presented the design to a congressional hearing in 1961. US Army.

it had created made the Corps capable of "meeting the need of all military and nonmilitary construction in support of space." Clearly, the Corps had the ability to do as much. Indeed, the space program would require a global network of facilities including radar sites, communication links, and stations from which the astronauts would be retrieved upon returning to Earth. For example, NASA created an extensive network of antennae around the world. Armstrong's first comments from the surface of the Moon were picked up by an eighty-five-foot antenna at Honeysuckle Creek, near the city of Canberra in Australia.[5]

Toward the end of his allotted time, Itschner pivoted to the Corps' ability to project into the cosmos. He unveiled an artist's concept of an "early Lunar base" (fig. 15.1). The Army's familiar-looking lunar base is a combination of a Bonestell landscape painting and a Camp Century blueprint. The horizon is ringed with peaks; Earth hangs in the dark sky.

Itschner explained that prefabricated tubes would be sent to the Moon and buried partially into the regolith for protection from the harsh elements (this was the same architecture as for the radar sites in Greenland, though to poor effect). Similar to Camp Century, the Moon base would be powered by a portable nuclear plant. We have "a vast store of knowledge and experience which can be utilized to assist the national space program in the field of lunar construction," Itschner said. The Corps had lab space already devoted to the extremes and experience in power generation, soil trafficability, and construction "under severe conditions here on Earth." That very summer, in fact, the Corps would be building a city under the ice, in a place "warm and friendly" compared to the Moon.[6]

The Corps created the Space Sciences Section precisely to channel its existing competencies toward the extraterrestrial.[7] This included a proposal for the Extraterrestrial Engineering Center, where the particular difficulties of building in space could be addressed (fig. 15.2). In doing this, the Corps was following a long-standing strategy for dealing with extreme environments on Earth: create a special research division devoted to them. In the late 1940s and early 1950s, the Corps had set up the Yuma Test Station for desert work, and the Snow, Ice, and Permafrost Research Establishment for cold and snow. At a space center, it would be able to more holistically address basic questions of extraterrestrial construction. Would the wonder of modern global construction—the basic bulldozer—work in the cosmos? Would space dust behave similarly to terrestrial materials such as sand or ice? How would lack of gravity change materials?

The organizational systems and processes might be familiar, but the engineers working in this new hyperenvironment clearly understood that space upped the ante. Collaborative, interdisciplinary teams were going to be the norm. The Space Science Section team included not only engineers but also chemists, physicists, and an astronomer.[8]

Just as new weapons and technologies had opened the world up after World War II, fostering new ways to imagine things, space travel threw open even more possibilities. But reliance on existing pathways, institutions, and questions led not to new and stunning ways of

LUNAR ENVIRONMENT SIMULATOR

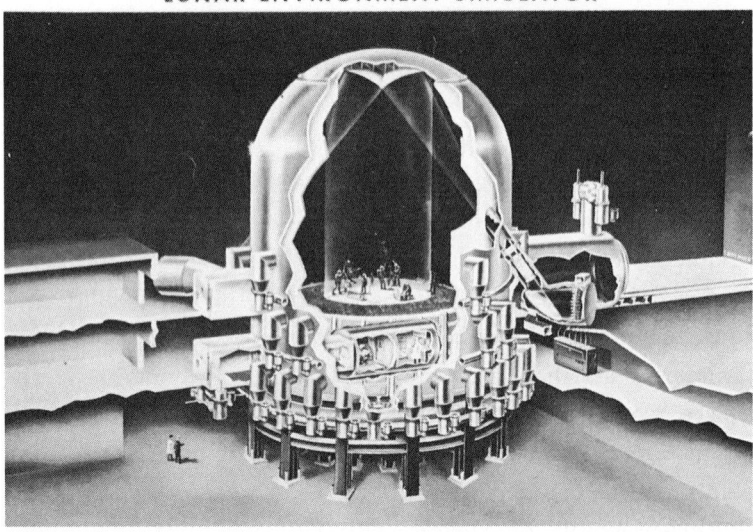

FIGURE 15.2 Lunar environment simulator, the Extraterrestrial Engineering Center. The Corps proposed to build a space simulator as part of a much larger research center, but it was never authorized. As with climate chambers used for terrestrial extremes, the engineers wanted to simulate all aspects of work on the Moon, including moondust and rocks. From, *Lunar Construction*, US Army Corps of Engineers.

imagining Earth and the cosmos; rather, it replicated familiar tropes about power and access. There were glimmers of different ways of seeing the Moon. But the reality of who could make space exploration happen—the US military, despite the creation of NASA—meant that the project would be constrained by strategic visions. How else to explain that the United States sent men to the Moon six times and then simply stopped. Tied to the defense infrastructure, space exploration would remain tethered to earthly ideas about prestige.

The man put in charge of the Corps of Engineers' program evaluating cosmic construction does not seem to have been a classic space enthusiast. In 1962, Bruce M. Hall was called to Washington to run the Space Sciences Section. Hall grew up in a military family and dreamed of being a doctor like his father, but financial constraints kept him from enrolling in medical school. Instead, after World War II, he received his bachelor of science in geology and soon thereafter joined the Corps. As an engineering geologist in California, he rose rapidly through the ranks and eventually oversaw the construction of major dams, such as the Isabella, Success, Terminus, and New Hogan facilities. In 1962, he was recruited to the Corps' Extraterrestrial Research Agency. Clearly his expertise with major earth-moving projects had appointed him well. He moved across the country to Washington, where he began work on the engineering model for the Moon.[9]

Hall understood that in addition to all the other variables that space threw at him, the entire idea of a lunar construction program depended on the grains of sand or stardust or ash (who knew?) that covered the lunar surface. Indeed, to be able to create a viable construction model, Hall's team needed the equivalent of a bucket of moondust to sift through their fingers and run through various soil tests. Engineering reports to date had been bearish on the Moon precisely because so little was known about it. Information remained "incomplete" and "sketchy" at best. With "limited hard data," the engineers had to rely on "extensive assumptions" about the surface characteristics of the lunar environment. As the various theories of the lunar surface swirled, the engineers admitted that too much information about the Moon was "subject to interpretation and controversy." They could use some cold, hard facts. In the 1963 report *Lunar Construction*, Hall admitted that "a single precise measurement of any of a number of lunar environmental factors" would be far more useful and valuable than "all the theories and debates about moon dust, rock, volcanism, and moonquakes that have persisted for years."[10]

What the Moon is made of was tied up into all sorts of other crucial questions that scientists and engineers wanted answered. Not only for construction purposes but also for landing, moving, and understanding the solar system itself. How had the Moon formed? Today the dominant hypothesis is that the Moon is likely a chunk of the same mass that became Earth. Astronomers had long seen a surface dominated by craters, but how were they made? Were they caused by long-ago volcanic activity, which would have been suggestive of a molten inner core, or by perpetual storms of meteors hitting the surface? By 1960, most North American researchers favored the latter theory. But at the time all ideas were still hypotheses.[11]

Through the early 1960s, while most of the nation focused on the more glamorous side of space exploration—rockets and orbiters, astronauts and space suits—construction engineers had to get dirty. Questions about materiality—what a place was made of and how to work the materials found there—continued to link engineering plans and practices across the extremes. In fact, the more fantastical a potential construction site became, the more the engineers had to reach across the world for useful terrestrial examples. Which of their existing Earth-based experiences might help them best prepare for outer space? How could desert sand and the behavior of granular materials be helpful when thinking about the Moon? Would granular materials behave the same without moisture of any kind? What about the effect of gravity? Was building in ice a good way to plan for lunar dust, a material that soon became known as *regolith* to distinguish extraterrestrial dirt from that on Earth?

The ultimate destination was certainly new, but the practices and processes that had to be thought through and mapped out were familiar. And yet space construction was like taking the difficulties of the desert, the Arctic, and the deep ocean and jamming them all together, making something new and even more challenging.[12] Building in space was a ratcheting up after all.

Even men in spaceships needed good terrain maps. That is why, in the late 1950s, the Department of Defense threw its support behind several parallel programs, including a US Geological Survey (USGS) lunar mapping project. Indeed, USGS and US military interests had a long been aligned. During World War II, it was the USGS that crafted terrain assessments of global battlefields. There was thus no question in the late 1950s that the level of detail required of strategic space maps went far beyond landing bombers on ice or sand. They could not have a space vehicle land on a crevasse or a cliff, or hit the ground at an odd angle and tip over. The Air Force Aeronautical Chart and Information Center created the Lunar Aeronautical Chart, including artist-drawn maps and images. At the same time, the Corps of Engineers collaborated with the USGS's Military Geology Unit to create the 1960 *Engineer Special Study of the Surface of the Moon*. As

Lieutenant General Itschner explained to Congress, men from the Corps had conducted a remote survey of the Moon (a process he called "selenodesy") with an eye toward future landing and identifying potential construction sites.[13]

Not surprisingly, the sorts of terrain assessments and mapping techniques used to try and fill in the blanks in hard-to-reach terrestrial places were redeployed in space. For example, aerial photography and terrain assessment used in Greenland to help site research and military stations could be reframed for lunar landing sites. In the Sahara, photographs taken from the air were used to consider the morphology of desert dunes so engineers could best locate emergency runways such as El Uotia and consider how landforms might best reveal other natural resources, such as oil deposits. Even if the Corps' Military Evaluation of Geographic Areas analog mapping program had not proved as useful in creating comparative desert areas, it had taught folks how to read Earth for signs of what was there. The tools developed and practiced in the terrestrial extremes were redeployed in space, where there was no chance of setting feet down until good reconnaissance had been done.[14]

Hall and the engineers could learn a lot from these ever-more-sophisticated maps and theories, but they drew most from their own cadre of experts. The bibliography of the two-part report *Lunar Construction* that the Corps submitted in 1963 highlights the tight links that stitched together various engineering centers. Dozens of men attached to Cold Regions Research and Engineering Laboratory (previously known as the Snow, Ice, and Permafrost Research Establishment) were consulted, as were a handful from the US Engineer Research and Development Laboratory.[15] Their expertise was augmented by industry and academic experts who also moved in and out of government work.

Lunar Construction was truly a sophisticated amalgam of previous plans and ideas, many of which had ties to some of the terrestrial extremes. Others were linked to the early work of the space boosters like Willy Ley and Wernher von Braun. Hall and his peers described the ways that the first six "pioneers" (projected to arrive in 1970) would set up facilities for more. In 1971, eight more men would arrive for three-month tours and continue to conduct reconnaissance, recover materials, create construction sites, scout for resources, and so on. In 1972, eight more men would arrive to create a space port and continue testing and conditioning construction equipment. Finally, they would identify the best location for the semipermanent base that would be built in 1973. The men would also set up a nuclear-powered generator, similar to the one at Camp Century. It might be possible, the engineers suggested, that two men could "overwinter" on the Moon in 1972–1973.[16]

The parallels to Arctic construction were obvious. "Probably the closest terrestrial approach was found in the severe challenges which were successfully met in our polar bases," the engineers wrote. But even there it had taken approximately ten years. Way back in 1952, Henri Bader had been digging the test pits that would lead to bases and research camps, places like Camp Fistclench and Camp Century. Those men had years of study, months of constant trial and error, water to drink and air to breathe. Far more time would be needed for the Moon.[17] Camp Century was a stepping stone, but a cautionary tale rather than a success story.

Supply was one area in which they had to find synergies. The logistic networks that the Corps of Engineers and broader military had put in place to provide reasonably steady goods to men in remote locations had to be refigured for outer space. "No plan for lunar construction will be meaningful without a definition of the logistical support available," the engineers explained.[18] Even if they could get a man to the middle of the desert, the top of the ice, or the surface of the Moon, they had to be able to supply him again and again. A significant amount of time and money would be spent stockpiling resources. This is where the logistic power of the US military came into play—moving things, packing things, storing things—competencies built on a massive scale since the 1940s. On the Moon, everything would need backup. If Camp Century had sixty days' worth of extras, and Arctic weather stations had four hundred, what would a base in space use? (Certainly more than the thirty days of excess allocated in the 2015 movie *The Martian*.) But what were the engineers to do about the supply glitches that had plagued Greenland? If bad winter weather kept them from getting fuel to Camp Century, leaving GIs without hot showers for weeks, which obstacles might keep them from reaching men in space?

Lunar Construction tackled the most important ingredient of all: water. The engineers identified the daily quantity needed by humans in space from careful assessments going back to wartime laboratory and physiology tests. During the first thirty days, each man would need 0.71 gallons per day for drinking and food prep; there was zero allowance made for hygiene and cleaning. Those extras would have to be added over time; after a year, once they had figured out how to recycle what they had, the men could use 2 gallons per day, and as that technology was perfected, they could expect 3.4 gallons for consumption and care.[19]

No matter the amount of analogous planning and designing, at the end of the day, the engineers wanted their Moon rocks. Since World War II, acquiring samples and testing materials had been the linchpin to successful engineering in the extremes. But until a man landed on the lunar surface, bent down to scoop some moondust, and returned safely home, there would no "real" stuff to play with. In the end, of course, no astronaut would actually bend down to scoop the dust; they would use specially made shovels. No one wanted an astronaut's hands or gloves to contaminate the collected samples.

Absent good firsthand data and material, Hall and his team had to contemplate how to fill the void. They needed chemical analysis and information on the mechanics of lunar regolith. Hall wanted to know the ground's bearing strength and the makeup of its grains. How would the lunar surface behave? Casagrande's studies of materials needed a cosmic update, and Hall and his team were the men to do it. Would moondust support the weight of structures? How could they be sure? An engineer model had to take into account basic construction, to be sure, but also the "tools, equipment, techniques and knowledge required" to do all things on the Moon.[20] Hall wanted to be able to create the sorts of charts and tables that Bader had made for snow, or Bagnold for sand, and classification systems for the grains and slivers the men would find when they arrived.

Training was equally as important. As *Military Engineer* explained, "Experience with construction in hostile environments" on Earth had shown that outer space, too, would require "a long program of engineering research and development." "It would be foolhardy," said Itschner, "to try to send anybody to the moon without having a laboratory . . . in which to test man against moon conditions."[21] The lunar environment was surely going to constrain how a person could act. Drilling might be harder. Even walking might prove awkward with gravity one-sixth of that on Earth. Certainly moving materials and rearranging terrain would require special care. All this meant the engineers had to fill their lunar simulators with materials that could mimic the Moon. Not real moondust, but as close a fake as the engineers could make.

The development and manufacture of fake lunar regolith was integral to *Lunar Construction*'s grand plan. At the proposed Extraterrestrial Engineering Center, "a wide range of manned and unmanned engineering activities will be conducted," with fake lunar soil, including "excavation by hand and by machines," and "assembly of equipment, welding and cutting metals, tunnelling, drilling, coring, quarrying without explosives, and testing of equipment and systems including construction machinery and vehicles." All of that, of course, required the most accurate version of the "lunar surface and subsurface environments."[22] A variety of agencies and companies were already creating lunar simulants. At the Extraterrestrial Engineering Center, teams would test various derivatives and keep track of simulants, cataloging fakes in the same way the Corps had cataloged Earth's sand and snow.

But how to make the best fake regolith? Enter the geologists. Their studies and conclusions about the makeup of the lunar surface would feed directly into any terrestrial formulas. Combinations of terrestrial materials could be identified and pulverized, then recombined into mixes that might act like moondust. One Apollo-era simulant was a mixture of "sand and kaolinite," subsequently shown to be entirely inadequate. Others consisted of crushed basalt mixed with crushed glass (MLS-1 lunar soil simulant). Slivers of silica were considered essential properties, sharp jagged grains that were very unlike those on Earth. Indeed, the Moon is made of slivers: the regolith on the Moon would prove sharp and angled, as the Moon lacked erosion and weathering, which turned terrestrial grains into rounded particles.[23]

And yet no simulant could perfectly represent the real thing. Just like the sands of one desert could never really mimic the grains of another. That was clear as soon as astronauts landed on the Moon. Teams had practiced digging and drilling in the fake stuff, but astronauts found that the real stuff would not give below ten centimeters. "It's a very soft surface," Armstrong had reported when he was digging his first samples. "But here and there where I plug with the contingency sample collector," he added, "I run into a very hard surface." Drilling failed. They had a hard time planting their flags. Excavation often led to severe abrasion. Dust accumulated in the Lunar Roving Vehicles radiator and required maintenance. It covered the astronauts and filled the lunar module when they went inside. It left some of them coughing and sneezing. In some photos, the astronauts are covered in the stuff, smudged with a chalky gray substance that almost looks like soot. The engineers had underestimated the effect of impact on the Moon. Each small object that hurled into it created finer and finer shards on the surface. These fine, sharp particles tore through the layers of fabric that made up Moon boots and space suits. Even packaging samples became compromised as lunar dust managed to leach into the seals. Moondust was perhaps more like sand than anyone realized.[24] These were the very sorts of intimate, in-person details— and problems—that the engineers tended to fret about.

The mysteries of moondust continue to animate twenty-first-century conversations about NASA's plans to return to the Moon. The Artemis mission hopes to land women and men on the lunar surface by 2030. But according to Dr. Erica Montbach, manager of NASA's Planetary Exploration Science Technology Office, moondust "is very damaging in ways that we don't see on Earth." As a result, study continues on its effect on humans and machines, and on ways to mitigate its most damaging qualities. Not only can the material be abrasive and damage machinery and tools; it leads to serious health effects. When the astronaut and geologist Harrison Schmitt returned from the Moon in 1972, he reported having suffered a sort of lunar hay fever as the particles settled in his lungs. "One of the most aggravating, restricting facets of lunar surface exploration is the dust and its adherence to everything no matter what kind of material," Apollo 12 commander Pete Conrad explained. "Whether it be skin, suit material, metal, no matter what it be and its restrictive friction-like action to everything it gets on."[25] What the Apollo missions did was show how the smallest of particles could prove to be one of the most befuddling problems, able to harm astronauts, surface systems, tools, and machines in ways that have to be addressed if longer-term missions are to be carried out.

It is not just the effect of lunar dust on astronauts and materials that is of concern. Scientists remain uncertain of how granular materials will behave in outer space, largely because they remain uncertain of how such materials behave on Earth. Using moondust or Martian regolith in construction or in the manufacture of needed materials will hinge on how the stuff responds to digging, drilling, moving, and piling. Landing and taking off from the extraterrestrial surfaces, too, depends on how the material functions in relation to terrestrial designs. But the physics of granular materials remains poorly understood. No one can explain with total certainty how sand flows through an hourglass, why grain silos can spontaneously collapse, or where and when a landslide or avalanche might take place. Taking things to more extreme environments only ups the ante. "The potential for major catastrophe is high," a 2005 NASA study warned, "if human lives will depend on the granular behavior of extraterrestrial soils and regolith."[26] Changing variables like humidity, temperature, or surface conditions can drastically alter how equipment and materials respond. One result has been more intense focus on granular physics on earth to better grapple with what astronauts will find in the cosmos.

FIGURE 15.3 The surface of the Moon. NASA's unmanned Surveyor III soil mechanics surface sampler scoops from the Moon's surface. The Surveyor landed in 1967. This image was taken in 1969 by Apollo 12 astronauts who visited the probe. From 1969 to 1972, the Apollo missions brought more than eight hundred pounds of lunar material—rocks, dust, pebbles, and core samples—back to Earth. Many of those rocks are still being studied for information about the Moon. AS12-48-7129, NASA.

The first scoop of moondust was made not by a human, but by a small metal shovel attached via an accordion-style arm to the Surveyor 3 probe (fig. 15.3). It was the culmination of a handful of NASA missions to better understand the lunar surface before Apollo would land men there. The first missions were called Ranger, and those vehicles were aimed at taking photos and smashing into the lunar surface. The Surveyor program, however, was to conduct "soft" landings, showing the possibility of setting a machine down on the Moon. In 1966, the first Surveyor, had—at long last—achieved that aim, providing a useful midcentury selfie: a photo of its metal foot gently pressing into the very solid lunar surface. This was outstanding evidence that the surface could hold the weight of men and machine.

Geologists and engineers pored over images of the small holes and slashes that Surveyor 3 made in April 1967. The largest trench was fifteen inches long and five inches deep; the trench's sides held up well, and the regolith stuck together in small clods. "The soil," declared the chief scientist for the digging program, "behaves in perhaps a disappointingly ordinary way." NASA investigators explained that the ground consisted of a "lumpy, granular soil" that looked somewhat "like beach sand after the tide goes out."[27] The Surveyor 5 launched later that year carried a soil-sampling kit that gave even greater detail about the chemical makeup of the Moon. Engineers ran the data against their classification systems, hoping to find potential matches.

These images and data transformed understanding of the lunar surface. It was not jagged and mountainous; it was smooth and old. Bonestell's *A Lunar Landscape* mural was quietly removed from the wall of the Boston Science Museum, rolled up, and put in storage. He had painted the Moon as "it ought to be," not how it actually was.[28]

By the time the geologists and engineers got their Moon samples in 1969 (the Apollo 11 astronauts returned with eighty pounds of the stuff), the US government had decided not to pursue lunar construction. It was too costly and, frankly, unnecessary. Moreover, international conversations about the peaceful uses of space meant that the military should not be involved in Moon landing. As a result, few of the engineering dreams ever came to life. *Lunar Construction* was written, studies undertaken, and detailed, beautiful maps of the Moon created. But a Moon base never came. As a result, it is easy to see why discussion about lunar construction rarely figures into discussion of the space history today. But stories of regolith and rocks, engineering centers and base blueprints, are crucial in understanding how seamlessly ideas about extreme environments traveled. This was how knowledge was made and structured. Attention to the details—the dirt and stardust and ice and sand—provided continuities for engineers working rapidly across extremes. Indeed, these connections also make legible the ways that plans that might seem far-fetched today seemed more plausible in the 1960s: they were part of a continuum of work. The expansion of the US defense apparatus across all terrestrial environments had necessitated a military engineering apparatus capable of managing and imagining what seemed utterly impossible. Space was merely the next "wilderness" frontier.

Importantly, space was a frontier that was envisioned in strategic terms, normalized and described in the same terms that US military programs used to discuss extreme environments on Earth. This helps explain and show the narrowness of space visions in the 1960s (at least government-funded ones). The push into new places and new environments did not translate into new ways of thinking or new strategies; it merely replicated old ones.

There is much more we know about the Moon today, of course. It is not merely a ball of rocks and dust; it is integral to life on Earth as we know it. Whether Earth's ocean tides, tilt, or magnetic field, terrestrial processes depend on the Moon's quiet presence. So much so that as astrobiologists search for life on other planets, they must contend with the possibility that a moon like ours might be a prerequisite for it.[29]

★ 16 ★
ANALOG EARTH

"Mare Tranquillitatis, Flagstaff, Arizona," was the setting of a 1967 film about the space program (fig. 16.1). It was there, at a place locals called Cinder Hills, that the US Geological Survey and NASA had decided to re-create a small patch of the Moon. Because "there's no place on Earth that offers a topography quite like" the lunar surface, the narrator explained, it had to be faked. Using increasingly detailed photographs of the lunar surface, a common bulldozer, and a lot of TNT, the geologists excavated 143 craters, as close in depth, size, and placement as they could gather from the images returned from unmanned rovers. The sort of photographic forensics that had been perfected over years of studying the Earth's far-flung areas was being brought to bear ferreting out the secrets of the Moon.

Cinder Lake Crater Field No. 1 had a simple purpose: astronaut training (fig. 16.2). Years before Neil Armstrong took that famous first step in space, he had to practice what to do and what to look for. His description of the "very fine grained powder" came not from some innate sense of regolith, but from training. If astronauts were going to be the only eyes on outer space, the only men to intimately interact with an alien world, they had to know how to observe, describe, sample, and catalog what they found.

Crater Field No. 1 was just one of many training sites created for the Apollo program. Field No. 2 was created the next year, and more after that. *Military Engineer* reported on a fake field replete

FIGURE 16.1 Apollo 17 astronaut and geologist Harrison H. Schmitt on the lunar surface with a scoop used to collect samples (1972). AS17-145-22165, NASA.

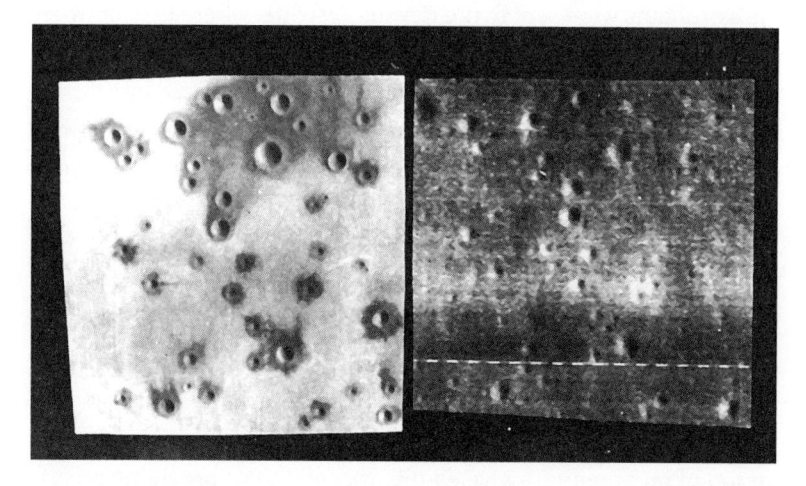

FIGURE 16.2 The Cinder Lake Crater Field (right) and the proposed lunar landing site to be simulated (left). In the late 1960s, the USGS and NASA created a crater field in the Arizona desert to mimic potential landing sites for Apollo 11 missions. P421, F867187, USGS.

with craters created on Long Island. In these analog landscapes, astronauts-in-training bumped over volcanic gravel in vehicles with tires and parts being tested; they maneuvered scoops and shovels in their bulky space suits; and they practiced the geologists' parlance for describing a cratered landscape. What were the best words for talking about ejecta, lava flows, and cascading edges? Which observations should they make and how? At Cinder Lake, the trainees walked across a terrain made distinctly unlike Earth even as they could look up and see a horizon crowded in ponderosa pine.[1]

Simulated spaces—fakes, copycats, virtual, stand-ins, analogs—were a requirement of the space age. But they were not new. In creating the labs, climate chambers, and analog environments for space programs, NASA drew heavily on work that the defense establishment had been doing for decades for extreme environments on Earth.[2] Not only were labs and climate chambers repurposed; the extremes themselves became simulacra for space. What better way to practice for a hyperenvironment than to get comfortable in the merely extreme places nearer to home?

Not all such places had to be remade for space. On the contrary, Cold War militarism had already pockmarked swaths of the desert with nuclear and chemically created craters. And from 1970 to 1972, the astronauts were sent to the Nevada Test Site, where they crawled across a nuclear wasteland that was suddenly made useful again for mimicking the Moon.[3] The Schooner crater was produced by a thirty-kiloton blast that created a hole 200 feet deep and 850 feet wide, and geologists thought it was an ideal place for astronauts to learn about the types of formations they might see in space.

Blasting craters into the desert floor was just one way to turn Earth into the Moon. Across the globe, a wide range of landscapes and labs became lunar, and later Martian, replicas. The best-known sites were the chambers where men and technologies were tested. Astronauts practiced landing, launching, and maneuvering in the Lunar Module Mission Simulator. The Rendezvous Docking Simulator, a large ring on top of an antenna, swiveled this way and that, providing an opportunity for men to try and dock one vehicle to another. Inside the Reduced Gravity Waking Simulator, the astronauts mimicked space walks. To test weightlessness, they flew a C-131 transport (nicknamed the "vomit comet") to thirty thousand feet and let it drop. The trainees and their lunch floated about.[4]

Beyond the well-photographed and well-documented chambers, every place on Earth became a potential space analog.[5] Astronaut training and materials testing took place outdoors, in deserts and polar regions where the extremes could begin to approximate the discomfort of the intemperate and the disorientation of being in a place without trees. Trainees became world travelers well before they ever reached a launch pad. Just as the Sonora Desert became a stand-in for deserts everywhere, the world became a mirror of the solar system.

But it was a mirror with a strategic point of view, one that reflected ideas rooted in military capabilities and planning. The terrestrial training areas that the astronauts visited were all strung along the routes of US strategic power. Just as the decade before Wilkins had traveled on US aircraft, the astronaut programs used the same machines and routes to move about the world. In Iceland, they stayed at the air base's officers' club; in Arizona, they traveled on military craft. The astronauts themselves would have found this familiar enough, because—with one later exception—they were all military pilots. NASA was a civilian agency, but its guts and infrastructure—the pathways that made its missions possible—were born in the military. The result was that even as scientists tried to insert more and more science into the Apollo and lunar missions, their ideas were often subordinate to the simple Cold War mission of getting men to space first and bringing them home again.[6]

There were many reasons the desert became a space-age destination. Percival Lowell knew a few of them: clear skies and thin air made for good stargazing. Observatories proliferated across the high desert of Arizona. The Air Force, too, liked the desert because the weather was good for flying. That, and there were few people around to get in the way. The Air Force created testing ranges in California, Nevada, and Utah. In 2022, NASA described Arizona as an ideal training site for future space missions precisely because it was "analogous to a lunar

FIGURE 16.3 Eugene Shoemaker. The father of astrogeology is shown here in a space suit collecting rock samples in a lunar simulator. Shoemaker was vital to NASA's plans to acquire information on the Moon. USGS.

environment." Today, the Atacama Desert in Chile is home to nearly half of the world's astronomical infrastructure.[7]

Lack of stuff was the main reason the desert could become the Moon. Not just lack of people, but the absence of organic life altogether. Stripped of grass and dirt and trees and shrubs, the desert is a geologist's playground. But—as Lowell had intuited—there was also something sublime about watching the cosmos from a place that seemed most like it.

The men of the USGS's Astrogeological Branch agreed. In 1963, Eugene Shoemaker, the group's creator, moved his team from California to Arizona to better take advantage of observatories and proximity to craters (fig. 16.3). In 1964, the branch used the base to start astronauts' geological training.[8]

Shoemaker's trajectory was not uncommon for space enthusiasts at midcentury. He had long harbored a fascination with the Moon, but professionally it was considered a dead end. As a kid, Shoemaker had been drawn to minerals and rocks: he collected them and read about them. Ultimately, he earned a PhD in geology. In 1950, he joined the USGS and began mapping uranium deposits in the American West. When the Army began experimenting with captured V-2 German rockets, Shoemaker realized that one day those rockets would make it to the Moon. He wanted to be there.[9]

The young geologist's interest in space and rocks collided serendipitously in Arizona, at a place called the Meteor Crater. Shoemaker's work with the USGS had been mapping uranium deposits. The Cold War requirement for uranium, and later plutonium, kept geologists busy for some time. As part of this work, Shoemaker went to the Nevada Test Site, where he investigated craters created by atomic bomb tests. The experience there gave him new data and ideas for how to see craters elsewhere, namely in Arizona. Shoemaker explained that Meteor Crater was created not by a volcanic eruption, as generally believed, but by impact. An asteroid had slammed into Earth, turning sandstone into glass and creating the bowl that people could look down into. Studying impact craters at home went a long way toward helping Shoemaker and others explain what was going on above. The Moon, too, was pockmarked with rims and depressions. As a geologist, Shoemaker understood that the best way to test out the theories, to really know about the Moon, was going to be to get up there and collect samples. The Apollo program was what would make that possible. Health issues disqualified him from the pool, but he came tightly linked to astronomical studies. In fact, the comet Shoemaker-Levy 9 bears his name: he and his wife, Carolyn S. Shoemaker, discovered it along with Dave Levy.[10]

Where Bonestell and Ley had painted visions of a Technicolor solar system, Shoemaker was at ease with what he saw: a monochromatic Moon, contoured in shades of gray (although astronauts would eventually report hints of red and green, yellow and brown). The Moon was a rock that needed explaining. Like the glaciologists who would soon scale the Greenland ice cap, hoping to drill down to read the climate archive, Shoemaker understood that the lunar crust was its own sort of archive. The layers of the Moon could be read like the layers of Earth for clues about the deep history of the solar system.[11] But the Moon would be ever better since it had not been scraped and rubbed away through processes of erosion and the plate tectonics that kept Earth's crust folding over itself. Shoemaker was committed to using photographs and topographical features to create accurate and usable geological maps of the Moon. His interest was in the pursuit of both science and exploration, although it would take strategic interest in space to get such efforts supported.

In the early 1960s, Shoemaker joined a new arm of the USGS, the Astrogeological Branch, which had a bold task: to study, map, and deeply describe a place none of the employees had been to. As the amateur astronomer Davy Levy explained, because Shoemaker could not go to the Moon to map it, he "proceeded with his fieldwork in much the same way as he [would have] out on the Colorado Plateau." Instead of shovels and a compass, he used his eyes, intently scanning high-quality photographs of the Moon taken forty years earlier. "From these photographs, his team made the first geologic map of a lunar feature."[12]

Shoemaker had pressed for scientists, namely a geologist, to go into space. He believed that only a trained geologist could make the most sense of what they saw on the Moon. He wanted NASA to look beyond test pilots and Air Force officers. They might be fit and capable, cool under pressure, and able to withstand stunning environmental stresses, but they lacked the scientific training and imagination that the moment so clearly demanded. Anyone could scoop dust and dirt into bags, but it took a trained eye to know what to look for, to get the best scoops and samples.

NASA was not interested. For the first few rounds of training and selection, NASA shut down all other avenues. Even a rigorous private screening program for women—known now as Mercury 13— was never seriously considered despite the women's accomplished records. The Soviet cosmonaut Valentina Tereshkova was the first woman in space in 1963; it took twenty years for an American, Sally Ride, to get there. In the 1960s, the American astronaut was white, Protestant, and male. As Roger Launius, a former chief historian of NASA, explained: "They seemed to embody the personal qualities in which Americans of that era wanted to believe: bravery, honesty, love of God and country, and family devotion." They were thus also the embodiment of the ideal man the military had long been planning for, the model man they tested extreme clothing and gear on, the specifications that went into "clo" measurements and into copper-plated thermal manikins that could stand in for human beings.[13] The tension between what scientists wanted and what the military and NASA were willing to consider highlighted the narrowness of a strategic visions, one that constrained options even for the most dramatic program of all: sending humans to outer space.

The Grand Canyon was the first stop on Shoemaker's astronaut geology training program. This was not because the red, pink, and orange walls of the Grand Canyon looked anything like the Moon, but because it was a remarkable place to gain an appreciation for geology. Geologists knew that they could read the past through rock layers and that similar clues were likely to be found in space. Nearly two billion years of Earth's geological past are exposed where the Colorado River channels down through layer after layer of stuff. If they were going to teach someone to read rocks, to understand the importance of reading rocks, why not start with a place where the rocks speak for themselves? It was like geology kindergarten: get to know the basics.

Geologists such as Shoemaker must have spent a lot of time on Earth imagining how to map the Moon. For one, they had to make a lot of educated best guesses about the terrain the astronauts would find. They had their pictures and increasingly good imaging from the various Surveyor missions, but guesstimates were the norm.[14] At the same time, they wanted to identify as many analog sites as possible to match the hypothetical moonscape.

Shoemaker and his fellow teachers had a long list of potential geological features and formations that might (or might not) resemble features on the Moon. In addition to evaluating basic igneous and sedimentary rock formations, the astronauts needed to study all nature of flows (ash, lava, and obsidian) and cones (cinder, spatter, and pumice). There were broad, shallow craters called maars to see and tuff rings, like a maar but with higher edges, to identify. Men like Buzz Aldrin, used to manning flight controls, needed to be trained in making stratigraphic observations and scratching down field notes. Aldrin and his fellow astronauts practiced mapping various features and evaluating aerial photographs. They were advised on what to look for as their spaceships descended to the Moon. For example, they needed to distinguish between impact craters, caused by meteors smashing into the surface, and calderas, formed through volcanic activity. Shoemaker led field trips to Arizona's Meteor Crater to explain how to spot impact ejecta, another sign of meteor activity. There were a lot of things to know—data sets to master, structures to memorize—but the ultimate goal was for the material to cohere somehow into a geologist's instinct. Charts, graphs, maps, and manuals were all good sources of information, but as aviation engineers had been showing for decades, knowledge came by doing. The hope was that the astronauts might get to a point where they could look at a cliffside and figure out rather quickly which section of the cliff would yield the best samples.[15]

It is almost possible to imagine Arthur Casagrande, of the wartime engineering soil school, asking the men to rub materials through their fingers and perhaps take a small bite. Although Apollo 17 astronauts returned with a capsule that smelled like burned gunpowder, and Neil Armstrong described the odor of moondust as being like "wet ashes."[16]

The men participated in a "Moon game" to test their abilities. Teams of two were sent out to an unidentified location and told to complete a set of tasks. The teams had to plan a traverse across alien terrain and collect samples along the way. But these were to be "important" and "representative" samples of area rocks and soil, without the aid of the geologists. The conceit of the game was simple but totalizing: a man and his partner were the only two people who had ever been on—and might likely ever be on—that stretch of terrain. You might be collecting all that was ever to be known. What if they collected the wrong rock?[17]

The stakes were considerable. The men were to be the interpreters for all mankind. Their eyes were the amateur lenses through which humanity would be able to see and begin knowing the extraterrestrial. Their recognition (or failure to recognize) certain rocks and formations could be the difference between finding minerals on the Moon or seeing nothing. Their reading of craters could even help determine the origins of the planets in the solar system. And they would not have a lot of time or space to do it in; the Apollo program was aimed at getting to the Moon and back, not at how many rock samples could be collected and stored.

In this light, the transcripts of the Apollo missions take on new meaning. They are not just men in space suits marveling over a new place; they are looking for things. The astronauts spent an awful lot of their time on the surface of the Moon talking about rock formations and samples, pondering a particular boulder or—as in the case of Commander David Scott during the Apollo 15 mission—whooping in delight. "Oh boy!" he exclaimed. "Guess what we just found?" It was something he went on to describe as a "plag," or plagioclase, "something close to anorthosite, because its crystalline." Scott scooped it up and put the sample in a specially marked bag, number 196, to be shuttled back down to Earth.[18] Scott apparently came to like geology so much that his wife later took it up as a hobby so they would have something to do together.

That sample quickly became known as the "Genesis Rock" because it was believed to be the oldest specimen recovered from the Moon, dating back some four billion years.[19] It turned out not to be the oldest rock, but because all rocks from the Moon are older than those found on Earth, it was plenty old. In any case, it was a piece of the Moon's primordial crust and would hopefully help astrogeologists learn more about the origins of the Moon.

The first round of studies, maps, and images were aimed at identifying lunar landing sites, but the longer-term need was to identify potential lunar resources (fig. 16.4). This was also central to the sort of geological formations that the astronauts could observe, describe, and sample on the Moon. Ongoing lunar exploration and the very future of space exploration depended on the ability to locate materials. The more they wanted to dig down and in, to stay a while, the more they had to find local materials to make that happen. If engineers could dig a well in the middle of the Sahara, they did not have

FIGURE 16.4 Astronaut geology school. Apollo astronauts were trained to observe and identify rocks and rock formations so that they could properly interpret the surface of the moon. Astronauts training at Philmont Ranch, NM, 1964. S64-238880, NASA.

to carry in water; if they could use snow as a construction material, they could drag fewer metal beams and wooden rooms across the ice. What might the Moon hold?

Knowing how to find and use "indigenous materials" would help shorten the logistic bridge that the Corps had to build to space. Could lunar dust be turned to bricks? After all engineering teams in Greenland had successful molded chairs out of perma-crete. Were there chemicals in the lunar surface that might be turned to fuel? What about water? Could the building blocks of life be extracted from Moon rocks? Were there chemicals on the Moon that could be used as propellant? How would moondust react with Earth metals and materials? Rumors abounded about water on the Moon, perhaps stored in a layer like permafrost or as a "lunar iceberg."[20]

These questions and possibilities were at the core of the creation of the Working Group on Extraterrestrial Resources, created by Bruce Hall, the engineer who wrote *Lunar Construction*. Shoemaker was also a member. The working group remained relatively small; there were still not a lot of space "experts" around and because the grand plan was to get a man to and from the Moon as easily and quickly as possible. Mirroring the laments of the Research Development Board in the late 1940s as its panels complained about the lack of US expertise on deserts and Arctic environments, space enthusiasts continued to voice concern about general lack of scientific and engineering engagement with planetary sciences as the space age began in earnest.

The working group focused on identifying resources. "We have reason to expect," Shoemaker explained during its second meeting, "that we may be able to identify different minerals on the Moon."[21] Those minerals would be central to plans for building and living in space. Good maps and reconnaissance of geological features might be clues to underlying resources (think of Ralph Bagnold's work reading the morphology of desert dunes), which could help determine where bases might best be established. At the time, this thinking was novel, but given the limited parameters of the space program—prestige built around strategy—it was never fully integrated into Apollo or lunar missions.

The quest for lunar resources clearly upped the stakes in what the astronauts collected and how they understood what they were seeing. An awful lot went into making sure they could collect properly. The *Catalog of Apollo Lunar Surface Geological Sampling Tools and Containers* (1989) is thick with information and drawings of the various devices designed for the use of a few men on a few trips to the Moon. There were special scoops and drills, sampling pads of cloth and velvet. Core tubes of various sizes, along with tongs and handles; hammers and rakes; brushes, gnomons, scales, bags, labels, and special tool carriers.[22] The processes and tools changed for each mission as new things were learned and problems encountered. No one had anticipated just how profoundly the grit of the Moon would mess with the materials from Earth.

The bags sent to the Moon were prelabeled and numbered so there would be no confusion. (Much like the process of viewing ice cores, the samples were useless if mislabeled or poorly packed). Most of the rocks and fluff the astronauts packaged in bags would be squirreled away in labs and storage facilities, highly secured. Today most of the 842 pounds of rocks, dust, sand, and pebbles returned by the Apollo missions are stored at the Lyndon B. Johnson Space Center in Houston. Some of it is displayed in museums. Some of it has gone missing. President Nixon gave out small samples to every nation as a sign of goodwill. In 2002, two NASA interns were arrested for stealing a safe containing some of the treasure. The Soviet Union, too, collected lunar samples during its unmanned Luna probe program, a few fragments of which were sold in auction for $855,000 in 2018. Most recently China was able to bring samples back through the Change 5 mission.[23]

Not unlike the ice cores drilled in Greenland in the 1960s, the Moon rocks collected in the 1960s and 1970s became more scientifically valuable as new tools were developed to analyze them. Indeed, as NASA plans for the Artemis return trips to the Moon in the 2020s or 2030s, the Moon rocks collected decades ago are the basis for what it knows and what can be done. Research teams from around the world continue to apply for samples from the Johnson Space Center, where they are carefully parceled out. In the 2010s, for example, scientists learned that—after years of assuming the opposite—the Moon does, indeed, contain water, and that the Moon and Earth are, in fact, made from the same stuff. New tools, new questions, new people. If twenty-first-century plans for space exploration are to succeed, even more about lunar resources will have to be known, more Moon rocks collected and studied. Lunar resources will be required to create a feasible base for space exploration.[24]

In many ways, the engineering teams of the 1950s understood that they were working ahead of their own time. The technologies to do the extracting, mixing, combining, and processing were not yet in their tool kits. But their job was to imagine and study what might be possible. Little did they realize it would take until the twenty-first century for some of their ideas to be considered viable. That is a good reminder that discussions about lunar bases and the colonization of Mars are hardly new. Nor are many of the problems and solutions being procured to cope with the extremes of space. Twenty-first-century tools, plans, and visions are rooted firmly in US government plans, military and civilian, to reach the stars. There is still a need for better lunar simulants and testing centers, for places where tons of fake Moon rocks can be shoveled and prodded. In fact, a 2016 report from NASA's Lunar Exploration Advisory Group recommended a comprehensive review of simulants and their uses. The field was a mess, full of bad copycats and poor science. No one had bothered to do what the Corps once proposed doing: cataloging them. Although part of the problem was lack of central planning, so was secrecy. Research into extraterrestrial soil simulants was contained in "gray" or "dark-gray" defense publications and not shared openly or widely reviewed. Like information on the terrestrial extremes that had been carefully crafted but kept largely inaccessible, information on space dust was shrouded in the haze of defense imperatives. Not exactly secret, but not easy to find.[25]

Desert and Arctic areas, too, remain central to NASA plans. That is because Earth's most extreme environments are the best stand-ins for outer space. Although some of the locations are similar, the analog missions of the twenty-first century are aimed at preparing for longer deep space missions, such as to Mars. Field tests of different machines, materials, habitats, communications, storage, and behavior can all be conducted at planetary analog locations. Analog sites exist in Antarctica, the Arctic, Utah, Hawai'i, the Atacama Desert in Chile, and more. The Johnson Space Center has a simulation chamber, and the International Space Station is often considered the best analog around. In October 2022, NASA began using a portion of the Arizona desert to practice moonwalks. The desert, NASA has declared, "possesses many characteristics that are analogous to a lunar environment including challenging terrain, interesting geology, and minimal communications infrastructure, all of which astronauts will experience near the lunar South Pole during Artemis missions."[26] It is possible to imagine that someday, perhaps the surface of the Moon might become an analog for all that has yet to be experienced.

It is not a stretch to say that in January 1978, Ralph Bagnold's career came full circle. The former British military engineer turned

FIGURE 16.5 NASA Mars conference in the Mojave, 1978. Held in Palm Springs, California, the conference focused on how Earth's dune formations and desert land-forms could help scientists better understand the surface of the Red Planet. Ralph Bagnold gave the keynote. From "Learning about Planets through Studying Wind-Related Process on Earth," *Journal of Geophysical Research: Planets*, April 19, 2018. Reprinted by permission of Wiley.

amateur scientist and then global sand consultant was invited to give the keynote address at a NASA conference on Mars. Bagnold would speak specifically on the topic of aeolian processes, the movement of particles in a fluid—in this case, sand in wind. Since Bagnold's foundational studies on the topic in the 1930s and 1940s, very little had been studied or published about the interplay of wind and desert sands, the formation of dunes, or the mechanisms of a desert environment. The conference was meant to change that, to leverage the enthusiasm over the Red Planet to boost terrestrial studies of the world's deserts.[27]

The conference took place in Palm Springs, California, perched between Pasadena, headquarters of the key NASA contractor the Jet Propulsion Laboratory, and the Mojave (fig. 16.5). The location was no mere coincidence; part of the program included field trips organized by area geologists to the lands that Mary Hunter Austin and

Minerva Hamilton Hoyt had once walked through. The NASA conference participants were not there for the plants or animals, however, but to look at dunes. As geologists had long argued, it was best to visit the sorts of landforms one was talking about. Conference attendees went out to explore the Coachella Valley, the dunes near the Salton Sea, and the Kelso Dunes.[28]

Bagnold was eighty-two years old, no longer the young man who risked everything to see what was over the next dune, but he had hardly slowed down. In fact, other attendees noted that he could outwalk just about anyone there. Since retiring from the British military in the 1940s, Bagnold had stayed active the world of eolian processes. He also traveled widely, including a trip to the Arctic. All the while he clipped files of people who achieved great exploratory feats, not just in desert areas. The papers in Bagnold's archive show that he continued to work, write, and evaluate until the end of his life. It was his lifetime of learning and observation, then, that NASA hoped to draw on.

That NASA invited Bagnold to provide expertise on images from far away was emblematic of both the towering importance Bagnold held in the field of eolian processes and to the field's rather narrow bench. In the late 1970s, there were not a lot of people interested in studying the way particles moved in fluids, such as sand in wind or in water, although interest in Mars in the 1980s and 1990s would ultimately change that.[29] To be sure, those earliest Mariner images from 1965 of the Red Planet had proved disappointing. Mars looked far more like the Moon than the lichen- or vegetation-covered terrain long anticipated. In the mid-1960s, the dream of a habitable Mars had been dashed. Mariner 9 images from 1971, however, were more promising. Mars was still dead and barren, but the landscape was far more varied than earlier images had implied: there were mountains and deep valleys, areas that looked as if they had been carved out by liquid. Just five years later, the Viking missions set down, snapping the first photos from the planet's surface. Humanity was getting its first real glimpses of a planet that turned out to have been weathered and altered by processes not dissimilar to those on Earth: wind, water, time.

The combined data and images led to an explosion in thinking about how various dunes and terrain features had been created on Mars, which led to renewed interest in deserts closer to home. The questions about how eolian processes on Earth might inform what was being seen on Mars were vast and potentially illuminating. What clues about Mars could dune formation provide? Did the dune and rock formations hint at a watery past? Could the geological formations provide clues of the climate? The wind? Indeed, studying dune patterns can not only reveal insights into a planet's surface composition but also hint at its atmospheric history. Large dunes seemed to suggest that Mars's atmosphere was once far thicker than today.

The line from Bagnold's wind tunnel experiments and equations about sand grains in the desert and grains on Mars thus seemed straight and—to many—far clearer than Bagnold was willing accede. To be sure, in 1978, the Sandman was flattered that his fundamental work on movement of particles on Earth by wind could be applied to other planetary bodies with atmospheres, but he was wary of making predictions based on incomplete information. "Things may not be what they appear," he cautioned. After all he had driven across what seemed like stable sand, only to sink into a sand pit. He had once been in a small plane that needed to make a forced landing in the middle of the desert, but the entire surface looked to be covered in large dunes. They landed anyway, steeling themselves for impact. Instead, they landed on a flat, smooth surface. "The dunes had marched away," he wrote, "but had left their outlines clearly mosaicked in the flat surface of fine gravel."[30] Even for a man who had spent decades traveling through and studying the extremes, the materials could still deceive.

Bagnold was thus cautious about using too many desert analogies to fill in the blank spots on Mars. Photographs might reveal "apparent sand dune forms like those here on Earth," he explained, but little else was certain. It is but "mere guesswork," he admitted, "as to what the stuff of the so-called dunes really is." Bagnold wondered whether it was "granular like sand" or "fluffy like snowflakes." Extremely dry ice and snow, Bagnold knew, acted like sand. It formed into dunes. But did stardust?

From decades spent traveling the most desolate deserts in the world, Bagnold knew that the extremes were capricious. Only the foolish made bets on what might be. He told his audience that without samples of Martian rock and soil, it was going to be difficult to properly read the images from space. Get the rocks, he advised, to understand and ultimately build on Mars.[31] Even in an age of remote sensing, distant orbiters, and increasingly powerful camera lenses, simulations and extrapolation can get one only so far. Technology cannot erase what is the rather prosaic need for boots on the ground and human eyes to do the interpreting.

Bagnold was repeating an oft-lamented reality about the extremes: they are places with an impossibly thin margin between what could go right and what could go wrong, with such scarce background information and detail that some bit of guesswork becomes necessary. But as this book has shown, the more extreme the environment got—from deserts to the ice cap and then up to the Moon—the more carefully one had to plan and the more intimate the environmental intelligence had to be. Recall Brigadier General Donald Davison—the aviation engineer of the war—standing in mud pits of Tunisia, damning the midget bulldozers they had packed away on the transatlantic crossing. The big bulldozers they used at home were far away, hard to access—the tangled logistic network was out of grasp. Or remember Henri Bader, stuck on top of the ice cap for ten days because the airplane that delivered him, rather impossibly to the top of the ice, was too heavy to take off again. Or the engineer Paul Troxler, who moved sand off his runways one night, only to discover more the next morning. The things they did not know—sometimes they got lucky, sometimes not. To be sure, the logistic networks that the engineers were putting into place helped stack their odds a bit, but nothing was foolproof. Bagnold knew it. He had cheated death and disaster himself many times, but others had not been so lucky.

CONCLUSION
EVERYDAY EXTREMES

Since I started tracking grains of sand and snow through military engineering reports that were created in far-off, little-thought-of places, the environments at the center of this book have become increasingly part of our everyday. Global warming has pulled the extremes from the margins into the mainstream. Newspapers frequently carry stories about melting ice and rising sea levels, about expanding deserts, and—most intimately felt—extreme weather events in places accustomed to temperate climatic cycles. Every year seems to be the hottest year on record; indeed, the ten warmest years since 1850 have all been since 2014.[1] Today, hurricanes are stronger, wildfires are more frequent and intense, and rainstorms bring flash floods. These are all the utterly predictable effects of anthropogenic climate change. Experts project things will only get more severe unless humanity finds a way to come together and reduce greenhouse gas emissions.

If the everyday has become more extreme, what does that mean for the places that already were extreme, where life already exists on a razor-thin margin of what is and what is not? The prognosis is grim. By 2100, it is estimated that Joshua Tree National Park will no longer house groves of the namesake trees. Hotter, drier deserts are also more susceptible to wildfires, a condition made worse by the spread of invasive plants that thrive in the damaged ecosystem.[2]

What of Greenland? The ice is melting. That is certain. How much and how fast remains the trillion-dollar question. Scientists are still uncovering new glacial systems and feedback loops that shift how well they understand the ice cap. Much of the new information makes melt projections worse. The Intergovernmental Panel on Climate Change estimates that by 2100, meltwater from Greenland will contribute up to 10.6 inches to global sea levels.[3] Other estimates are much higher. That means unprecedented flooding, worldwide.

Outer space is not immune from the effects of human-induced climate change. In some circles, Mars has emerged as an escape hatch from the troubles of Earth. But exporting the tendency to abuse environments doesn't seem like much of a solution. When we think about how midcentury planners saw this world's deserts and ice caps, it suggests that similar environmental damage would happen to the Red Planet. Already space exploration has left a trail of broken items in Earth's orbit, on the Moon, and—even though no human has yet been there—on Mars. The Moon is littered with the wreckage of exploration, including old spacecraft, American flags and the canisters that carried them to space, a pair of boots, and ninety-six bags of human waste.[4] Old rovers and landers sit on the Martian surface, baking in the sun, being smothered by dust. "Pack in, pack out" is a wilderness mantra that has not yet entered the realm of extraterrestrial planning. Planning to escape this planet does nothing to solve the problems that are on it—and it might even make them worse.[5]

I suggested at the outset that looking at the extremes may change the way we see the world. Writers have long suggested that such places harbor the sublime, a potential for deeper knowing. Barry Lopez found both the Arctic and the desert to be "spare, balanced, extended, and quiet." Places where one could gain some perspective. "Motionless and silent," Edward Abbey wrote in *Desert Solitaire*. The desert "evokes in us an elusive hint of something unknown, unknowable, about to be revealed."[6] Today, our thinking about the extremes is far less introspective and placid. Collapsing ice and burning sand highlight a collective human failure. As the conservationist Minerva Hamilton Hoyt realized while trying to save desert plants in the interwar period, what happens in the desert does not stay in the desert. She didn't have the language or the science to back her up, but she was right. Earth consists of interlocking systems and processes that—in concert—have led to life as we know it. When the system is out of whack, we do not know quite what might happen.

The military engineers at the center of this book were not thinking about Earth systems, climate change, or their own impact on the world's environments. They were steeped in wartime planning and then strategic Cold War sensibilities that suggested that the whole Earth had to be monitored and mastered as best as possible, environments manipulated to achieve certain ends. But their work amassing environmental intelligence, rearranging places, and defining environmental categories set the foundation for how we understand these places even now. It is a way of seeing and knowing that we must grapple with if we are to understand why we know what we know and what we should do with that information going forward.

It would be most convenient if I could report that strategic thinking about global environments—the type of environmental intelligence the military gathers today and for what purposes—has changed along with the climate crisis. But that is not the case. Although there are individuals in the strategic community who are undoubtedly concerned about the military's role in contributing to global warming, the Department of Defense's broad approach to the problems has not been to account for the military's emissions. Rather, climate change is seen as one of many potential threats to national security. In 2018, General Joseph F. Dunford Jr., chairman of the Joint Chiefs of Staff, explained that climate change was "in the category of sources of conflict around the world and things we have to respond to."[7] Rising temperatures will create new instabilities—resource scarcity that could lead to conflict, refugees and migration, flooding and famine—that may require military responses. This is particularly true because the US military often responds in times of humanitarian crisis precisely because of the vast strategic infrastructure the engineers still maintain. Although its network of facilities is not as vast as its Cold War footprint, the US military still maintains as many as eight hundred overseas facilities.[8]

Sand, Snow, and Stardust has shown that the United States became a planetary power because of the acquisition of environmental intelligence. Indeed, it was this very intelligence that allowed for the construction and maintenance of a global base structure that—though dynamic in its geography—has been consistent in its reach and power. Around the world, US military engineers still acquire data, witness change, study how to shore up their runways and barracks. To be a planetary presence, US strategic thinkers have realized, the military has had to continually demonstrate some degree of mastery over every possible environment, even those long dismissed as no-man's-lands, perhaps those particularly that are no-man's-lands. Climate change has amplified strategic interest in the extremes, not to stop the changes but to cope with the aftermath.

Since the 1940s, the US military has accumulated a stunning amount of data and expertise about the world's environments. Though impressive in its magnitude, the environmental intelligence acquired has been limited and utilitarian: how to best use what was available, how to move, how to ensure the operation of weapons and machines. The extremes, in this view, became useful wastelands that could be bombed, strafed, dug into, treated as trash dumps, all without seeming consequence.

In the process, military planners and engineers created new global categories to simplify and streamline how all of Earth could be administered. Local conditions were to be measured, charted, quantified, and then turned into well-researched categories. "Desert" and "Arctic," as we have seen, became containers for the sort of features that enhanced military maneuvering, if little else. Sir Hubert Wilkins took hundreds of pictures and collected dozens of vials of materials from the Arctic and desert regions, but his aim was to find reductive commonalties, not diversity. Mapping projects did the same thing on a grander scale, picking apart places for characteristics like slope, vegetation, or rockiness that could be matched across environments to inform a Platonic ideal of "desert," or "Arctic." Even the engineers sent to far-flung bases, the men who experienced local places and materials, ended up planning across broader categories. How else to order a wild world?

Here was a vision with power: environmental categories that looped around Earth, erasing political boundaries. Imagine a globe painted in blobs of color—like Wilkins's maps, but three dimensional—each color representing an environmental category: yellow for "desert," cyan for "ocean," perhaps deep green for "rainforest," a lighter green for "alpine forest," white for "Arctic," and so on. A world of vibrant color swatches but not politics.

It could almost seem utopian. In fact, for plenty of people, the view of planet Earth—the *whole* Earth—as an orb floating in a black abyss, elicited not strategic visions but ecological ones. The view from space afforded by the early Moon shots is often credited with having animated both the environmental movement and the rise of Earth systems science. People for the first time saw Earth not as a puzzle of nations, but as a palette of blues, greens, cloud formations, systems. After the actor William Shatner was hurled into space in 2021 on a private Blue Origin rocket, he used terrestrial analogies to describe the grandeur of space: "I saw how small the earth is," he said. "It's a particle of sand in an endless desert."[9]

This was hardly the view that strategic planners and military engineers were going for. But it highlights the possibility of divergent thinking about environmental categories. The way that the strategic community defined *environment* in operational terms matters not because it was most accurate, but because it was so powerful. Precisely when interest in the environmental sciences was growing, it was the US defense apparatus that picked up the bill. The result is a strategic infrastructure that remains deeply entrenched around the world; it is still the most efficient way for research teams to travel to and from the extremes. It is also how US military engineers continue to study material realities. Permafrost is being studied; ice and snow, measured; the growth of sand, seas, and dune fields, monitored.

In the early 2000s, Greenland's Thule Air Base emerged as a crucial site for civilian scientific projects to better understand the ice at the top of the world. From 2009 to 2019, NASA's Operation IceBridge used Thule for its efforts to measure the ice sheet. The National Science Foundation, too, leases space there, for offices, for cold storage, and for a heated warehouse.[10] For decades, international research teams have moved to and from the base to study the ice, the ocean, animals, and even black holes, all maneuvering on the backbone of US military power.

During the April 6, 2023, renaming ceremony when Thule became Pituffik, the US ambassador to Copenhagen noted the role the base had long played in security and science. He hoped friendship could be added to the mix. The new name, he said, could be "a symbol of our cooperation in science, climate, and space research, the common defense of our countries, and the stability of this amazing part of the world that is so vital for our survival, the Arctic."[11] The ambassador melded the local and the global, the past and the future, all the while describing a strategic vision that was cooperative, science-based, security-driven, and environmentally sensitive. Yet the ceremony avoided addressing the two elephants in the room: Pituffik Space Base is on land forcibly taken from the Inughuit and the US military has not changed its own interactions with local and global environments.

Nor did the ceremony address the limitations the strategic infrastructure has placed on the types of knowledge acquired. I wish I could draw a direct line between exclusion and failure; the "if only they had listened" story can be a powerful antidote to power. There is no proof that had the US military attended to the existence of the Inughuit in northern Greenland, US strategic planning for the ice cap would have been measurably different. I cannot pretend that had the US Army invited Minerva Hamilton Hoyt to the Desert Training Center that military ideas about arid ecosystems would be different or the Joshua tree not endangered today. But more fruitful than trying to prove counterfactuals is to use the past to think through alternate futures. Given what has been known, what might still be?

Today, responses to climate change in global terms—big data, big plans, big programs—are aimed at tackling Earth systems. These are the same sorts of categories that strategic thinkers liked not only because they seemed to make the planet simpler to manage and imagine but also because they erased the local, where violence and dispossession were taking place. We would do well to hold some suspicion of big solutions and plans. I do not mean to cast doubt on the models and data that warn us of our climate catastrophe. Human-induced climate change is not only happening; it is the crisis of our time. But in thinking through what that might mean and how we might come to terms with the consequences, the places matter most of all. It is through intimate stories that we can find the effects of planetary changes as well as unexpected responses and adaptations. The stories won't all fit into big-picture models of what should be, but they go far in explaining what might become.

ACKNOWLEDGMENTS

I did not set out to write a book about extreme environments, or granular materials, and certainly not about outer space. No one would call me a space enthusiast. And while I like to travel, I prefer beaches and mountain vistas to hard-to-reach places. But as a historian I tend to be drawn to unexpected stories, strange places, and details that—when added up—create particular ways of seeing and knowing. I like to tell stories about things and people that are generally overlooked. I like to tell stories about things that might at first appear of little interest. And so when I found grains of sand appearing across archival collections, I took notice. From there the materials ran away, moving to snow and ice, to moondust and Martian regolith. I hope I have been able to bring some of these materials to life in this book.

There are a lot of people to thank. Some for helping me track down materials; others for helping me make choices and cuts. Far more for just being around to offer support. I am fortunate that I have been able to move through remarkably kind and generous communities of scholars, colleagues, neighbors, friends, and family.

Over the past eight years, I have accumulated many debts to archivists and librarians. I have always enjoyed my trips to the US Army Corps of Engineers library and history center in Alexandria, Virginia. Douglas Wilson and John Lonnquest have let me peruse the files in the basement reading room for days on end, often helping me locate alternate materials and images. The staff at the National Archives and Record Administration in College Park, Maryland, are always courteous and helpful even as they are grossly understaffed. Laura Kissel at the Byrd Polar and Climate Research Archive at Ohio State University helped me gather materials during a quick trip. The folks at the Minnesota Historical Society tracked down key images. Trips across the pond allowed me to check out Bagnold's archives, and frequent email exchanges since have unearthed some good photos. Thanks to Charlie Jeffries, a doctoral student in Cambridge, who took time to photograph many of Bagnold's papers. Kylie Neal at the NASA Jet Propulsion Laboratory helped me navigate NASA items. Appreciation, too, for the archivists at Harvard's Countway Library and the folks at the General Patton Memorial Museum in California. I am deeply grateful to scholars willing to share ideas and documents, including Matt Bischoff, Kristian Nielsen, and Paul Landsberg. I am also appreciative of the creative minds who generated maps out of vague data sets: Alejandro Gonzalez and Heather Dart.

My writing group deserves a special shout-out. Megan Black, Daniel Immerwahr, Julia Irwin, and David Milne have been kind but firm readers, reminding me that sometimes I must just say what I mean. Plus, they are all just good fun. Jumping on to morning Zoom calls with Charles Peterson's morning writing group was a great way to start my writing. Thanks to all who were involved in that for providing a place to remember we do not have to work alone. I also am deeply indebted to plenty of other folks who read this manuscript in various forms, including Lisa Brady, Ron Doel, Stephen Macekura, and anonymous reviewers. Many others have provided valuable feedback along the way. There are far too many to mention here, but the participants at the Society for Historians of American Foreign Relations summer second-book workshop in 2022 helped keep me working. Petra Goedde, Penny Von Eschen, and Susan Ferber provided valuable insight, and Nicole Anslover brought us all together. Feedback I receive from talks and workshops always serve as a lifeline. Just when I think I am done, someone helps me find a new connection or reason to keep working. The chance to share my work at places such as the Rothermere American Institute at the University of Oxford, the Munk School of Global Affairs and Public Policy at the

University of Toronto, Columbia University's International History Workshop, the Massachusetts Historical Society's Environmental History Seminar, Dartmouth College's Institute of Arctic Studies, and Massachusetts Institute of Technology's Seminar on Environmental and Agricultural History have been generative. Thank you to the conveners, the participants, and the folks who have taken the time to break bread in the aftermath.

It is fortunate indeed to find an academic community. I found mine years ago in the Society for Historians of American Foreign Relations, where every year I meet new and exceptional people and scholars. More recently, I have been welcomed into the world of the American Society of Environmental Historians. While I still feel a bit of an interloper, I am grateful for the opportunities. The chance to share ideas in these venues is unparalleled. A shout-out to fellow panelists, commentators, and attendees at these meetings over the years. My ideas have only been enriched and challenged through our conversations and collaborations. The same goes for the editors, readers, and reviewers who have been involved in the articles I have written related to this material. Essays for *Diplomatic History*, *Endeavour*, *Environmental History*, *Military and the Market*, and *Cambridge History of America in the World* have all helped me hone certain ideas or look at materials in new ways. Although none of those articles is reprinted here, some of the ideas or background from them will be familiar. Indeed, writing and processing is an iterative process. I have enjoyed the ride as ideas have morphed with the addition of new materials and input.

It was during a blissful year in 2018–2019 as a faculty fellow at Harvard University's Charles Warren Center for Studies in American History that this project really turned from a book about engineering into a book about materials and environments. A big thanks to the center and to my colleagues that year who pushed me in some new directions. A huge thanks to Monnikue McCall for letting me stay connected to the amazing resources at the Harvard libraries. It was also the year that I met Jeffrey Nesbitt, whose interest in technical lands and landscapes has been a great introduction to interdisciplinary thinking on these topics. Matthew Farish, too, has helped me think of these things in new, interdisciplinary ways. Financial support from the American Philosophical Society and Northeastern University's College of Social Science and Humanities allowed me to conduct important archival work.

As my work moved into the extraterrestrial, I was fortunate to intersect with folks also heading for the cosmos. Neil Maher let me be an observer at a conference on NASA history in 2023, where I was able to learn a great deal from a lot of very smart people. Thanks to Teasel Muir-Harmony for the great tour of the Moon exhibit at the National Air and Space Museum, and to John McNeill for taking some time to chat about environmental history. Ron Doel nudged me toward the Shoemakers. Thanks to Andrew Preston and Susan Carruthers for more than one conversation about writing and publishing.

The material for this book has always threatened to race away from me, and I am grateful to my agent, Farley Chase, for helping me squinch material for three books into one. I have appreciated his patience and his interest. The folks at University of Chicago Press have been a delight to work with; thank you to everyone who has helped shaped this final product, including Andrea Blatz, Katherine Faydash, Daniel Lewis, Adriana Smith, and Anne T. Strother. Tim Mennel is an amazing sounding board and has pulled me back from far too many tangents.

Northeastern University has been home base while I have written this book. My colleagues in the department and across the college have been unflappable supporters of my work as a teacher and scholar. The university had provided ample support and encouragement as I have tacked this way and that. A special note of gratitude to the History Department's faculty, staff, and students. Since I arrived in 2013, Victoria Cain, Chris Parsons, and Philip Thai have been my comrades. Louise Walker was my ally during the promotion process. Two dear colleagues and friends passed away suddenly while the book was being finished, Angel David Nieves and Jeffrey Burds. I wish they were here to celebrate its completion. Liza Weinstein dared cross disciplinary borders to become my writing partner; our weekend in the Berkshires helped solidify some of my ideas. Thanks to the Roth family for the writing retreat and for thirty-plus years of friendship.

I have finally reached that point in my career when I realize why graduate students are so important—they keep me on my toes! Working with the innovative students at Northeastern, both grad and undergrad, has been constantly refreshing, particularly because so many of them understand the fierce urgency of thinking about the here and now: how our work excavating the past might help us create better futures. Thank you in particular to my advisees Jeff Lamson, Chris McNulty, Aaron Peterka, and Cassie Tanks. Chris deserves special shout-out for being research assistant extraordinaire. At times he has been editor, fact-checker, and sounding board.

I am able to do any work at all because of the support of friends near and far. This is especially true due to the fierce women in my life who everyday show me how to live with grace and compassion. It is hard to balance all of the things—work, family, friends, activism, the messiness of the world, a global pandemic that shut down much of what we knew—but done together, it is less terrifying and far more gratifying. And, of course, when I stumble you have all helped me pick up the slack. I am especially fortunate to have a cadre of confidants from my high school days back in Minnesota. Our text chain is a source of entertainment and often solace. My college friends remain my rock. You all know who you are. I was lucky to be able to spend a few days in Columbus, Ohio, with my dear friend Jason Krumm and his family. They fed me after long days at the archive. Thank you, Midge and Tally, for letting me spend some time with your son. Neighborhood friends have reminded me to take walks, stay involved with politics, share good food, watch bad 1980s movies, volunteer for school plays, and enjoy the day-to-day even when I am under tight deadline. Our group, "The Stand," that started as a daily pandemic-era social hour, has turned into a family of sorts. I suspect that most of these folks don't even remember what I am working on—it has been so long!—but they always ask, somewhat sheepishly, "How's the writing going?" I can at last report that it is done.

Finally, this book would not have happened without the support of my family. Thank you to the Bergsteins for putting up with my odd research interests. My mom is a great travel partner and research assistant. Her willingness to do just about anything—even to howl at the desert moon—is contagious. She is the fiercest woman around. I miss my dad every single day. We lost him in 2019, when I was still gathering materials, but I know he would have loved to see this book in print—whether he would have read it or turned it into a lovely household prop is up for debate. Never was there a bigger fan. My brother and sister, nieces, and nephew are a constant source of joy and strength. Thank you to Eleanor and Owen who have grown up with this research and put up with my odd hours. Eleanor has kept me energized with homemade cookies, and Owen distracts me with homemade crosswords. Every day I look forward to what I will learn from you. And every day you remind me that I study the past in order to envision—and work toward—more just futures. This book is for you. There is always something worth fighting for. To Brian, my best friend and confidant, I can't believe you are still willing to edit my words, staying up late into the night because I have not yet learned to get things done before the last possible minute. I love you so much I can hardly believe it. Onward to our next adventure!

ABBREVIATIONS

ACEHQ US Army Corps of Engineers Headquarters, Historical Research Office, Alexandria, VA

ADTIC Arctic, Desert, and Tropics Information Center, US Army Air Forces/US Air Force

AFHRA Air Force Historical Research Agency, Maxwell AFB

BAGNOLD PAPERS Papers of Brigadier Ralph Alger Bagnold, GBR/0014/BGND, Churchill Archives Center, University of Cambridge, Cambridge, UK

CRREL US Army Corps of Engineers, Cold Regions Research Engineering Laboratory (formerly US Army Corps of Engineers Snow, Ice, and Permafrost Research Establishment)

DTIC Defense Technical Information Center, http://discover.dtic.mil

ERDC US Army Engineer Research and Development Center's online knowledge core http://erdc-library.erdc.dren.mil/jspui/

FRUS Foreign Relations of the United States (Washington, DC)

HFL/COUNTWAY Records of the Harvard Fatigue Laboratory, Center for the History of Medicine, Francis A. Countway Library of Medicine, Harvard University

JPL ARCHIVES Jet Propulsion Laboratory Archives, California Institute of Technology, Pasadena, CA

NACP US National Archives, College Park, MD

NAUK National Archives of the United Kingdom, Kew Gardens

PATTON MUSEUM General Patton Memorial Museum, Chiriaco Summit, CA
RDB Research and Development Board
RG 27 Records of the Weather Bureau, NACP
RG 57 Records of the Geological Survey, NACP
RG 59 General Records of the Department of State, NACP
RG 77 Records of the Chief of Engineers, NACP
RG 165 Records of the War Department General and Special Staffs
RG 330 Records of the Secretary of Defense, NACP
RG 342 Records of the US Air Force Commands, NACP
RG 377 Records of HQ Army Ground Forces, NACP
SIPRE US Army Corps of Engineers Snow, Ice, and Permafrost Research Establishment
WILKINS PAPERS Sir George Hubert Wilkins Papers, SPEC.PA.56.0006, Byrd Polar and Climate Research Center Archival Program, Ohio State University
USACMH US Army Center for Military History, http://history.army.mil

NOTES

INTRODUCTION

1. William Woodward, "Space Logistics from Earth to Mars," *Military Engineer* 54, no. 360 (1962): 264.
2. United States Army Research and Development Board, Progress Report No. 6, video from Camp Century, "The Story of Camp Century: The City under the Ice" (Department of Defense, 1964), Atomic Heritage Foundation, July 19, 2018, https://ahf.nuclearmuseum.org/ahf/history/camp-century/.
3. Henry Serrano Villard, *Libya: The New Arab Kingdom of North Africa* (Ithaca, NY: Cornell University Press, 1956), 33–34.
4. Clyde Reedy, "Engineering Problems of Lunar Exploration," *Military Engineer* 53, no. 352 (1961): 107–9. Willy Ley, *Engineers' Dreams: Great Projects That Could Come True* (New York: Viking, 1954).
5. Quoted in Paul H. Nesbitt, "A Brief History of the Arctic, Desert, and Tropic Information Center and Its Arctic Research Activities," in *United States Polar Exploration*, ed. Herman R. Friis and Shelby G. Bale (Athens: Ohio University Press, 1970), 135.
6. Although we have plenty of accounts of each of these environments on its own, this is the first to examine how they were linked in military plans and research programs. Much work on the extremes remains about explorers or adventure, often accounts of individuals who have experienced great difficulties. Nature writing, too, has focused on these areas as places of unexpected and sublime beauty, where the hidden processes of life are revealed. For example, see Barry Lopez, *Arctic Dreams: Imagination and Desire in a Northern Landscape* (New York: Charles Scribner's Sons, 1986); Edward Abbey, *Desert Solitaire: A Season in the Wilderness* (New York: McGraw-Hill, 1968). For noteworthy new scholarship on each of these environments, refer to the notes that follow.

7. The literature on the United States as a global power is voluminous. For the most contemporary scholarship, see essays in David C. Engerman, Max Paul Friedman, and Melani McAlister, eds., *The Cambridge History of America and the World*, vol. 4, *1945 to the Present* (Cambridge: Cambridge University Press, 2022). See also Daniel Immerwahr, *How to Hide an Empire: A History of the Greater United States* (New York: Farrar, Straus & Giroux, 2019); Melvin P. Leffler, *A Preponderance of Power: National Security, the Truman Administration, and the Cold War* (Stanford, CA: Stanford University Press, 1993); Alfred W. McCoy, *In the Shadows of the American Century: The Rise and Decline of US Global Power* (Chicago: Haymarket Books, 2017).

8. The concept of planetary power is frequently linked either to ideas about global governance in the latter half of the twentieth century or to the global environmental movement. In my reading, *planetary* is the physical reality of Earth as a planet. The materiality of the world today—amplified and made visible by climate change—is rooted in the defense establishment's conception of the world as a series of material processes. That our "environmental" sensibilities have somewhat ironic roots in an institution that has done more than almost any other to disrupt the world is taken up in the conclusion. See Joyce Chaplin, *Round about the Earth: Circumnavigation from Magellan to Orbit* (New York: Simon & Schuster, 2012); Dipesh Chakrabarty, *The Climate of History in a Planetary Age* (Chicago: University of Chicago Press, 2021). On the more social use of *planetary*, see Ben Huf, Glenda Sluga, and Sabine Selchow, "Business and the Planetary History of International Environmental Governance in the 1970s," *Contemporary European History* 31, no. 4 (2022): 553–69; Diane Labrosse and Damien Mahiet, eds., "From the Global to the Planetary: A Conversation with Glenda Sluga, Stephen Macekura, and Jonathan S. Blake," Robert Jervis International Security Studies Forum, June 1, 2023, https://issforum.org/ISSF/PDF/jrd-2023.pdf. On global environment, see Paul Warde, Libby Robin, and Sverker Sörlin, *The Environment: A History of the Idea* (Baltimore: Johns Hopkins University Press, 2021); Perrin Selcer, *The Postwar Origins of the Global Environment: How the United Nations Built Spaceship Earth* (New York: Columbia University Press, 2018).

9. For an overview of the literature on the global base network, see Gretchen Heefner, "Overseas Bases and the Expansion of U.S. Military Presence," in *The Cambridge History of America and the World*, vol. 4, *1945 to the Present*, ed. David C. Engerman, Max Paul Friedman, and Melani McAlister (Cambridge: Cambridge University Press, 2022), 55–79; Rachel Woodward, "From Military Geography to Militarism's Geographies: Disciplinary Engagements with the Geographies of Militarism and Military Activities,"

Progress in Human Geography 29, no. 6 (2005): 718–40; Mark L. Gillem, *America Town: Building the Outposts of Empire* (Minneapolis: University of Minnesota Press, 2007); Kent E. Calder, *Embattled Garrisons: Comparative Base Politics and American Globalism* (Princeton, NJ: Princeton University Press, 2007); Catherine Lutz, ed., *The Bases of Empire: The Global Struggle against U.S. Military Posts* (New York: New York University Press, 2009); Immerwahr, *How to Hide an Empire*.

10. Heefner, "'A Slice of their Sovereignty': Negotiating the U.S. Empire of Bases, Wheelus Field, Libya 1950–1954," *Diplomatic History* 49, no. 1 (2017): 50–77.

11. The loss of eggs was a costly mistake, estimated to account for $148,000 of waste in 1951. Nathaniel Finch, Auditor, "Audit and Inspection Report No. 2-52," April 16, 1952, Corps of Engineers, Military Overseas Ops, Box XII-23, ACEHQ.

12. For overview of local-military relations, Lutz, ed., *The Bases*; David Vine, *Base Nation: How U.S. Military Bases Abroad Harm American and the World* (New York: Metropolitan Books, 2015); Seungsook Moon and Maria Hahn, eds., *Over There: Living with the U.S. Military Empire from World War Two to the Present* (Durham, NC: Duke University Press, 2010); Colleen T. McCaffrey, *Military Power and Popular Protest: The U.S. Navy in Vieques, Puerto Rico* (New Brunswick, NJ: Rutgers University Press, 2012); Marc Becker, "Ecuador's Early No Foreign Bases Movement," *Diplomatic History* 41, no. 3 (2017): 518–42; Amy Austin Holmes, *Social Unrest and American Military Bases in Turkey and Germany since 1945* (Cambridge: Cambridge University Press, 2014).

13. Villard, *Libya*, 59.

14. Carol Clark, "Movers and Shakers: New Evidence for a Unifying Theory of Granular Materials," Emory University, January 6, 2020, https://news.emory.edu/features/2021/01/esc-granular-materials/index.html. On sand as granular material, see Michael Welland, *Sand: The Never-Ending Story* (Berkeley: University of California Press, 2009); Vince Beiser, *The World in a Grain: The Story of Sand and How It Transformed Civilization* (New York: Riverhead Books, 2018).

15. I am indebted to the environmental historians who have established tools and methods for better integrating the ways that human interaction with environments can shape how we view the past. Works particularly important to my thinking include Megan Black, *The Global Interior: Mineral Frontiers and American Power* (Cambridge, MA: Harvard University Press, 2018); Kate Brown, *Plutopia: Nuclear Families, Atomic Cities, and the Great Soviet and American Plutonium Disasters* (New York: Oxford University

Press, 2013); Kurk Dorsey, "Dealing with the Dinosaur (and Its Swamp): Putting the Environment in Diplomatic History," *Diplomatic History* 29, no. 4 (2005): 573–87; Kurk Dorsey and Mark Lytle, "Introduction," *Diplomatic History* 32, no. 4 (2008): 517–18; Jacob Darwin Hamblin, *Arming Mother Nature: The Birth of Catastrophic Environmentalism* (New York: Oxford University Press, 2013); Thomas Robertson, "Cold War Landscapes: Toward an Environmental History of U.S. Development Programs in the 1950s and 1960s," *Cold War History* 16, no. 4 (2016): 417–41; Lisa Brady, "War from the Ground Up: Integrating Military and Environmental Histories," in *A Field on Fire: The Future of Environmental History*, ed. Mark Hersey and Ted Steinberg (Tuscaloosa: University of Alabama Press, 2019), 250–62. Within the broader frame of environmental history, see Ellen Stroud, "Does Nature Always Matter? Following Dirt through History," *History and Theory* 42, no. 4 (2003): 75–81; J. R. McNeill and Corinna R. Unger, eds., *Environmental Histories of the Cold War* (New York: Cambridge University Press, 2010); William Cronon, "The Trouble with Wilderness; or, Getting Back to the Wrong Nature," in *Uncommon Ground: Rethinking the Human Place in Nature*, ed. William Cronon (New York: W. W. Norton, 1995), 69–90; Linda Nash, "Furthering the Environmental Turn," *Journal of American History* 100, no. 1 (2013): 131–35. For environmental histories of the Arctic and polar regions, see Bathsheba Demuth, *Floating Coast: An Environmental History of the Bering Strait* (New York: Norton, 2020); Julia Herzberg, Christian Kehrt, Franziska Torma, eds., *Ice and Snow in the Cold War: Histories of Extreme Climatic Environments* (New York: Berghahn Books, 2018); Adrian Howkins, *The Polar Regions: An Environmental History* (Cambridge, UK: Polity Press, 2016). For deserts, see Welland, *The Desert: Lands of Lost Borders* (Chicago: University of Chicago Press, 2015); Natalie Koch, *Arid Empire: Entangled Fates of Arizona and Arabia* (New York: Verso Books, 2023); Diana K. Davis, *The Arid Lands: History, Power, and Knowledge* (Cambridge, MA: MIT Press, 2016).

16. There are other extremes that could be considered in this book, such as the tropics, potential atomic landscapes, and oceans. These were all environments where the operation of the military's regular machines and men were thrown into doubt. But a person stranded on a tropical island without supplies for a week is much more likely to survive than a person stranded in the desert, the Arctic, or outer space.

17. On what Armstrong said, see Amy Stamm, "'One Small Step for Man' or 'a Man'?," National Air and Space Museum, June 17, 2019, https://airandspace.si.edu/stories/editorial/one-small-step-man-or-man. On the Moon landing, see "Apollo 11 Transcripts," NASA History Division, https://history.nasa.gov/alsj/a11/a11trans.html; "The Apollo 11 Mission Was Also a Global

Media Sensation," *New York Times*, July 15, 2019; David Meerman Scott and Richard Jurek, *Marketing the Moon: The Selling of the Apollo Lunar Program* (Cambridge, MA: MIT Press, 2014).

18. On links between the space race and environmental thinking, see Neil Maher, *Apollo in the Age of Aquarius* (Cambridge, MA: Harvard University Press, 2017); Peder Anker, "The Ecological Colonization of Space," *Environmental History* 10, no. 2 (2005): 239–68; Lisa Ruth Rand, "Falling Cosmos: Nuclear Reentry Vehicles and the Environmental History of Earth Orbit," *Environmental History* 24, no. 1 (2019): 78–103; Roger Launius, "Writing the History of Space's Extreme Environment," *Environmental History* 15, no. 3 (2010): 526–32; Black, *The Global Interior*. On the space race and global perspectives, see Denis Cosgrove, "Contested Global Visions: One-World, Whole-Earth, and the Apollo Space Photographs," *Annals of the Association of American Geographers* 84, no. 2 (1994): 270–94.

19. Steve Pyne, "Extreme Environments," *Environmental History* 15, no. 3 (2010): 509–13; Lopez, *Arctic Dreams*; Mary Hunter Austin, *Lost Borders, the People of the Desert* (New York: Harper & Bros, 1909), 3.

20. Colin Woodward, "US Cold-War Waste Irks Greenland," *Christian Science Monitor*, August 22, 2008, https://www.csmonitor.com/Environment/Living-Green/2008/0822/us-cold-war-waste-irks-greenland; Julia Rosen, "Mysterious, Ice-Buried Cold War Military Base May Be Unearthed by Climate Change," *Science*, August 4, 2016, https://www.sciencemag.org/news/2016/08/mysterious-ice-buried-cold-war-military-base-may-be-unearthed-climate-change.

21. Neta C. Crawford, "Pentagon Fuel Use, Climate Change, and the Costs of War," Watson Institute for International and Public Affairs, Brown University, November 2019, https://watson.brown.edu/costsofwar/files/cow/imce/papers/Pentagon%20Fuel%20Use%2C%20Climate%20Change%20and%20the%20Costs%20of%20War%20Revised%20November%202019%20Crawford.pdf. On great acceleration and anthropogenic climate change, see J. R. McNeill and Peter Engelke, *The Great Acceleration: An Environmental History of the Anthropocene since 1945* (Cambridge, MA: Harvard University Press, 2016); Stephen Macekura, "Environment, Climate, and Global Disorder," in *The Cambridge History of America and the World*, vol. 4, *1945 to the Present*, ed. Engerman, Friedman, and McAlister, 488–511. For war and the environment, Richard P. Tucker and Edmund Russell, eds., *Natural Enemy, Natural Ally: Toward an Environmental History of War* (Corvallis: University of Oregon Press, 2004); Edwin Martini, ed., *Proving Grounds: Militarized Landscapes, Weapons, Testing, and the Environmental Impact of U.S. Bases* (Seattle: University of Washington Press, 2017); Akino Oshiro, "From 'Footprint' to Relationships: Impacts of U.S. Military Base on

Okinawa," *Sociology Compass* 18, no. 1 (2024): http://compass.onlinelibrary.wiley.com/doi/ful/10.111/soc4.13099.

22. "NASA Explores a Winter Wonderland on Mars," NASA, December 22, 2022, https://mars.nasa.gov/news/9326/nasa-explores-a-winter-wonderland-on-mars/. See also NASA's fantastic image gallery from the Webb Space Telescope at http://webbtelescope/org/news/first-images/gallery.

23. Ronald E. Doel, "Constituting the Postwar Earth Sciences: The Military's Influence on the Environmental Sciences in the USA after 1945," *Social Studies of Science* 33, no. 5 (2003): 636; Hamblin, *Arming Mother Nature*; Stuart W. Leslie, *The Cold War and American Science: The Military-Industrial-Academic Complex at MIT and Stanford* (New York: Columbia University Press, 1994); Aaron L. Friedberg, *In the Shadow of the Garrison State: America's Anti-Statism and Its Cold War Grand Strategy* (Princeton, NJ: Princeton University Press, 2000). For the latest thinking on Cold War and the economy, see Jennifer Mittelstadt and Mark R. Wilson, eds., *The Military and the Market* (Philadelphia: University of Pennsylvania Press, 2022); Jennifer Mittelstadt, *The Rise of the Military Welfare State* (Cambridge, MA: Harvard University Press, 2015); Mark R. Wilson, *Destructive Creation: American Business and the Winning of World War II* (Philadelphia: University of Pennsylvania Press, 2016). Pentagon study, quoted in Doel, "Constituting the Postwar Earth Sciences," 636.

24. Richard Flint, "Snow, Ice, and Permafrost in Military Operations," SIPRE Report 15, 1953. SIPRE reports are available at the ERDC Knowledge Core online library, https://erdc-library.erdc.dren.mil/jspui/handle/11681/6043.

25. Naomi Oreskes, *Science on a Mission: How Military Funding Shaped What We Do and Don't Know about the Ocean* (Chicago: University of Chicago Press, 2021).

26. Michael Adas, *Dominance by Design: Technological Imperatives and America's Civilizing Mission* (Cambridge, MA: Harvard University Press, 2006); David E. Nye, *American as Second Creation: Technology and Narratives of New Beginnings* (Cambridge, MA: MIT Press, 2004), 9–20; Linda Nash, "Traveling Technology? American Water Engineers in the Columbia Basin and the Helmand Valley," in *Where Minds and Matters Meet: Technology in California and the West*, ed. Volker Janssen (Berkeley: Huntington Library and University of California Press, 2012), 135–58. On the complexities of relocating technologies, see Bruno Latour, *Science in Action: How to Follow Scientists and Engineers through Society* (Cambridge, MA: Harvard University Press, 1987); Vincanne Adams and Warwick Anderson, "Pramoedya's Chickens: Postcolonial Studies of Techno-science," in *The Handbook of Science and Technology Studies*, 3rd ed., ed. Edward J. Hackett, Olga Amsterdamska,

Michael E. Lynch, and Judy Wajcman (Cambridge, MA: MIT Press, 2007), 181–204; David Arnold, "Europe, Technology, and Colonialism in the 20th Century," *History and Technology* 21, no. 1 (2005): 85–106.
27. See especially Linda Nash, "The Agency of Nature or the Nature of Agency?," *Environmental History* 10, no. 1 (2005): 67–69. Environmental historians have long looked at the hybridity of nature and the ways that nature and culture are tightly intertwined. In addition to books already cited, see Richard White, *The Organic Machine: The Remaking of the Columbia River* (New York: Hill & Wang, 1996). See also essays from "The State of the Field: American Environmental History," *Journal of American History* 100, no. 1 (2013): 94–148, specifically Paul S. Sutter, "The World with Us," at 94–119; Linda Nash, "Furthering the Environmental Turn," at 131–35; and David Igler, "On Vital Areas, Categories, and New Opportunities," at 120–23.
28. Nash, "Agency of Nature."

PART ONE

1. On weather, see Col. John S. Allard, interview, March 1943, Archive Reel No. A1272, frame 115, AFHRA; "The Battle of Sidi Bou Zid, February 1943" (CSI Battlebook, 1984), 4-D, available at the Defense Technical Information Center (DTIC), https://apps.dtic.mil/sti/tr/pdf/ADA151626.pdf; Ernie Pyle, *Here Is Your War: Story of GI Joe* (New York: Henry Holt, 2004), 48; Dwight D. Eisenhower, "Eisenhower Report on Torch" (Department of the Army, 1944), Ike Skelton Combined Arms Research Library Digital Library, https://cgsc.contentdm.oclc.org; Stuart C. Godfrey, "Airdromes Overseas," *Military Engineer* 25, no. 211 (1943): 213–17.
2. Pyle, *Here Is Your War*, 48; Rick Atkinson, *An Army at Dawn: The War in North Africa, 1942–1943* (New York: Henry Holt & Co., 2002); Robert Walker, *The Namesake: The Biography of Theodore Roosevelt Jr.* (London: Brick Tower Press, 2014), 257.
3. *A Pocket Guide to North Africa*, Special Service Division, Army Service Forces, US Army, 1942, Patton Museum; also available at the Internet Archive, https://archive.org/details/APocketGuideToNorthAfrica_879.
4. Pyle, *Here Is Your War*, 136–37; Atkinson, *Army at Dawn*; Walker, *Namesake*.
5. US War Department, General Staff, "Logistical History of NATOUSA, MTOUSA, 1942–1945" (Naples, Italy, Mediterranean Theater of Operations General Staff, 1945), 53.
6. Donald Davison, interview, March 1943, Air Force Archive Reel No. A1272, frame 964, AFHRA. On the pilot's quote, see Charles Messenger, *The Tunisian Campaign* (n.p.: Allan, 1982), 13.

7. *Foreign Relations of the United States*, The Conferences in Washington, 1941–1942, and Casablanca Jan. 1943. Eds., Frederick Aandahl, William M. Franklin, and William Slany (US Government Printing Office, 1948), Proceedings of the Conference, Friday January 15, pp. 568–569 (Document 342), https://history.state.gov/historicaldocuments/frus1941-43.

CHAPTER ONE

1. FDR, "February 23, 1942: Fireside Chat 20: On the Progress of the War," Miller Center, University of Virginia, https://millercenter.org/the-presidency/presidential-speeches/february-23-1942-fireside-chat-20-progress-war. For background, see Doris Kearns Goodwin, *No Ordinary Time: Franklin & Eleanor Roosevelt: The Home Front in World War II* (New York: Simon & Schuster, 1994), 319–20; "Maps to Use Tonight . . ." *Boston Globe*, February 23, 1942, 22; "Clip and Save This Map," *New York Times*, February 23, 1942; "People Urged To Use Maps," *Atlanta Constitution*, February 21, 1942, 1; "President's Plug Booms Map Trade," *New York Times*, February 22, 1942, 15.
2. FDR, "Fireside Chat 20: On the Progress of the War."
3. Daniel Immerwahr notes that "global" was not new, but its reference to US positioning in the world was. FDR first used it in September 1942; every president has used it since. See *How to Hide an Empire*, 222–23.
4. On lack of intelligence, see Stephen Ambrose, "Eisenhower and the Intelligence Community in World War II," *Journal of Contemporary History* 16, no. 1 (1981): 153–66. Thomas M. Pitkin, *Quartermaster Equipment for Special Forces* (Washington, DC: Historical Section, Office of Quartermaster General, 1944), 1 and 10.
5. Quoted in Matt Bischoff, *Preparing for Combat Overseas: Patton's Desert Training Center* (n.p.: Lulu.com, 2016), 21; "War in the Sand, Jungle, Snow," *New York Times*, January 4, 1942, SM5.
6. Brooke Blower, "From Isolationism to Neutrality: A New Framework for Understanding American Political Culture, 1919–1941," *Diplomatic History* 38, no. 2 (2014): 345–76.
7. Atkinson, *Army at Dawn*, 8; *Logistics in World War II: Final Report of the Army Service Forces*, vol. 7 (Washington, DC: Government Publishing Office, 1948), 1–3. In 1939, when Hitler and the Germans invaded Poland, the US Army consisted of just 227,000 troops. By June 1941 the US Army had 1.5 million enlisted, and by the end of 1942, it had 5.4 million. See George C. Marshall, *Biennial Reports of the Chief of Staff of the United States Army to the Secretary of War, 1 July 1939–30 June 1945* (Washington, DC: Center of Military History, US Army, 1996), USACMH. On poorly equipped military, see Adas, *Dominance by Design*, 224.

NOTES TO CHAPTER ONE ★ 323

8. Maj. Gen. George S. Patton Jr., to Gen. Jacob L. Denvers, March 13, 1942, quoted in Martin Blumenson, *The Patton Papers: 1940–1945* (Boston: Houghton Mifflin, 1974), 60; Maj. Gen. George S. Patton Jr., "The Desert Training Corps," *Cavalry Journal*, September–October 1942; Patton, "Notes on Tactics and Techniques of Desert Warfare, 1942," both reprinted by the Patton Society, http://www.pattonhq.com/textfiles/desert.html.
9. See Sidney L. Meller, *The Army Ground Forces: The Desert Training Center and C-AMA, Study No. 15* (Washington, DC: Historical Section, US Army Ground Forces, 1946), preface, DTIC (hereafter *Study No. 15*).
10. The Western Desert was named for its relation to the British Empire in Egypt, not its continental geography. For example, "Lawrence of Libya," *Time*, February 24, 1941, 28; "Sahara War," *Saturday Evening Post*, May 23, 1942, 26–61; "Fighting in—and against—the Desert," *New York Times*, February 16, 1941, SM7; "British in Desert Make Some Gains," *New York Times*, July 18, 1942, 1; "Fighters of the Great Sand Sea," *New York Times*, November 20, 1941, SM4.
11. Clayton R. Newell, *Egypt-Libya* (Washington, DC: Government Printing Office, 1993), https://history.army.mil/brochures/egypt/egypt.html.
12. "Explorers and Their Work," *Saturday Evening Post*, August 22, 1931, 6–85. For the cover, see *Time*, October 29, 1923. For details on Andrews, see "Who was Roy Chapman Andrews?," Roy Chapman Andrews Society, https://roychapmanandrewssociety.org/roy-chapman-andrews/.
13. See Bischoff, *Preparing for Combat Overseas*, 21; Meller, *Study No. 15*, 12.
14. Pitkin, *Quartermaster Equipment for Special Forces*, 10 and 1.
15. Report of the Quartermaster-General (US Army to the Secretary of War, War Department, fiscal year ended June 30, 1919), 58; Pitkin, *Quartermaster Equipment for Special Forces*, 7.
16. US intelligence officers accumulated materials from construction companies, consulates, geologists, and even people sending postcards. For descriptions of roads, bridges, and water sources, see Blanche D. Coll, Jean E. Keith, and Herbert H. Rosenthal, *United States Army in World War II, the Technical Services, the Corps of Engineers: Troops and Equipment* (Washington, DC: Center of Military History, US Army, 1988), 450, USACMH.
17. Frank Norris, "WWII Cold Weather Training in Denali," Administrative History of Denali National Park, 2006, https://www.nps.gov/articles/dena-wwii-training.htm. See Pitkin, *Quartermaster Equipment for Special Forces*, for organizational structure and various boards and units.
18. Lt. Col. H. O. Russell, "Information, If You Please!," *Air Force Magazine*, March 1943, 25, https://archive.org/details/AirForceMagazine194301/page/n43/mode/2up; Col. Russell, "Report on a Trip to the ADTIC, New York," February 11, 1944, Box 3165, Sarah Clark Correspondence Files, RG 342, NACP.

19. Harry A. Greveris, "Desert Environmental Handbook" (US Army, Yuma Proving Ground, November 1977), DTIC, https://apps.dtic.mil/sti/pdfs/ADA048608.pdf.
20. On Wilkins, see R. A. Swan, "Sir George Hubert Wilkins (1888–1958)," *Australian Dictionary of Biography*, https://adb.anu.edu.au/biography/wilkins-sir-george-hubert-9099; Simon Nasht, *The Last Explorer: Hubert Wilkins–Australia's Unknown Hero* (Sydney: Hodder, 2005).
21. Patton quoted in John S. Lynch, John W. Kennedy, Robert L. Wolley, "Patton's Desert Training Center" (Fort Myer, VA,: Council on America's Military Past, 1984), 5.
22. George W. Howard, "Desert Training Center/California-Desert Maneuver Area," *Journal of Arizona History* 26, no. 3 (1985): 273–94. Within the year, a number of additional camps were created in California and Arizona. Ground like "flour dough," see interview with Perrotti, in Roger M. Baty and Eddie Maddox Jr., eds., *Where Heroes Trained* (New York: Fenestra Books, 2004), 161. On general response to environment, Jesse B. Cress and James H. Sawyer Jr., "Desert Roads," *Military Engineer* 36, no. 223 (1944): 162–63; "Spearhead in the West: The Third Armored Division: 1941–45," 45–46, manuscript on file, General Patton Memorial Museum, Chiriaco Summit, California; "World War II Desert Training Center," pamphlet, Bureau of Land Management, Needles Field Office, https://www.blm.gov/sites/default/files/documents/files/media-center-public-room-california-desert-training-center-brochure.pdf.
23. John C. Coveney was interviewed in 2003. See Baty and Maddox, *Where Heroes Trained*, 136. See also Meller, *Study No. 15*, 9. On the Ninety-Third and dog tags, see "93rd Division Trains in Intense Desert Heat," *Pittsburgh Courier*, September 11, 1943, 12.
24. Lynch, Kennedy, and Wooley, "Patton's Desert Training Center"; "Letters from a Solider at Desert Training Center" binder, Patton Museum.
25. Mary Hunter Austin, *The Land of Little Rain* (Boston: Houghton Mifflin, 1903), 3. On Americans' ideas about the desert, see Patricia Nelson Limerick, *Desert Passages: Encounters with the American Deserts* (Albuquerque: University of New Mexico Press, 1985); Gerald Nash, *The Federal Landscape: An Economic History of the 20th Century West* (Tucson: University of Arizona Press, 1999). For militarization of US deserts, see Maria E. Montoya, "Landscapes of the Cold War West," in *The Cold War American West*, ed. Kevin Fernlund (Albuquerque: University of Mexico Press, 1998), 12–15; Valerie Kuletz, *The Tainted Desert: Environmental and Social Ruin in the American West* (New York: Routledge, 1998); Ryan Edginton, *Range Wars: The Environmental Contest for White Sands Missile Range* (Lincoln: University of Nebraska Press,

2014). For scholarship on the ways Europeans created an "environmental imaginary" about deserts that provided a logic for continued colonial exploitation and violence, see esp. Diana K. Davis, *Resurrecting the Granary of Rome: Environmental History and French Colonial Expansion in North Africa* (Athens: University of Ohio Press, 2007); Diana Davis and Edmund Burke III, eds., *Environmental Imaginaries of the Middle East and North Africa* (Athens: Ohio University Press, 2011).

26. Maj. Gen. Patton to Gen. Devers, March 13, 1942, in Blumenson, *The Patton Papers*, 60; *Basic Field Manual for Desert Operations, FM 31–25* (Washington, DC: Government Printing Office, 1942).

27. "Southland Rivals Sahara Desert," *Los Angeles Times*, April 19, 1940, 12; "Tank Army Turns Desert into Training Ground," *Los Angeles Times*, April 21, 1932, A1; *Basic Field Manual for Desert Operations*.

28. Quoted on placard, "Leadership Exhibit," Patton Museum.

29. Pyle, *Here Is Your War*, 190–91.

CHAPTER TWO

1. Donald A. Davison, "The Aviation Engineers in North Africa," *Aviation Engineer Notes*, August 1943, no. 14, Box V-25: World War II, General Files, Folder: "Aviation Engineer Notes," ACEHQ.

2. "President Sees Allies Holding Initiative Now," *Stars and Stripes*, January 8, 1943, 4.

3. Howe notes military reluctance to initiate Torch and slow planning process. George F. Howe, *Northwest Africa: Seizing the Initiative in the West* (Washington, DC: US Army Center of Military History, 1993), 12–14. On various aspects of the problems and lack of preparation, see Atkinson, *An Army at Dawn*, esp. 34 (on the chaos of loading), 53–54 (on the lack of training), 170 and 283 (on the lack of a plan), and 184 (on rivalry with allies). For a more recent assessment of the failures in North Africa, see Daniel R. Mortensen, *A Pattern for Joint Operations: World War II Close Air Support, North Africa* (Washington, DC: Office of Air Force History and US Army Center of Military History, 1987), USACMH. On Ike blaming the weather, rising enemy air and ground strength, and logistic shortfalls, see Jeffrey Bryan Mullins, "For the Want of a Nail: The Western Allies' Quest to Synchronize Maneuver and Logistics during Operations Torch and Overlord" (PhD diss., Kansas State University, 2020). On learning experience, see Col. F. Randall Starbuck, "Air Power in North Africa, 1942–43: An Additional Perspective," US Army War College, Carlisle, PA, 1992, DTIC, https://apps.dtic.mil/sti/tr/pdf/ADA251886.pdf.

4. US War Department, General Staff, "Logistical History of NATOUSA, MTOUSA, 1942–1945" (Naples, Italy, Mediterranean Theater of Operations General Staff, 1945), 485.
5. On location, see *Eisenhower Report on Torch* (Washington, DC: Department of the Army, 1944), DTIC; Robert Murphy, *Diplomat among Warriors* (New York: Doubleday & Co., 1964); Atkinson, *Army at Dawn*, 46; Corroboration can be found in Thomas W. Dorrel Jr., "The Role of the Office of Strategic Services in Operation Torch" (master's thesis, University of Minnesota, 1995).
6. From the memoir of Ike's aide Harry C. Butcher: *My Three Years With Eisenhower: The Personal Diary of Captain Harry C. Butcher, USNR, Naval Aide to General Eisenhower, 1942 to 1945* (New York: Smith & Schuster, 1946), 105–10.
7. Murphy, *Diplomat among Warriors*, 103.
8. Mabel H. Cabot, *Vanished Kingdoms: A Woman Explorer in Tibet, China, and Mongolia, 1921–1925* (New York: Aperture, 2005).
9. Pitkin, *Quartermaster Equipment for Special Forces*, 168; "Extract from Military Attaché Report, Cairo," October 12, 1941, reprinted in *Aviation Engineer Notes*, April 1942, Box V-25: World War II, General Files, Folder: "Aviation Engineer Notes," ACEHQ.
10. Pyle, *Here Is Your War*, 59, 60. On Pyle piling clothes on, see James Tobin, *Ernie Pyle's War: America's Eyewitness to World War II* (New York: Free Press, 2006), 77. For Roosevelt, see Atkinson, *Army at Dawn*, 277. See also "Pilots Brave Ice, Storms to Hit Axis," *Stars and Stripes*, December 31, 1942, 1; Atkinson, *Army at Dawn*, 184–85; "More Planes Reach Allies for Big Push," *Stars and Stripes*, December 11, 1942, 1; "Nazis Retiring on Coast Road towards Tripoli," *Stars and Stripes*, December 24, 1942, 1; "Allies Seek New Battle in Tunisia," *Stars and Stripes*, January 7, 1943, 1.
11. Pyle, *Here Is your War*. On muddy conditions, see also "The Other North African Battle," *Air Force Magazine* 26, no. 1 (January 1943): 11–12; "Notes from North Africa," *Air Force Magazine* 26, no. 2 (February 1943): 4–6.
12. Weather and terrain have long been factors in military strategy, although, as with many factors of the North Africa campaigns, the Germans seemed to grasp the subtleties of them a bit more clearly.
13. "Log of the Trip of the President to the Casablanca Conference, 9–31 January, 1943," Naval History and Heritage Command, July 25, 2019, https://www.history.navy.mil/research/library/online-reading-room/title-list-alphabetically/l/log-of-the-the-trip-of-the-president-to-the-casablanca-conference-9-31-january-1943.html.
14. "Runway Construction in War," *Military Engineer* 34, no. 204 (1942): 473–77; Stuart C. Godfrey, "Airdromes Overseas," *Military Engineer* 25, no. 211

(1943): 213–17. On aviation and war, see Jennifer Van Vleck, *Empire of the Air: Aviation and the American Ascendancy* (Cambridge, MA: Harvard University Press, 2013).

15. George Mayo, "Airfields 'Custom Built' by Aviation Engineers," *Civil Engineering* (April 1945), Box V-1: World War II Airfields, Folder: "Articles and News Clippings, ACEHQ.
16. Harold J. McKeever, "Where We Are in Airport Design," *American Road Builders' Association* (January 1945), Box V-1: World War II Airfields, Folder: "Articles and News Clippings," ACEHQ. Ben H. Fatherree, *The History of Geotechnical Engineering at the Waterways Experiment Station, 1932–2000* (Vicksburg, MS: US Army Engineer Research and Development Center, 2006), 35.
17. See "Resume of Investigation and Development of Pavement Design Procedures and Temporary Airfield Landing Surfaces," Box V-1: Air fields, Folder: "airfields-runways," ACEHQ.
18. On the general history of soil mechanics and Casagrande's role in the twentieth century, see Philotheos Lokkas et al., "Historical Background and Evolution of Soil Mechanics," *WSEAS Transactions on Advances in Engineering Education* 18 (2021): 96–113.
19. *The Unified Soil Classification System* (Vicksburg, MS: US Army Engineer Waterways Experiment Station, 1960), ii.
20. General information on Casagrande comes from "Arthur Casagrande," American Society of Civil Engineers, https://www.asce.org/about-civil-engineering/history-and-heritage/notable-civil-engineers/arthur-casagrande; Rubén Galindo-Aires, Antonio Lara-Galera, and Gonzalo Guillán-Llorente, "Contribution to the Knowledge of Early Geotechnics during the Twentieth Century: Arthur Casagrande," *History of Geo- and Space Sciences* 9, no. 2 (2018): 107–23. For his airport system and the fact that it is still in use, see Gilson Co., "Soil Classification: Foundation and Pavement Design Starts Here," https://www.globalgilson.com/blog/soil-classification-foundation-and-pavement-design-starts-here; Geoffrey C. Bowker and Susan Leigh Star, *Sorting Things Out: Classification and Its Consequences* (Cambridge, MA: MIT Press, 2000); Benno P. Warkentin, *Footprints in the Soil: People and Ideas in Soil History* (Amsterdam: Elsevier, 2006). On World War II and standardization along US lines, see Immerwahr, *How to Hide an Empire*.
21. Edwin F. Clements, "Soil Tests for Military Construction," *Military Engineer* 34, no. 206 (1942): 604–8. See also Fatherree, *History of Geotechnical Engineering*, 47; D. R. Casagrande, "Memorial to Arthur Casagrande, 1902–1981," Geological Society of America, May 1982, https://www.geosociety.org/documents/gsa/memorials/v13/Casagrande-A.pdf; Galindo-Aires

et al., "Contribution to the Knowledge of Early Geotechnics." On the visual and sensory reading of soils, see also Gilson Co., "Soil Classification."

22. Richard H. Anderson, "Special Observers: A History of SPOBS and USAFBI, 1941–1942" (PhD diss., University of Kansas, 2016); Alfred Beck, Abe Bortz, Charles W. Lynch, Lida Mayo, and Ralph F. Weld, *The Corps of Engineers: The War against Germany* (Washington, DC: Center of Military History, US Army, 1985), 8, USACMH; Report of Reconnaissance to Iceland, Annex 1, Engineer, WPD 4493-20, War Plans Division (WPD) General Correspondence, 1920–1942, RG 165, NACP.

23. Report of Reconnaissance to Iceland; Ronald B. Hartzer, Lois E. Walker, Rebecca Gatewood, Katherine Grandine, and Kathryn M. Kuranda, *Leading the Way: The History of Air Force Civil Engineers, 1907–2012* (Washington, DC: General Printing Office, 2015), 33.

24. Beck, Bortz, Lynch, Mayo, and Weld, *The Corps of Engineers: The War against Germany*, 63, USACMH; Coll, Keith, and Rosenthal, *United States Army in World War II, the Technical Services, the Corps of Engineers: Troops and Equipment*, 450, USACMH. During World War I, commanders got used to 1:20,000 maps, which were very close. Strategic planning usually requires maps smaller than 1:100,000,00; ideal tactical maps are at a scale of 1:100,000. Maps of Morocco were also compiled from Michelin guides. See Atkinson, *Army at Dawn*, 34. On changes to mapping during World War II, see Bill Rankin, *After the Map: Cartography, Navigation, and the Transformation of Territory in the Twentieth Century* (Chicago: University of Chicago Press, 2016); Welland, *The Desert*, 15.

25. Tebesse Field took fifteen days. See *History of Aviation Engineers in the Mediterranean Theater of Operations* (US Army Air Forces, 1946), 17, Box X-39, Folder: "History Division April–May 1943," ACEHQ. On the lack of advanced information, see also "Airborne Missions in the Mediterranean, 1942–1945" US Air Force Historical Studies No. 74, USAF Historical Division, Air University, September 1955, DTIC, https://apps.dtic.mil/sti/tr/pdf/ADA522511.pdf; Beck et al., *Corps of Engineers: The War against Germany*, 97; Coll, Keith, and Rosenthal, *United States Army in World War II, the Technical Services, the Corps of Engineers: Troops and Equipment*, 56–57; Stuart Godfrey, in "Airdromes Overseas," *Military Engineer* 35, no. 211 (1943): 213–17, also mentions small equipment brought for airborne aviation engineering units and the need to get them on planes. Davison notes that they didn't have the right stuff. See interview with Davison, Air Force Archive Reel No. A1272, AFHRA. On equipment being "too light," see also Lt. Robert P. Young, "Letter from North Africa," in *Aviation Engineer Notes*, February–March 1943, Box V-25, Folder: "Aviation Engineer Notes," ACEHQ. On asphalt, US War Department, General Staff, "Logistical History of NATOUSA, MTOUSA,

1942–1945" (Naples, Italy, Mediterranean Theater of Operations General Staff, 1945), 208. On the type of advanced information that engineers increasingly realized they needed, see Albert Land, "Engineering Operations on Advanced Bases," *Military Engineer* 37, no. 238 (1945): 299–302.

26. "Airborne Engineers Speed Construction of Vital Airdromes in North Africa," January 11, 1943, news release, War Department, Box V-1, Folder: "articles and news clippings," ACEHQ.

27. Sgt. E. K. Stiles, "Those Aviation Engineers!," *Air Force Engineer*, June–July 1945, Box V-25, Folder: "Aviation Engineering News," ACEHQ.

28. Frederick Aandahl, William M. Franklin, and William Slany, eds., *Foreign Relations of the United States, The Conferences in Washington, 1941–1942, and Casablanca Jan. 1943* (US Government Printing Office, 1948), Proceedings of the Conference, January 15, p. 568–69 (doc. 342), https://history.state.gov/historicaldocuments/frus1941-43.

29. For numbers of airdromes, see the interview with Davison, Air Force Archive Reel No. A1272, AFHRA. For an engineer's critique of work done, see *History of Aviation Engineers in the Mediterranean Theater of Operations* (US Army Air Forces, 1946), 20, Box X-39, ACEHQ.

30. See, e.g., "Mat, Airplane, Landing, Aluminum Alloy, Pierced Plank Type," "Paving with Concrete," and "Resume of Investigation and Development of Pavement Design Procedures and Temporary Airfield Landing Surfaces," Box V-1, Folder: "airfields-runways," ACEHQ.

31. Fatherree, *History of Geotechnical Engineering*.

CHAPTER THREE

1. "Equip-Engr" file, Box V-8, ACEHQ.
2. Examples include George Howard, "Desert Effects on Engineer Equipment," *Military Engineer* 43, no. 285 (1951): 362–64; George Howard, "The Yuma Test Branch of the Engineer Board," *Military Engineer* 39, no. 263 (1947): 384–85; R. C. Hannum and H. A. McKim, "Desert Motor Transport," *Military Engineer* 42, no. 290 (1950): 435–37; and W. R. Shuler, "Major Test Runway Built for Research," *Military Engineer* 47, no. 318 (1955): 265–67.
3. Pitkin, *Quartermaster Equipment for Special Forces*, 163. The War Department also established cold weather and high-altitude boards and facilities. For quote, see Meller, *Study No. 15*, 12.
4. Lawrence R. Walker and Frederick H. Landau, *A Natural History of the Mojave Desert* (Tucson: University of Arizona Press, 2027), 182–85.
5. Austin, *Land of Little Rain*, sec. 1 "The Land of Little Rain." On the desert as a place for seekers, see Limerick, *Desert Passages*; Maria E. Montoya, "Landscapes of the Cold War West," 12–15.

6. Frank E. Egler, "Forrest Shreve and the Sonoran Desert," *Geographical Review* 44, no. 1 (1954): 137–41.
7. On desert studies at the time, see George E. Webb, *Science in the American Southwest: A Topical History* (Tucson: University of Arizona Press, 2002).
8. Forrest Shreve, "The Edge of the Desert," *Yearbook of the Association of Pacific Coast Geographers* 6 (1940): 6–11; Shreve, *The Cactus and Its Home* (Philadelphia: Williams and Wilkins Co., 1931); Forrest Shreve and Ira L. Wiggins, *Vegetation and Flora of the Sonoran Desert* (Stanford, CA: Stanford University Press, 1964).
9. Nick Middleton, *Deserts: A Very Short Introduction* (New York: Oxford University Press, 2009), 66.
10. "The Desert: Lecture I," *ADTIC: Pertinent Data on Air Forces Activities in Arctic, Desert, and Tropics Areas: Nine School Lectures* (New York: Office of Assistant Chief of Air Staff, Intelligence, 1943), 94.
11. "The Desert: Lecture I," 94.
12. "The Desert: Lecture 1," 142.
13. Mary Hunter Austin, *The Land of Little Rain* (1903), chap. 1, "The Land of Little Rain." On King Clone, see Kim Stringfellow, "King Clone Creosote," August 2014, Mojave Project, http://mojaveproject.org.
14. Webb, *Science in the American Southwest*, 17; Gary Nabhan, *Gathering the Desert* (Tucson: University of Arizona Press, 1986), 11–16. Nabhan is an agricultural ecologist, ethnobotanist, ecumenical Franciscan friar and an author whose work has focused on the interaction of biodiversity and cultural diversity in the arid binational Southwest. See also Austin, *The Land of Little Rain*, 4.
15. Damon Akins and William Bauer, *We Are the Land: A History of Native California* (Berkeley: University of California Press, 2022), 103; Catherine Fowler, "Reconstructing Southern Paiute—Chemeheuvi Trails in the Mojave Desert of Southern Nevada and California," in *Landscapes of Movement: Trails, Paths, and Roads in Anthropological Perspective*, ed. James E. Snead, Clark L. Erickson, and J. Andrew Darling (Philadelphia: University of Pennsylvania Press, 2009), 88.
16. Pvt. Horace Barrett letter home, March 12, 1943, "Letters from a Solider at Desert Training Center" binder, Patton Museum Archive; Bischoff, *Preparing for Combat*, 40; "Three Soldiers Die of Thirst," *Chicago Tribune*, July 28, 1943.
17. "Designated Historic Civil Engineering Landmarks: Colorado River Aqueduct, Blythe, California, USA," American Society of Civil Engineers, 2009, https://web.archive.org/web/20060625150733/http://www.asce.org/history/landmark/projects.cfm?menu=name. On the camps' setup, see Bischoff, *Preparing for Combat Overseas*.

18. J. W. Kennedy, "The CAMA World War II, Some Medical and Nonmedical Notes," *Arizona Medicine* 40, no. 7 (1983): 484–87; Lynch, Kennedy, and Wooley, "Patton's Desert Training Center"; Howard, "The Desert Training Center/California-Arizona Maneuver Area," 290; Weldon F. Heald, "With Patton on Desert Maneuvers," *Desert Magazine*, July 1960, 24.
19. See "Biographical Note" to the Phyllis and Weldon Heald Papers, Special Collections, University of Arizona, https://archives.library.arizona.edu/repositories/2/resources/1405.
20. Lynch, Kennedy, and Wooley, "Patton's Desert Training Center," 26–27; Heald, "With Patton on Desert Maneuvers." See also "Prevention, Symptoms, Signs and Emergency Treatment of the Three Chief Ills Associated with Hot Weather," December 1942, Series 00302, Box 24, Folder 3, HFL/Countway.
21. *Basic Field Manual Desert Operations* (Washington, DC: Government Printing Office, 1942), 13.
22. R. A. Bagnold, "A Lost World Refound," *Scientific American* 161, no. 5 (1939): 261–63.
23. Austin, *Land of Little Rain*.
24. Glucozade (British), a citrus-flavored sugar water, was invented in 1927. Gatorade was invented in the 1950s by a Florida University football coach, in efforts to prevent his players from flaking out in the summer heat.
25. Ernest A. Davidson, Regional Chief of Planning, "Notes and Pictures Relative to proposed army maneuver road at the Joshua tree national monument," Interior Department, February 1943, available online through the National Park Service, https://npshistory.com/publications/jotr/index.htm
26. Quotes from Chris Clarke, "The Joshua Tree: Myth, Mutualism, and Survival," Mojave Project, https://mojaveproject.org/dispatches-item/the-joshua-tree-myth-mutualism-and-survival/original; John C. Frémont and James Hall, "Report of the Exploring Expedition to the Rocky Mountains in the Year 1842, and to Oregon and North California in the Years 1843–44" (Washington, United States: Gales and Seaton, Printers, 1845); Joseph Smeaton Chase, *California Desert Trails* (Boston: Houghton Mifflin, 1919), 49–50; Austin, *Land of Little Rain*.
27. "Strange California 'Trees,'" *Christian Science Monitor*, November 28, 1921, 3. Sam Schipani, "How a Tree and Its Moth Shaped the Mojave Desert," *Smithsonian Magazine*, August 10, 2017, https://www.smithsonianmag.com/science-nature/how-tree-and-its-moth-shaped-mojave-desert-180964452/.
28. Erin Rode, "Joshua Trees Rejected for Protection under the Federal Endangered Species Act," *Desert Sun*, March 10, 2023, https://www.desertsun.com/story/news/environment/2023/03/10/joshua-trees-rejected-for-federal-endangered-species-act-listing/69989860007/#.

29. Minerva Hamilton Hoyt, "The International Deserts League," *Americana* (July 1931): 315-325.
30. "Desert Vandals," *Los Angeles Times*, July 22 1930, A4. Hoyt even started offering $100 rewards for information on the vandals.
31. "New York Has Tiny Desert," *Los Angeles Times*, May 15, 1928, 2; On the Bronx, see "Mrs. Albert S. Hoyt: Philanthropist, Ex-Head of the Desert Conservation League," *New York Times*, December 19, 1945, 25.
32. Hoyt, "The International Deserts League."
33. "Minerva Hamilton Hoyt," National Park Service, https://www.nps.gov/people/minerva-hamilton-hoyt.htm. In addition to Joshua Tree National Monument, she was instrumental in writing reports for and getting support for parks in Death Valley, the Anza-Borrego Desert, and in the Joshua tree forests of the Little San Bernardino Mountains north of Palm Springs.
34. "Desert Troops Train in California's 'Little Libya,'" *Christian Science Monitor*, May 7, 1942, 13; Patton, "The Desert Training Corps," *Cavalry Journal*, September–October 1942.
35. On remnants and the Bureau of Land Management's study of camps, see Bischoff, *The Desert Training Center/California-Arizona Maneuver Area, 1942–1944* (Redlands, CA: US Bureau of Land Management, 2000). For an aerial view of wounds from the 1940s through 1960s, see D. V. Prose, "Map Showing Areas of Visible Land Disturbances Caused by Two Military Training Operations in the Mojave Desert," USGS Publications Warehouse, https://doi.org/10.3133/mf1855.
36. California Park Commission "Report of State Park Survey of California," 1929, 51–52. Olmsted used Hoyt as a consultant to identify desert areas in need of preservation.

CHAPTER FOUR

1. Nick Cullather, "The Foreign Policy of the Calorie," *American Historical Review* 112, no. 2 (2007): 337-64; Timothy Mitchell, *Rule of Experts: Egypt, Techno-Politics, Modernity* (Berkeley: University of California Press, 2002), 82; Harwood Belding "Physiological Testing of Quartermaster Items," in *Symposium on Military Physiology* (Washington, DC: US Army Chemical Corps, 1947), 45, https://collections.nlm.nih.gov/bookviewer?PID=nlm:nlmuid-06830900R-bk. On laboratory spaces especially in World War II and Cold War, see Matthew Farish, "The Lab and the Land: Overcoming the Arctic in Cold War Alaska," *Isis* 104, no. 1 (2013): 1–29.
2. In 1947, the Air Force created its own indoor climate chamber to test planes and parts at Eglin Air Force Base in Florida. Today the chamber is called the McKinley Climatic Laboratory.

3. Quoted in Andi Johnson, "'They Sweat for Science': The Harvard Fatigue Laboratory and Self-Experimentation in American Exercise Physiology," *Journal of the History of Biology* 48, no. 3 (2015): 425–54.
4. B. L. Bennett, "David Bruce Dill: A Man of Many Seasons and Environments—Committed to Life, Heat, and Altitude," *Wilderness and Environmental Medicine* 17, no. 2 (2006): e10–e13; Steven Horvath and Elizabeth Horvath, *The Harvard Fatigue Laboratory: Its History and Contributions* (New York: Prentice-Hall, 1973), 26. Most men made it only nine minutes before uncontrollable shivering made them leave.
5. Heald, "With Patton on Desert Maneuvers." See also "Preselection for work in Hot Climates," Harvard Fatigue Laboratory, Series 302, Box 22, Book 2, HFL/Countway; Committee on Medical Research of the National Research Council, Harvard Fatigue Laboratory, "Report No 6, Fall Report on Physical Fitness of Soldiers and Observations of Their Clothing;" and Sid Robinson, E. S. Turrell, H. S. Belding, and S. M. Borvath, "Rapid Acclimatization to Work in Hot Climates," *American Journal of Pathology*, November 2, 1943, both in Series 302, Box 22, Book 1, HFL/Countway.
6. Dill, introduction to *Military Physiology: Reports from the Fatigue Laboratory*, vol. 1, July 1940–June 1941, Series 302, Box 22, Book 1, HFL/Countway.
7. For a list with names, see Horvath and Horvath, *Harvard Fatigue Laboratory*, 56; Arlie V. Bock to Dr. Howard Means, October 1942, Series 300, Box 1, Folder 10, HFL/Countway. The men's correspondence is abundant and includes handwritten notes about the smallest details relating to experiments. See Forbes to Dill, January 14, 1943, Series 00300, Box 1, Folder 3b. See also Dill, "The Harvard Fatigue Laboratory," Series 132, Box 30, Folder 14; both HFL/Countway; Pitkin, *Quartermaster Equipment for Special Forces*, 166.
8. Ancel Keys, "The Problem of Increasing Human Efficiency in the Field," March 1948, Series 302, Box 26, Folder 28, HFL/Countway; Todd Tucker, *The Great Starvation Experiment* (Minneapolis: University of Minnesota 2007). Keys pitched his work toward the need for postwar rebuilding efforts because he estimated that so many people in war-torn areas would be starving and in need of aid: what was the minimum caloric intake that could be supplied to the largest amount of people in areas where resources were limited? Efficiency was measured in survival, not satisfaction.
9. Minutes of a Conference on Fitness, December 11, 1942, National Research Council, Division of Medical Sciences, Series 302, Box 24, HFL/Countway.
10. Bennett, "David Bruce Dill: A Man of Many Seasons and Environments"; G. Edgar Folk, "The Harvard Fatigue Laboratory: Contributions to World War II," *Advances in Physiology Education* 43, no. 3 (2010): 119–62. See also

Sid Robinson, H. S. Belding, and S. M. Horvath, "Rapid Acclimatization to Work in Hot Climates," May 1942, Series 302, Box 22, Book 2, HFL/Countway.

11. On creation of the clo, Lyman Fourt and Normal Hollies, "The Comfort and Function of Clothing," US Army Natick Laboratories, June 1969, 9, DTIC, https://apps.dtic.mil/sti/pdfs/AD0703143.pdf. On studies reifying racial differences, Folk, "Harvard Fatigue Laboratory." The clo was first presented in 1941 in "A Practical System of Units for the Description of the Heat Exchange of Man with His Environment," *Science*, n.s., 94, no. 2445 (1941): 428–30. Vanessa Heggie calls this faith in environmental determinism "both incorrect, and somewhat lazy." See "Blood, Race, and Indigenous Peoples in Twentieth Century Extreme Physiology," *History and Philosophy of Life Sciences* 41, no. 2 (2019): 26. See also Susan L. Smith, "Mustard Gas and American Race-Based Human Experimentation in World War II," *Journal of Law, Medicine, and Ethics* 36 (2008): 517–21.

12. Fourt and Hollies, "The Comfort and Function of Clothing," 1–34, 71.

13. On wartime experiments, Susan L. Smith, *Toxic Exposures: Mustard Gas and the Health Consequences of World War II in the United States* (New Brunswick, NJ: Rutgers University Press, 2019); Jordan Goodman, Anthony McElligott, and Lara Marks, eds., *Useful Bodies: Humans in the Services of Medical Science in the Twentieth Century* (Baltimore: Johns Hopkins University Press, 2008). Ancel Keys's son later apologized for his father's work. See Tucker, *Great Starvation Experiment*. On Alaska experiments, see Farish, "The Lab and the Land."

14. Richard A. Howard, *Sand, Sun, and Survival* (New York: Desert Publications, 1971), 1–3.

15. Pitkin, *Quartermaster Equipment for Special Forces*, 166. For "Desert Warfare Board," see p. 163.

16. Pitkin, *Quartermaster Equipment for Special Forces*, 185–86.

17. Pitkin, *Quartermaster Equipment for Special Forces* 24–25.

18. Pitkin, *Quartermaster Equipment for Special Forces*, 5. Rachel S. Gross writes about the ways that military clothing and testing practices made way for the postwar boom in outdoor clothing. Gross explicitly links government wartime work with private clothing manufacturing companies well after. "Layering for a Cold War: The M-1943 Combat System, Military Testing, and Clothing as Technology," *Technology and Culture* 60, no. 2 (2019): 378–408.

19. See Folk, "Harvard Fatigue Laboratory"; Thomas L. Endrusick, Leander A. Stroschein, and Richard R. Gonzalez, "US Military Use of Thermal Manikins in Protective Clothing Research," (Natick, MA: Army Research Institute of Environmental Medicine, 2002), DTIC, https://apps.dtic.mil/sti/citations/tr/ADP012410; "Fatigue Lab Scientists Drop Mercury to 40 Below Zero to Test Effects of Arctic on Army Men and Equipment," *Harvard*

Crimson, February 2, 1945, https://www.thecrimson.com/article/1945/2/2/fatigue-lab-scientists-drop-mercury-to/.
20. "McKinley Climatic Laboratory Eglin Air Force Base, Florida," National Historic Mechanical Engineering Landmark pamphlet, 1987, https://www.asme.org/wwwasmeorg/media/resourcefiles/aboutasme/who%20we%20are/engineering%20history/landmarks/116-mckinley-climatic-laboratory-1944.pdf.
21. Quoted in Paul Landsberg, "Deploy Globally, Train Locally: the US Army and the Global Environment" (PhD diss., Lawrence, KS, University of Kansas, 2022), 29; "Minutes of Meeting, National Research Council, Committee on Quartermaster Problems, Subcommittee on Environmental Protection at National Academy of Sciences," October 16, 1947, 3, Consultant Work-US Military: National Research Council: 1947–1947, Box 3, Folder 34, Wilkins Papers.

CHAPTER FIVE

1. "Integration of the Fatigue Laboratory with Industry," Box 26, Folder 4, Series 00303: IV Administrative Records, 1939–1950, Harvard Fatigue Lab Administrative Records, HFL/Countway. On closure and funding issues, G. Edgar Folk, "The Harvard Fatigue Laboratory: contributions to World War II," *Advances in Physiology Education* 43, no. 3 (2010): 119–62.
2. Forbes to Dill, December 15, 1942, Series 00300, Box 1, Folder 3, HFL/Countway.
3. "Integration of the Fatigue Laboratory with Industry," Series 303, Box 26, Folder 4; and "War Work and Peace Plans of the Fatigue Laboratory," *Harvard Business School Alumni Bulletin*, Series 303, Box 25, Folder 15, both HFL/Countway.
4. "Engineer News: Authoritative Information from the Chief of Engineers," *Military Engineer* 38, no. 243 (1946): 24–29. The next chief seconded this in his article in *Military Engineer*, saying that the war accelerated research and development like nothing else. See also R. A. Wheeler, "The Corps of Engineers' Extensive Peacetime Program," *Military Engineer* 40, no. 273 (1948): 297–300.
5. "Both survival and progress . . ." from Lt. Gen. Wilson, "The Army Engineers' Role in Space Construction," July 13, 1962, Box XVIII-30, ACEHQ. See also Wheeler, "Corps of Engineers' Extensive Peacetime Program." The literature on postwar roles is voluminous. See Leffler, *Preponderance of Power;* Adas, *Dominance by Design*, 221–29; Michael J. Hogan, *The Marshall Plan: America, Britain, and the Reconstruction of Western Europe, 1947–1952* (Cambridge University Press, 1987); Wheeler, "The Corps of Engineers' Extensive Peacetime Program"; Heefner, "Building the Bases of Empire," in

The Military and the Market, ed. Mittelstadt and Wilson, 105–19. On postwar rebuilding, see Robert P. Grathwol and Donita M. Moorhus, *Bricks, Sand, and Marble: U.S. Army Corps of Engineers Construction in the Mediterranean and Middle East, 1947–1991* (Alexandria, VA: US Army Corps of Engineers, 2010); Grathwol and Moorhus, *Building for Peace: U.S. Army Engineers in Europe, 1945–1991* (Washington, DC: US Army Center of Military History, 2005).

6. J. J. Manning, "Future Trends in Military Engineering," *Military Engineer* 41, no. 283 (1949): 333–35.
7. Wheeler, "Corps of Engineers' Extensive Peacetime Program."
8. Doel, "Constituting the Postwar Earth Sciences," 635–66.
9. Oreskes, *Science on a Mission*. On weather, see Kristine Harper, *Make It Rain: State Control of the Atmosphere in Twentieth-Century America* (Chicago: University of Chicago Press, 2017).
10. For example, the medical Department Field Research Lab in Fort Knox shrank at the end of the war, but in 1946, RDB decided it should stay open. See *Symposium on Military Physiology* (Washington, DC: Department of the Army, 1947), 27–28.
11. Scholarship on the RDB is surprisingly thin. See Michael Aaron Dennis, "Reconstructing Sociotechnical Order: Vannevar Bush and US Science Policy," in *States of Knowledge*, ed. Sheila Jasanoff (London: Routledge, 2004); "Joint Research and Development Board," Committee on Geographical Exploration, 1947, Box 160, Records of the Office of the Secretary of Defense, RG330, NACP. The RDB was created in June 1946 to coordinate all research and development activities of joint interest to the War and Navy Departments. Bush was head of the RDB in October 1947. Hamblin, *Arming Mother Nature*, 19–36, 43.
12. For "not normally accessible," see "Budget Statement from the Committee on Geographical Exploration," September 30, 1947. For a similar sentiment ("the global aspect of modern warfare demands an intense, progressive, continuing study of environmental factors"), see JRDB Committee on Geographical Exploration Annex "E," to minutes of third meeting, "report of conference on performance of material under extreme environmental conditions," March 17, 1947; "Intimate knowledge" in Sidney Paige memo, JRDB Committee on Geographical Exploration, n.d.—all in Box 160, RG 330, NACP. For more general info on RDB organization, see "Research and Development Board: Structure and Functions," RDB, Washington, DC, August 5, 1952, https://www.osti.gov/opennet/servlets/purl/16006145.pdf.
13. Hamblin, *Arming Mother Nature*; Simone Turchetti and Peder Roberts, eds., *The Surveillance Imperative: Geosciences during the Cold War and Beyond* (New York: Palgrave Macmillan, 2014).

14. Allan Needell has called these groups part of a "powerful coalition of interest groups that included military officers, government officials, and national political leaders," each of which valued the role of science in dealing with a host of domestic and geopolitical challenges. See Allan A. Needell, *Science, Cold War, and the American State: Lloyd V. Berkner and the Balance of Professional Ideals* (Amsterdam: Harwood, 2000); Annual Report of the Committee on Exploration, September 30, 1947–June 30, 1948, Box 26, Folder 25, Series 00303, HFL/Countway.
15. On Siple, see "Paul Siple," American Polar Society, https://americanpolar.org/about/polar-luminaries/paul-siple/.
16. For these efforts, see Lemons to Roger Prior, March 17, 1952, and May 28, 1952; "British Desert Warfare Experience, May 13, 1952, London; and "List of Persons and Offices for Contact on Hot Desert Conditions," all in Box 449, Folder 2, RG330, NACP.
17. Committee on Geographical Exploration, "Annex F to Agenda of Second Meeting," January 20, 1947, Box 160, RG 330, NACP; "Budget Statement from the Committee on Geographical Exploration," Joint Research and Development Board, Committee on Geographical Exploration, September 30, 1947, Box 160, RG 330, NACP. To be sure, the group noted, there "are several agencies with the War and Navy Departments engaged in gathering and assembling data on environmental factors." But they were handicapped by lack of personnel and expertise in the very areas they were trying to study September 30, 1947, "Budget Statement from the Committee on Geographical Exploration, Box 160, RG 330, NACP.
18. Pitkin, *Quartermaster Equipment for Special Forces*, 21. The speech was to the Navy Medical Center on February 19, 1948.
19. Dr. M. C. Shelesnyak to Lt. Col. Robert B. Simpson, memo, "Stefansson's Letter to Dr. V. Bush," May 20, 1948; Paul Siple to Lt. Col. Robert B. Simpson, memo, "Answers to Questions Regarding the V. Stefansson Letter of 4 March 1948 for Dr. Vannevar Bush." For clo values, Hoyt Lemmons to Lt. Col. R. B. Simpson, April 22, 1948. All are in Box 452, Folder 1, RG 330, NACP.
20. Memorandum for Dr. Vannevar Bush, Chairman, RDB, "Correspondence with Dr. Stefansson," n.d.; and Memo for Lt. Col. Robert B. Simpson, 14 April 1948, both Box 452, Folder 1, RG330, NACP. For wartime study, see Pitkin, *Quartermaster Equipment for Special Forces*.
21. "Program Guidance Report for FY 1952 in the Field of Desert and Tropical Environments," December 12, 1949, Box 174, RG 330, NACP.
22. Committee on Geophysics and Geography, annual report, July 1950, Binder 3, RG330, NACP. See also Sidney Paige, Memo: Members of the Committee on Geographical Exploration, SUBJ: Establishment of Committee on Geophysics and Geography, November 5, 1948, Series 302, Box 26,

Folder 28: III. Reports, 1925–1947. For three subpanels, see "Annual Report of the Committee on Exploration," both in HFL/Countway. On concerns about Quartermaster funding, see Memo for Vannevar Bush, October 27, 1947, Records concerning Organization, Budget, and the Allocation of Research and Development, 1946–1953, RDB, NM-12, Entry 341, RG 330, NACP.

23. On lab work, see Naomi Oreskes and Ronald E. Doel, "The Physics and Chemistry of the Earth," in *The Cambridge History of Science: Part V—Mathematics, Astronomy, and Cosmology since the Eighteenth Century*, ed. Mary Jo Nye (Cambridge: Cambridge University Press, 2002), 538–58. On regions, see Hamblin, *Arming Mother Nature*; Farish, "The Lab and the Land."

24. "Annual Report to the Executive Secretary," "Agendas, Meeting Minutes, and Other Records of Research Panels, Committees, and Working Groups, 1949–1953," UD-UP 23, Research and Development Board (RDB), and "Annual Report to the Executive Secretary," Committee on Geophysics and Geography, 21 July 1950, in Binder 3, RG 330, NACP.

25. "Annual Report to the Executive Secretary," "Agendas, Meeting Minutes, and Other Records of Research Panels, Committees, and Working Groups, 1949–1953," UD-UP 23, Research and Development Board (RDB), and "Annual Report to the Executive Secretary," Committee on Geophysics and Geography, 21 July 1950, in Binder 3, RG 330, NACP. On "mathematical terms," see "Program Guidance Report for FY 1952 in the Field of Desert and Tropical Environments," Box 174, RG 330, NA. On weather collection, see "Annual Report to the Executive Secretary," Committee on Geophysics and Geography, 21 July 1950, Box 174, Binder 3, RG 330, NACP.

26. "Panel on Arctic Environments, Unsolved Problems," May 1951, Binder 6, RG 330, NACP; Reybold "Authoritative Information from the Chief of Engineers."

27. Hamblin, *Arming Mother Nature*.

28. "Program Guidance Report for FY 1952 in the Field of Desert and Tropical Environments," December 12, 1949, Records concerning Organization, Budget, and the Allocation of Research and Development 1946–1953, NM-12, Entry 341, RG 330, NACP.

29. Memo: Department of the Army Requirements for Desert Testing, April 15, 1949, Binder 6, RG 330, NACP; Memo for Lt. Col. Simpson, RDB, from Dept of Army, April 15, 1949, Binder 6, RG 330, NACP. Wilkins quoted in "Contract for Desert Ground Surface Study," Notes: "Desert Ground Surface Study," 1950–1952, Box 4, Folder 11, Series 2: Consultant Work, Wilkins Papers. ADTIC collections described in ADTIC to RDB Panel on Desert and Tropical Environments, memo, May 25, 1951, Box 390, RG 330, NACP.

30. Memo for Dr. Helmut Landsberg, "Joint Army Desert Research, Development, and Test Facility," January 29, 1951, Binder 6, RG 330; Hoyt Lemmons, "Desert Research and test Sites," Special Report No. 47, November 27, 1950,

Binder 6, RG330; "Allocation of Responsibility for Research and Development in the Field of Snow, Ice and Permafrost," Box 452, Folder 5; and "Program Guidance for FY 1952," Panel on Desert and Tropical Environments, December 12, 1949, Box 174, RG 330, all NACP. On the history of Yuma Test Station (now called the Yuma Proving Ground), see "Yuma Proving Ground," n.d., courtesy of the Yuma Public Library. By 1955, it was a $20 million test center, with family housing.

31. Siple report from Korea, Panel on Arctic Environments, Minutes, Binder 4, RG 330, NACP.

CHAPTER SIX

1. On uses of sand, see Welland, *Sand*. For a more modern take on the problem of what running out of sand might look like, see Beiser, *The World in a Grain*.
2. Material from this chapter comes from Heefner, "'A Slice of their Sovereignty,'" 50–77; "A Tract That Is Wholly Sand: Engineering Military Environments in Libya," *Endeavour* 40, no. 1 (2019): 38–47; Villard, *Libya*, 59; British Administration, Tripolitania, "Annual Report of the Chief Administrator," December 31, 1949, 55, FO 1015/593, NAUK. See also "The Italians Are Unwanted by the Arabs," July 2, 1946, 2, FO 1015/118, NAUK; "History of Mellaha Air Base, 1946–1947" and "History of Wheelus Field, 1948," both , Air Force Archive Reel No. A0083, AFHRA. For the "sandman" cocktail that was popular on the base, see "The Flying Pan Revue," Officers Wives Club of Wheelus Field, Tripoli, Libya, 1954. The recipe was as follows: six dashes of maraschino liqueur, six sprigs mint, half a jigger lemon juice, one and a half jiggers of gin. Shake well in ice, drain into cups. For information on social life at Wheelus see the online community of folks who were once stationed there, see the Libyan Heritage House, https://libyanheritagehouse.org/wheelus-a-microcosm-of-american-life/beach-club, and a Facebook group devoted to Wheelus stories, https://www.facebook.com/WheelusAirBaseTripoliLibya.
3. Richard White, *It's Your Misfortune and none of my Own: A New History of the American West* (Norman: University of Oklahoma Press, 1993), 497; Michael Welsch, "The Legacy of Containment," 87–100; Maria Montoya, "Landscapes of the Cold War West," 9–28, both in Kevin Fernlund, ed., *The Cold War American West*.
4. Grathwol and Moorhus, *Bricks, Sand, and Marble*, chap. 2.
5. Tom Clancy, with Chuck Horner, *Every Man a Tiger: The Gulf War Air Campaign* (New York: Penguin 2008); Robin Olds, *Fighter Pilot: The Memoirs of Legendary Ace Robin Olds* (New York: St. Martin's Griffin, 2011), 216.

Background on bombing and weather also from General Merrill McPeak, *Hangar Flying* (n.p., Lost Wingman Press, 2012), http://generalmcpeak.com; Humphrey Wynn, "Shores of Tripoli," *International Flight*, 1961.
6. US Army Corps of Engineers, "Middle East District Report, 1954," Military Construction Files, American Overseas Bases, OPS Box XII-30, ACEHQ.
7. Grathwol and Moorhus, *Bricks, Sand, and Marble*, 30, 50. For Morocco the joint venture was called Atlas construction and it was made up of Morrison-Knudsen, Nello L. Tear Co., Ralph E. Mills Co., Blythe Bros. Co., and Bates & Rogers Construction Corp.
8. Heefner, "Building the Bases of Empire," 105–19.
9. Heefner, 105–19; Grathwol and Moorhus, *Bricks, Sand, and Marble*, 12.
10. For example, in 1953, the Air Force approved the new B-52s, which would cost $3.6 million each (although the first four cost $20 million each). See "Chronology," *Air and Space Forces Magazine*, November 24, 2018, https://www.airandspaceforces.com/chronology-1950-1959/.
11. In re-creating what Troxler may have seen and how he may have traveled, I have layered various memoirs, diplomatic reports, and Air Force documents about the types of planes and materials available and widely used at the time. The Air Force had C-47s and C-54s. "History of Wheelus 1952," Air Force Archive Reel No. K3601, AFHRA; Vandewalle, *A History of Modern Libya* (Cambridge: Cambridge University Press, 2012), 15; George E. Thompson, *Life in Tripoli: With a Peek at Ancient Carthage* (London: Simpkin, Marshall & Co., 1894), 59–61.
12. "Port Facilities in Tripoli," November 8, 1948, Box 1, US Consulate, Legation and Embassy, Tripoli, General Records in Tripoli, 1948–1961, Records of the Foreign Service Posts of the Department of State, RG 84, NACP; Villard, *Libya*, 9; Grathwol and Moorhus, *Bricks, Sand, and Marble*:, 34. On repairing the breakwater, see "Wheelus Field, Construction of AF Facilities, FY 1952," April 1952, memo, DOD Papers, Box 3, ACEHQ. See also Office of the Secretary of State to Villard, December 8, 1952, 711.56373/11–2252; Foreign Service Dispatch, "Presentation of Check for $100,000," April 23, 1953, 711.56373/4–2353, both in Decimal Files, 1950–1954, RG 59, NACP.
13. Villard, *Libya*, 3.
14. Information on local farms is drawn from a number of sources, including A. J W. Hornby, "Northern Tripolitania: A Dry Mediterranean Coastal Region," *Economic Geography* 21 (October 1945): 231–51; Graeme Barker, David Gilbertson, Barri Jones, and David Mattingly, *Farming the Desert: The UNESCO Libyan Valleys Archaeological Survey* (Paris: UNESCO Publishing, 1996); Benjamin Higgins, Report of the Mission to Libya (New York: UNESCO, 1952); British Administration, Tripolitania, "Annual Report, 1949," FO 1015/599, NAUK.

15. See "Expropriation of Land for American Air Base," 1944, FO160/99, NAUK. For the full story, see Heefner, "'A Slice of their Sovereignty': Negotiating the US Empire of Bases, Wheelus Field, Libya, 1959–1954," *Diplomatic History* 41, no. 1 (2017): 50–77; "Tract That Is Wholly Sand."
16. The geographer Michael Welland has explained that information was not shared across imperial borders. See his books *Sand* and *Desert*. On colonial ideas about deserts, Davis, *Resurrecting the Granary of Rome*; Davis, *The Arid Lands*.
17. Maurice J. Kenn, "Ralph Alger Bagnold," https://royalsocietypublishing.org/doi/pdf/10.1098/rsbm.1991.0003.
18. Ralph Bagnold, "The Last of the Zerzura Legend?," Zerzura Club, 1936, Expeditions and Research 1924–1990, Papers of Brigadier Ralph Alger Bagnold, GBR/0014/BGND B.18, Churchill Archives Centre, University of Cambridge. "Explores Mysterious Africa Plateau Like 'Lost World,' but Sees No 'Lost Oasis,'" *New York Times*, November 17, 1930, 3.
19. W. B. Kennedy Shaw, *The Long Range Desert Group* (London: Greenhill Books, 1945); Ralph Bagnold, "Early Days of the Long Range Desert Group," *Geographical Journal* 105, nos. 1–2 (January–February 1945).
20. Ralph Bagnold, *The Physics of Blown Sand and Desert Dunes* (London: Methuen & Co., 1941), xviii–xix. For more background on Bagnold, see Welland, *Sand* and *Desert*; Ralph Bagnold, *Sand, Sun and War: Memoirs of a Desert Explorer* (Tucson: University of Arizona Press, 2019), 17 and 68.
21. Bagnold, *Sand, Sun and War*, 17, 68; Welland, *Sand*, 5.
22. Welland, *Sand*, 32–35.
23. These values come from the Udden-Wentworth scale to classify sediments based on diameter.
24. For sampling, see Letter from British Petroleum, December 2, 1963, GBR/0014/BGND B.22. In 1966, he was invited to Iraq, and in 1968, he was to advise Abu Dhabi's government on sand. In 1970, he went to Tehran. All reported in his draft autobiography: Autobiographical and Biographical, 1917–1991, GBR/0014/BGND A.13, Churchill Archives Center.
25. Jack Volin, "Historical Record, Tripoli Field" February 1946, Air Force Archive Reel No. A0083, AFHRA. "Shot Down over Libya," *Saturday Evening Post*, August 1, 1942; "History of Wheelus Field, July–December 1952," Air Force Archive Reel No. K3601, AFHRA; Waterways Experiment Station, "Limited Reconnaissance for Pavement Evaluation and Soil Type" (US Corps of Engineers, Humphries Engineering Center, 1952).
26. "A Study of Windborne Sand and Dust in Desert Areas," Technical Report ES-8, Earth Sciences Division, US Army Natick Laboratories, Natick, MA, August 1963.
27. "A Study of Windborne Sand and Dust in Desert Areas," 20, 29; Hornby, "Northern Tripolitania," 233; "Tree Planting," May 1951, FO 1012/29, NAUK;

"History of the 17th Air Force, January–March 1954," 6–9, Air Force Reel No. N0908, AFHRA.

28. "Tree Planting," May 1951, FO 1012/29, NAUK; "A Study of Windborne Sand and Dust in Desert Areas," 32.
29. Grathwol and Moorhus, *Bricks, Sand, and Marble*, 37 and 83.
30. "Mediterranean Division Report, Corps of Engineers, US Army," April 1956, General Files, Box 2, ACEHQ.
31. "Historical Summary of Middle East District, 1950–1951" ACE general files, Box 2, ACEHQ; Grathwol and Moorhus, *Bricks, Sand and Marble*, 79.
32. Grathwol and Moorhus, *Bricks, Sand, and Marble*, 42, 79, 87–88; "Mediterranean Division Report, Corps of Engineers, US Army"; "Strategic Desert Areas of the World," *Military Engineer* 43, no. 295 (1951): 362.
33. For example, see June 7, 1951, "Request for Waiver" Corps of Engineers Memo, ACE OPS Box XII-30, ACEHQ; April 2, 1952 Telegram Tripoli to DOS, Central Decimal Files, 3195a, RG59, NACP.

CHAPTER SEVEN

1. "Desert Symposium" November 14, 1952, Box 4, Folder 11, and "Color for Desert Clothing," Environmental Protection Branch, Special Report No. 55, January 15, 1952, Box 6, Folder 20, Wilkins Papers.
2. Hubert Wilkins, "A Journey through Hot Dry Areas in Pakistan, Saudi Arabia, Iraq, Syria, Lebanon, Egypt, Libya, Algeria, Morocco," June–August 1952, Series 2: Consultant Work: United States Military, 1929–circa 1950s, Box 6, Folder 19, Wilkins Papers.
3. Wilkins, "A Journey through Hot Dry Areas in Pakistan . . ." See also his typed notes about each location. For Libya notes, see Box 4, Folder 6, and Box 4, Folder 4, Wilkins Papers.
4. "Saudi Arabia: Some General Comments," Box 6, Folder 19, Wilkins Papers.
5. Wilkins, "A Journey through Hot Dry Areas in Pakistan . . ."
6. Wilkins Greenland notebook, Box 4, Folder 2, Wilkins Papers.
7. "Desert Symposium," November 14, 1952, Box 4, Folder 8, Wilkins Papers.
8. George Howard, "Desert Effects on Engineer Equipment," *Military Engineer* 43, no. 285 (1951): 362–64; George Howard, "The Yuma Test Branch of the Engineer Board," *Military Engineer* 39, no. 263 (1947): 384–85; R. C. Hannum and H. A. McKim, "Desert Motor Transport," *Military Engineer* 42, no. 290 (1950): 435–37; W. R. Shuler, "Major Test Runway Built for Research," *Military Engineer* 47, no. 318 (1955): 265–67.
9. "Basic Plan of Investigation and Determination of Coloration for Desert Clothing, Tentage, and Equipment," Engineer Research and Development Lab, March 1951, Box 4, Folder 11, Wilkins Papers.

10. "Contract for Desert Ground Surface Study," Box 4, Folder 11, Wilkins Papers.
11. James R. Shepard, James C. Johnstone, Alton A. Lindsey, Robert D. Miles, and Robert F. Frost, "Terrain Study of the Yuma Test Station Area," contract with the Chief of Engineers, (Lafayette, Indiana, Purdue University, March 1955), DTIC, https://apps.dtic.mil/sti/pdfs/AD0626500.pdf
12. Nevada National Security Site, Nuclear Timeline, https://nnss.gov/wp-content/uploads/2023/04/DOENV_1243-1.pdf.
13. Military Evaluation of Geographic Areas files are available online through the Defense Technical Information Center https://apps.dtic.mil/sti/citations/AD0626500. For scholarly overview, see Paul Landsberg, "Deploy Globally, Train Locally: The US Army and the Global Environment" (PhD diss., University of Kansas, 2022). On comparing deserts and assuming they are all similar, see Davis and Burke, eds., *Environmental Imaginaries of the Middle East and North Africa*. Jack R. Van Lopik and Charles R. Kolb, *Handbook: A Technique for Preparing Desert Terrain Analogs* (Vicksburg, MS: Waterways Experiment Station, 1959).
14. Jack R. Van Lopik, Charles R. Kolb, and John R. Shamburger, *Analogs of Yuma Terrain in the Northwest African Desert: Report 6, Volume I* (Vicksburg, MS: Waterways Experiment Station, 1965).
15. Warren E. Grabau, "Concept and Status of Terrain research," in *Military Evaluation of Geographic Areas, Reports on Activities to April 1963* (Washington, DC: Corps of Engineers, December 1963).
16. Van Lopik and Kolb, *Handbook: A Technique for Preparing Desert Terrain Analogs*, https://digital.library.unt.edu/ark:/67531/metadc303897/m2/1/high_res_d/metadc303897.pdf.
17. Shepard et al., "Terrain Study of the Yuma Test Area."
18. Warren E. Grabau, "Concept and Status of Terrain Research," 3.
19. Farouk El-Baz, *The View from Space* (Washington, DC: Smithsonian Institution, 1982), https://repository.si.edu/handle/10088/6349; Farouk El-Baz, "Earth Observations and Photography Experiment: Summary of Significant Results," NASA contract NAS 9-13831, (Washington, DC: Smithsonian Institutions, June 1978), 12–13, available at NASA Technical Reports Server, doc. 19790002327, https://ntrs.nasa.gov/citations/19790002327.
20. El-Baz, "Earth Observations."

PART TWO

1. Basic background comes from Jon Gertner *The Ice at the End of the World* (London: Icon Books, 2019); Carl Benson, oral history (Karen Brewster interviewer), June 22, 2001, Byrd Polar and Climate Research Center Archive Program, Ohio State University; Kristian H. Nielsen and Henry Nielsen,

Camp Century: The Untold Story of America's Secret Arctic Military Base under the Greenland Ice (New York: Columbia University Press, 2021); Ronald Doel, Kristine C. Harper, Matthias Heymann, eds., *Exploring Greenland: Cold War Science and Technology on Ice* (New York: Palgrave Macmillan, 2016).
2. For siting, see Captain Clifton T, Ken and Lt. Addison D. Minott, "'Project Dog Sled': The Construction of AC&W Stations N-33 and N-34 on the Greenland Icecap," January 1954, Box XVII-9, ACEHQ.
3. Jean Malaurie, *The Last Kings of Thule: A Year among the Polar Eskimos*, trans. Gwendolen Freeman (New York: Crowell, 1956).
4. R. L. Schuster and G. P. Rigsby, "Preliminary Report on Crevasses, April 1954," SIPRE Special Report 11, ERDC.
5. Bernt Balchen, quoted in "Engineering Problems in the Arctic," *Military Engineer* 44, no. 302 (1952): 426–28.

CHAPTER EIGHT

1. District Engineer, Northeast District, Monthly Report, March 1951, Box XII-18, ACEHQ. On the worst weather, see Dodd, Military Engineering File, ACEHQ.
2. There is no "daylight" during this window, although there is still civil twilight during part of it. See "Thule Air Base, Greenland—Sunrise, Sunset, and Daylength," time and date, https://www.timeanddate.com/sun/greenland/thule-air-base.
3. This book refers to the people living in northwestern Greenland, near Thule, as Inughuit. In the 1950s and earlier, many officials and outsiders referred to the people of the Thule area as "Polar Eskimos" or "Inuit." According to the International Work Group for Indigenous Affairs, in Greenland "the population is 88% Greenlandic Inuit with a total of 56,367 inhabitants (July 2020). The majority of Greenlandic Inuit refer to themselves as Kalaallit. Ethnographically, they consist of three major groups: the Kalaallit of West Greenland . . .; the Tunumi-it of Tunu (East Greenland) . . . and the Inughuit/Avanersuarmiut of the north. The majority of the people of Greenland speak the Inuit language, Kalaallisut, which is the official language, while the second language of the country is Danish." From "Indigenous Peoples in Kalaallit Nunaat (Greenland)," https://www.iwgia.org/en/kalaallit-nunaat-greenland.html.
4. Walter Wager, *Camp Century: City under the Ice* (Boston: Chilton Books, 1961); Russell Owen, "The North Polar Region," *New York Times*, January 12, 1947, SM12. The German glaciologist Ernst Sorge called it the "white desert," quoted in Gertner *The Ice at the End of the World*, xix.

5. Fitzhugh Green, "Abode of the White Terror," *New York Times*, May 17, 1931, 80.
6. Malaurie, *The Last Kings of Thule* (1982), 20. On Malaurie, see "Bio," Malaurie Institute of Arctic Research, https://miarctic.org/jean-malaurie/bio. Malaurie had spent the previous two winters in the Sahara.
7. Nikolaj Petersen, "SAC at Thule: Greenland in the US Polar Strategy," *Journal of Cold War Studies* 13, no. 2 (2011): 90–115. See also Ronald E. Doel, "Defending the North American Continent: Why the Physical Environmental Sciences Mattered in Cold War Greenland," 25–46, and, on weather, Matthias Heymann, "In Search of Control: Arctic Weather Stations in the Early Cold War," 75–98, both in Doel, Harper, and Heymann, eds., *Exploring Greenland. Greenland during the Cold War: Danish and American Security Policy 1945–68*, trans. Henry Allen Myers (Copenhagen: Danish Institute of International Affairs, 1997).
8. Lewis A. Pick, "The Story of BLUE JAY," *Military Engineer* 45, no. 306 (1953): 278–86.
9. "We are lagging," in "Support of Action Taken by the Committee on Geographical Exploration," RDB, August 20, 1948, Box 452, Folder 5, RG330, NACP. Arnold said, "If there is a third war, its strategic center will be the north pole." Quoted in Doel, "Defending the North American Continent," 28. Doel points out that the United States had long had interest in Greenland because of its fisheries, cryolite supply, geography between North America and Europe. Arthur Trudeau, quoted in Janet Martin-Nielsen, *Eismitte in the Scientific Imagination: Knowledge and Politics at the Center of Greenland* (New York: Palgrave Macmillan, 2013), 62.
10. Memo for Lt. Col. Simpson, Subject: Answers to Questions, April 14, 1948, Folder 1, Box 452, RG 330, NACP. See also general comments in "Analysis of Snow, Ice and Permafrost Questionnaires," RDB Committee on G&G Panel on Arctic Environments, 6 April 1949, Binder 5, RG 330, NACP.
11. Brig. Gen. Richard Whitcomb, Memo for Chief of Transportation, March 19, 1951, Folder 41E, Box XII; Sofia Ribeiro et al, "Vulnerability of the North Water Ecosystem to Climate Change," *Nature Communications* 12, no. 4475 (2021): 1–12; "North Water Polynya," Oceans North, https://www.oceans north.org/en/where-we-work/north-water-polynya/; Pew Trusts, "North Water Polynya," September 1, 2013, https://www.pewtrusts.org/en/research-and-analysis/fact-sheets/0001/01/01/north-water-polynya.
12. Pikialasorsuaq Commission, *People of the Ice Bridge: The Future of the Pikialasorsuaq* (Ottawa: Inuit Circumpolar Council Canada, 2017). In recognition that the polynya is at risk due to climate change, the Inuit Circumpolar Council created in 2013 a special Pikialasorsuaq Commission to begin making plans for the area.

13. Pikialasorsuaq Commission, *People of the Ice Bridge*, A-11.
14. Gertner, *Ice at the End of the World*, 44–46. See also Børge Fristrup, *The Greenland Ice Cap* (Seattle: University of Washington Press, 1967), 66–68; Martin-Nielsen, *Eismitte*. These sources also provide the scaffolding for the general history in the rest of this chapter.
15. See Fristrup, *Greenland Ice Cap*, 142; "Thule Air Force Base," *Progressive Architecture* 34, no. 12 (1953): 107–11.
16. Bernt Balchen, *Come North With Me: An Autobiography* (New York: E. P Dutton, 1941).
17. On relative lack of importance of Thule vis-à-vis other Greenland facilities, see "Historical Monograph Greenland," 1946, Box 1B, ACEHQ. For a strategic overview, *Greenland during the Cold War: Danish and American Security Policy 1945–68*, trans. Henry Allen Myers (Copenhagen: Danish Institute of International Affairs, 1997).
18. William Herbert Hobbs, "Secret of the Winds Sought in Greenland," *New York Times*, July 3, 1927, XX2: "Arctops Project for Studies in Problems of Arctic Operations," MIT, Box 1, Folder Polar OPS ARCTOPS, RG 27, NACP; Heymann, "In Search of Control," 78. On weather and climate data, see Paul Edwards, *A Vast Machine* (Cambridge, MA: MIT Press, 2005). On security information, see "Greenland Fact Sheet," Box XXII23, ACEHQ.
19. "Polar Ops Thule Report 1946-7," Box 1, RG 27, NACP.
20. On data collection for monitoring radioactive particles and weather patterns, and the computer programs and models created to account for them, see Edwards, *Vast Machine*; Hamblin, *Arming Mother Nature*. See also Toshihiro Higuchi, *Political Fallout: Nuclear Weapons Testing and the Making of a Global Environmental Crisis* (Stanford, CA: Stanford University Press, 2020).
21. On supply from Thule to Eureka, see "Report on Inspection of Eureka Sound and Thule Weather Stations During April 1947," Records of the Polar Operations Project, Arctic Stations Reports, RG 27, NACP. Heymann notes that the Danish government was concerned about US arrivals and wanted to have more Danes there than Americans. "In Search of Control," *Exploring Greenland*, 75–98.
22. Melanie McGauran, "Tales from Thule, Greenland Part II," *Leaving the Door Open*, November 14, 2020, https://leavingthedooropen.com/2020/11/14/tales-from-thule-greenland-part-ii/. "Report on Inspection of Eureka Sound and Thule Weather Stations During April 1947," Records of the Polar Operations Project, Arctic Stations Reports, RG 27, NACP.
23. "Supply Narrative," November 30, 1949, and March 28, 1950, Records of the Polar Operations Project, Arctic Stations Reports, RG 27, NACP.

24. On cooperation with weather bureau, Memo to "District Engineer, from AF Liaison Officer, Monthly Report, March 1951," April 2, 1951, Box XII-18, ACEHQ.
25. Richard Morenus, *Dew Line: Distant Early Warning, the Miracle of America's First Line of Defense* (New York, Rank McNally, 1957), 96.
26. Pick, "Operation Blue Jay," *Military Engineer*; "Thule Air Force Base," *Progressive Architecture* 34, no. 12 (December 1953): 107–11.
27. Heefner, "Building the Bases of Empire," 109–110. The firm Metcalf and Eddy, for example, was hired to help with the construction of Thule Air Force Base and then continued to get design contracts for Greenland projects such as Camp Century.
28. "Historical Monograph Greenland," 1946, Box 1B, ACEHQ.
29. "Inspection Report, April 1952," Box XII-23b; and "Birth of a Base," *Life Magazine* (September 1952), Box XII-Thule articles; both in ACEHQ.

CHAPTER NINE

1. "Panel on Arctic Environments, Minutes, Eight Meeting," June 11, 1951, Binder 4, RG 330, NACP.
2. "Technical Estimate FY1951," March 15, 1951, Binder 4B, RG 330, NACP.
3. "Panel on Arctic Environments, Minutes, 11th meeting," November 7, 1952, Binder 4, RG 330, NACP.
4. Lewis M. Alexander, "Samuel Whittemore Boggs: An Appreciation," *Annals of the Association of American Geographers* 48, no. 3 (1958): 237–43.
5. S. Whittemore Boggs, "An Atlas of Ignorance: A Needed Stimulus to Honest Thinking and Hard Work," *Proceedings of the American Philosophical Society* 93, no. 3 (1949): 253–58.
6. Boggs, "An Atlas of Ignorance," 253.
7. Selcer, *The Postwar Origins of the Global Environment*; Nick Cullather, *The Hungry World: America's Cold War Battle against Poverty in Asia* (Cambridge, MA: Harvard University Press, 2010); Michael Latham, *Modernization as Ideology: American Social Science and "Nation Building" in the Kennedy Era* (Chapel Hill: University of North Carolina Press, 2003); Mitchell, *Rule of Experts*.
8. The need for definitions and terminology was everywhere apparent in meetings. See November 7, 1952, "Minutes of 11th meeting," held on September 25, 1952, Panel on Arctic Environments, Minutes, Binder 4, RG 330; "Analysis of Environmental Factors in Arctic Regions," July 22, 1948, Folder 2, Box 451, and Committee on Geophysics and Geography, Panel on Arctic Environments, June 1, 1951, Box 389, RG 330; both in NACP.

9. On various efforts to rename, see "Status of the Three Geographic Panels," November 13, 1950, and "Joint Panel on Cold Environments," May 10, 1950, both in Binder 3, RG 330, NACP; *Glossary of Arctic and Subarctic Terms*, (Maxwell AFB, ADTIC Research Studies Institute, 1955), iii.
10. Harley J. Walker, *Man in the Arctic* (Maxwell Air Force Base, Alabama, ADTIC Research Studies Institute, 1962), reprinted in Air Force Archive Reel No. K2566, frame 618, AFHRA.
11. "The War Department Arctic Programs 1947–1950," July 28, 1947, Folder 1, Box 451, RG 330, NACP; "Minutes of 11th meeting."
12. See Paul H. Nesbitt, "A Brief History of the Arctic, Desert, and Tropic Information Center and Its Arctic Research Activities," in *United States Polar Exploration*, ed. Herman R. Friis and Shelby G. Bale (Athens: Ohio University Press, 1970), 144. The work was crafted into the ADTIC's *Glossary of Arctic and Subarctic Terms* (Maxwell AFB, ADTIC Research Studies Institute, 1955).
13. Kirk Bryan, "Cryopedology: The study of Frozen Ground and Intensive Frost Action with Suggestions on Nomenclature," *American Journal of Science* 244 (1946): 622–42; Bryan, "The Study of Frozen Ground and Intensive Frost-Action," *Military Engineer* (1948): 304–5; "Analysis of Snow, Ice and Permafrost Questionnaires," April 6, 1949, Binder 5, G&G Panel on Arctic Environments, RG 330, NACP.
14. The answer was no, but that data came years later, once the Corps had mapped and photographed much of the far north.
15. Madeline Ostrander, "In a Tunnel beneath Alaska, Scientists Race to Understand Disappearing Permafrost," *Smithsonian Magazine*, May 4, 2020, https://www.smithsonianmag.com/science-nature/tunnel-beneath-alaska-180974804/.
16. A lot more get stuck down in the permafrost. Though not known in the 1950s when permafrost became a focus of study, scientists learned over time that permafrost is a carbon sink, meaning it stores the material. Organic matter that is frozen down below will decompose when thawed, releasing greenhouse gases.
17. E. A. G. Schuur et al., "Climate Change and the Permafrost Carbon Feedback," *Nature* 520 (2015): 171–79.
18. "Minutes of the 17th Meeting of the Army Committee on Environmental Factors and Control," November 22, 1948, Box 452, Folder 2, RG 330, NACP.
19. "Some Aspects of Snow, Ice, and Frozen Ground," SIPRE Report 10 (Wilmette, IL: SIPRE, August 1953), 16, ERDC.
20. "Geology of the USA CRREL Permafrost Tunnel," Technical Report No. 199, CRREL (July 1967), ERDC; Margaret Cysewski, Kevin Bjella, and Matthew Strum, "The History and Future of the Permafrost Tunnel near

Fox, Alaska," US Cold Regions Research and Engineering Laboratory, https://pubs.aina.ucalgary.ca/cpc/CPC6-1222.pdf; Donald Rausch, "Ice Tunnel, Tuto Area, Greenland, 1956," Technical Report No. 44 (Wilmette, IL: SIPRE, 1956), ERDC; Ostrander, "In a Tunnel beneath Alaska, Scientists Race to Understand Disappearing Permafrost."

21. "Terrain and Construction Materials, Denali Area, Alaska," Engineer Intelligence Study (Washington, DC: Department of the Army, September 1959), Box 27, RG 77, NACP; "Collection of Information on Underground Installations, 1959," Engineer Intelligence Study 30 (Washington, DC: Department of the Army, 1959), Box 20 NN3-077-097-005, RG 77, NACP; "Collection of Information on Coasts and Beaches, 1959," Engineer Intelligence Guide 15 (Washington, DC: Department of the Army, 1958), https://digitalcommons.unl.edu/dodmilintel/80/.

22. "Statement of Engineer Intelligence Interest (Revised)" (Washington, DC: Department of the Army, 1958), NN3-077-097-005-Box 20; and "Seward Peninsula, Alaska", Engineer Intelligence Study 185 (Washington, DC: Department of the Army, 1959), Box 25; both RG 77, NACP.

CHAPTER TEN

1. Malaurie used the spelling *Outak* or *Uutaaq*, but the name is more frequently spelled *Odaaq* and sometimes *Oodaaq*. Malaurie, *The Last Kings of Thule* (1982), 384; Malaurie, *Ultima Thulé* (Paris: Bordas, 1990), 381. Note that I have consulted two translations of Malaurie's *The Last Kings of Thule*, the 1956 and 1982 versions. Each is noted here by year of publication.
2. Jens Brøsted and Mads Faegteborg, "Expulsion of the Great People When US Air Force Came to Thule: An Analysis of Colonial Myths and Actual Incidents," in *Native Power: The Quest for Autonomy and Nationhood of Indigenous Peoples*, ed. Jens Brøsted, Jens Dahl, Andrew Gray, Hand Christian Gulløve, Georg Henriksen, Jørgen Brøchner Jørgensen, and Inge Kleivan (Oslo: Universitetsforlaget AS, 1985), 224. According to Malaurie, in *Ultima Thule*, in 1951, there were 302 indigenous Greenlanders living in the region spread across a number of camps and settlement. See Steen Wulff, "The Legal Bases for the Inughuit Claim to their Homeland," *International Journal on Minority and Group Rights* 12, no. 1 (2005): 63–91.
3. Whitecomb Memo, Box XII-41E, ACEHQ.
4. Malaurie, *Last Kings* (1956), 263, and (1982), 383.
5. Malaurie, *Last Kings* (1982), 383–85, 392–403.
6. Malaurie *Last Kings* (1956), 262.
7. There are many stories of how US militarism has displaced people around the world. See Lauren Hirshberg, "Navigating Sovereignty under a Cold

War Military Industrial Colonial Complex: US Military Empire and Marshallese Decolonization," *History and Technology* 31, no. 3 (2015): 259–74; David Vine, *Island of Shame: The Secret History of the U.S. Military Base on Diego Garcia* (Princeton, NJ: Princeton University Press, 2009).

8. David Christopher Arnold, "'How Far North Can a Guy Git?' Minnesotans at the Top of the World," *Minnesota History* 65, no. 1 (2016): 4–13; "Transcript of interview with Harold 'Oakie' Priebe, Chief, Heavy Construction Division, Greenland Contractors," September 3, 1959, Military Engineer Files, American Overseas Bases, OPS Box XII-24, ACEHQ.

9. Malaurie, *Last Kings* (1982), 389. See Arnold, "Minnesotans."

10. Malaurie, *Last Kings* (1982), 389.

11. Interview with Harold "Oakie" Priebe, Chief, Heavy Construction Division, Greenland contractors, September 3, 1959, in XII-24 "Personnel." Narrative Report on Status of Construction 30 June 1951 confirms that planes were delayed for two periods of time in June due to weather, Box XII-20, ACEHQ.

12. Topography from a number of sources, including Robert L. Anstey, *Handbook of Thule, Greenland, Environment* (Natick, MA: Quartermaster Research and Engineering Center, 1956).

13. Bjarne Grønnow, "Living at a High Arctic Polynya: Inughuit Settlement and Subsistence around the North Water during the Thule Station Period, 1910–53," *Arctic* 69, no. 1 (2016): 10.

14. "General Considerations of Arctic and Subarctic Environments," June 1, 1951, Box 389, RG 330, NACP.

15. Quote from Malaurie, *Last Kings* (1982), 383.

16. Project Blue Jay, Scope of Work, Long Range Plan, May 18, 1951, Box XII-20, ACEHQ. *Kon-Tiki* note reported by Malaurie, *Last Kings* (1982), 388.

17. Malaurie *Last Kings* (1956), 267; "Project Blue Jay-Logistical Operations 1951," memo from USS Shadwell to CTF 118, Box XII-21, ACEHQ.

18. Malaurie, *Last Kings* (1956), 257, 268, and (1982), 394.

19. Robert L. Anstey, *Handbook of Thule, Greenland, Environment* (Natick, MA: Quartermaster Research and Engineering Center, 1956), 12.

20. Grønnow, "Living at a High Arctic Polynya."

21. "Relations with the Native Population," July 7, 1951, Box XII-21, Military, American Overseas OPS, Greenland Memos, ACEHQ.

22. See "Rules Set by Governor of North Greenland," Box XII-41E, ACEHQ.

23. On negotiations, see Martin-Nielsen, *Eismitte in the Scientific Imagination*, 62.

24. Malaurie, *Last Kings* (1982), 394–95.

25. Malaurie, *Last Kings* (1956), 165–66; Brøsted and Faegteborg, "Expulsion of the Great People," 214; Malaurie, *Last Kings* (1982), 393.

26. "Note on this Narrative Report on Status of Construction, 30 June 1951," Office of the District Engineer, Military Construction Files, Box XII-20, ACEHQ.
27. Arnold, "Minnesotans."
28. Ole Spiermann, "Hingitaq 53, Qajutaq Petersen, and Others v. Prime Minister's Office (Qaanaaq Municipality and Greenland Home Rule Government Intervening in Support of the Appellants)," *American Journal of International Law* 98, no 3. (July 2004): 572–78. See Wulff, "Legal Bases for the Inughuit Claim." Other sources on the removal and court cases include DeNeen L. Brown, "Trail of Frozen Tears," *Washington Post*, October 22, 2002, https://www.washingtonpost.com/archive/lifestyle/2002/10/22/trail-of-frozen-tears/21add819-7dd6-46e8-9585-88b40f4bbbb3/; Marie-Louise Holle, "The Forced Relocation of Indigenous Peoples in Greenland—Repercussions in Tort Law and Beyond," Working Paper No. 1940, Copenhagen Business School, November 2019.
29. Holle, "Forced Relocation of Indigenous Peoples."
30. Quoted in Brøsted and Faegteborg, "Expulsion of the Great People," 213; "Relations with Native Population," Memo No. 1, July 7, 1951, Military Files, Box XII-21, ACEHQ.
31. "Greenland Town Wants to Move from Planes' Roar," *Washington Post*, May 12, 1953, 3. "Thule, Greenland, 1954—Arctic Adventure with Nate Galbreath," account printed in the Thule Forum, an online platform for people to share stories and images from their time stationed at Thule. The Thule Forum recently ceased publishing; I have Galbreath's story saved.
32. Brøsted and Faegteborg, "Expulsion of the Great People," 227, 214. Claims made in 1960 and 1985. See Holle, "Forced Relocation of Indigenous Peoples," 21; Brown, "Trail of Frozen Tears." See also Jane George, "ICC President Welcomes Danish Apology for Thule Relocation," *Nunatsiaq News*, September 10, 1999, https://nunatsiaq.com/stories/article/icc_president_welcomes_danish_apology_for_thule_relocation/.
33. Brøsted and Faegteborg, "Expulsion of the Great People," 224.
34. Malaurie, *Ultima Thule*, 382; In 1968 a B-52 crashed with nuclear warheads on board, spewing radioactive contamination over the region.
35. For stuff that has been left behind and current concerns about it, see Colin Woodward, "US Cold-War Waste Irks Greenland," *Christian Science Monitor*, August 22, 2008, https://www.csmonitor.com/Environment/Living-Green/2008/0822/us-cold-war-waste-irks-greenland; Julia Rosen, "Mysterious, Ice-Buried Cold War Military Base May Be Unearthed by Climate Change," *Science*, August 4, 2016, https://www.sciencemag.org/news/2016/08/mysterious-ice-buried-cold-war-military-base-may-be-unearthed-climate-change.

CHAPTER ELEVEN

1. Jason Box, Alun Hubbard, David B. Bahr, William T. Colgan, Xavier Fettweis, Kenneth D. Mankoff, Adrien Wehrlé, Brice Nöel, Michiel R. van den Broeke, Bert Wouters, Anders A. Bjørk, and Robert S. Fausto, "Greenland Ice Sheet Climate Disequilibrium and Committee Sea-Level Rise," *Nature Climate Change* 12 (August 2022): 808–13. The National Snow and Ice Data Center has up-to-date data on Greenland's daily melt. See "Greenland Ice Sheet Today," National Snow and Ice Data Center, https://nsidc.org/greenland-today/. For basics, see "Greenland's Ice Is Melting," University Corporation for Climate Research, https://scied.ucar.edu/learning-zone/climate-change-impacts/greenlands-ice-melting. For the latest news, see American Geophysical Union, "The Greenland Ice Sheet Is Close to a Melting Point of No Return, Says New Study," phys.org, March 27, 2023, https://phys.org/news/2023-03-greenland-ice-sheet.html.
2. Corps of Engineers, Research and Development Board, "Greenland Ice Cap Research Program, Studies Completed in 1954," (Fort Belvoir, VA: Corps of Engineers ERD Laboratories, February 1957), ERDC.
3. Memo to Director of R&D, General Staff, US Army, "Snow and Ice Mechanics Laboratory," October 23, 1947, appendix to "Interim Report to Snow, Ice, and Permafrost Research Establishment," SIPRE Report 1, January 1950, 7–9, ERDC. On the lack of info and Bader's role, see especially Janet Martin-Nielsen, "'An Orgy of Hypothesizing': The Construction of Glaciological Knowledge in Cold War America," in *Ice and Snow in the Cold War: Histories of Extreme Climatic Environments*, ed. Julia Herzberg, Christian Kehrt, and Franziska Torma (New York: Berghahn Books, 2018), 73.
4. Quoted in Martin-Nielsen, "'An Orgy of Hypothesizing,'" 69.
5. Gertner, *The Ice at the End of the World*, 166–67; Chester C. Langway Jr., "The History of Early Polar Ice Cores," *Cold Regions Science and Technology* 52, no. 2 (April 2008), i.
6. "Interim Report to Snow, Ice and Permafrost Research Establishment," SIPRE Report 1, January 1950, 7, ERDC.
7. "Interim Report to Snow, Ice, and Permafrost Research Establishment," SIPRE Report 1, January 1950, 7–9, ERDC.
8. Quoted in Martin-Nielsen, "'An Orgy of Hypothesizing,'" 73.
9. "First SIPRE Snow Compaction Conference, December 13–14, 1950," SIPRE Report 2, January 1951, ERDC.
10. Bader quoted in "Preliminary Investigations of Some Physical Properties of Snow," SIPRE Report 7, June 1951, ix, ERDC. On differences of materials and lack of good terminology, see "Some Aspects of Snow, Ice, and Frozen Ground," SIPRE Report 10, August 1953, 5, ERDC.

11. "First SIPRE Snow Compaction Conference, 13–14 December 1950," SIPRE Report 2, January 1951, ERDC.
12. "Formation of Snow Crystals," SIPRE Research Paper 3, January 1954, 32, ERDC; Martin-Nielsen, "'An Orgy of Hypothesizing'"; Henri Bader and Daisuke Kuroiwa, *The Physics and Mechanics of Snow as a Material* (Hanover, NH: US Army Cold Research and Engineering Laboratory, 1962).
13. "Agenda for Discussion on Proposed Engineer Activities in Greenland, 1954," Box XVII-9, Folder 6, ACEHQ.
14. "Agenda for Discussion in Proposed Engineer Activities in Greenland, 1954."
15. Lawrence W. Shanahan and Frank L. Robertson, "Army Engineer Climatic Tests," *Military Engineer* 45, no. 306 (1953): 264–68.
16. Shanahan and Robertson, "Army Engineer Climatic Tests," 254–267.
17. "Military Clothing for Use in Desert Areas," November 2, 1949, Folder 20, Box 6, Wilkins Papers.
18. Camp Century, the "City under the Ice," has been the most intensively covered. See especially Nielsen and Nielsen, *Camp Century*. See also Doel, Harper, and Heymann, eds., *Exploring Greenland*.
19. Air Force to Soil Laboratory, "Establishment of a Semi-Permanent Station on the Greenland Ice Cap," memo, January 25, 1952, Box XII-23, ACEHQ.
20. Minutes of 11th meeting held on September 25, 1952, Panel on Arctic Environments, Box 453, Binder 4, Folder 3, RG 330, NACP; "Project Dog Sled—Greenland Ice Cap," June–October 1953, ACEHQ.
21. "Guide for Greenland Duty, 1958," SIPRE Special Report 25, April 1958, ERDC.
22. "Mastering the Arctic," Box 4, Folder 5, Wilkins Papers.

CHAPTER TWELVE

1. Donald Rausch, "Ice Tunnel, Tuto Area, Greenland, 1956," Technical Report 44, 1956, ERDC.
2. Most scholarship on US engineering projects in Greenland focuses, perhaps understandably, on what was built and how it was done, which misses a great deal of other activity undertaken by the engineers. Much of the work on Camp Century, for example, is a hagiographic celebration of it as an engineering feat. More comprehensive scholarship views the programs as stepping stones to scientific discovery and understanding; most recently, Nielsen and Nielsen have crafted an outstanding overview of Camp Century that highlights the diplomatic negotiations and fallout of the installations. See Nielsen and Nielsen, *Camp Century*.
3. Malaurie, *Last Kings* (1956), xix.

4. Malcolm Mellor, "Undersnow Structures: N-34 Radar Station, Greenland," Technical Report No. 132, August 1964 (Hanover, NH: US Army Materiel Command), ERDC. See also Captain Clifton T. Ken and Lt. Addison D. Minott, "'Project Dog Sled': The Construction of AC&W Stations N-33 and N-34 on the Greenland Icecap," January 1954, Box XVII- 9, ACEHQ.
5. Basic background and history of these expeditions is drawn from Gertner, *The Ice at the End of the World*; Martin-Nielsen, *Eismitte in the Scientific*; Fristrup, *The Greenland Ice Cap*.
6. Gertner, *Ice at the End of the World*, 152–58; Martin-Nielsen, *Eismitte*.
7. B. Lyle Hansen, "Instrumentation of Ice-Cap Stations," SIPRE Report 23, 1955, ERDC; Bader, Robert W. Waterhouse, J. K. Landauer, B. Lyle Hansen, James A. Bender, Theodore R. Butkovich, "Excavations and Installations at SIPRE Test Site, Site 2, Greenland," SIPRE Report 20, April 1955, 7, ERDC.
8. For the original plans, see Captain Clifton T. Ken and Lt. Addison D. Minott, "'Project Dog Sled': The Construction of AC&W Stations N-33 and N-34 on the Greenland Icecap," January 1954, Box XVII-9, ACEHQ. For movement of glaciers, see Ken and Minott, as well as "Analysis of Speed of Greenland Glaciers Gives New Insight for Rising Sea Level," National Science Foundation, news release 12-008, May 4, 2012.
9. Malcom Mellor, "Undersnow Structures: N-34 Radar Station, Greenland" (Hanover, NH: CRREL Laboratory, 1963), ERDC.
10. "First SIPRE snow compaction conference 13–14 December 1950," SIPRE Report 2, January 1951, ERDC.
11. Bader et al., "Excavations and Installations," 7.
12. Bader et al., "Excavations and Installations," 7–9.
13. Bader et al., "Excavations and Installations," 4.
14. Elmer Clark, "Camp Century: Evolution of Concept and History of Design, Construction and Performance," Technical Report No. 174, US Army Materiel Command, October 1965.
15. Herbert O. Johansen, "US Army Builds a Fantastic City under Ice," *Popular Science*, February 1960, reposted as Bill Gourgey, "From the Archives: Inside the US Army's Plan to Build a Luxurious City under the Arctic," *Popular Science*, May 17, 2022, https://www.popsci.com/environment/us-army-arctic-city/.
16. Nielsen and Nielsen, *Camp Century*.
17. Nikolaj Petersen, "The Iceman That Never Came: 'Project Iceworm,' the Search for a NATO Deterrent, and Denmark, 1960–1962," *Scandinavian Journal of History* 33, no. 1 (2008): 75–98. For ICBM deployment and strategy, see Heefner, *The Missile Next Door: The Minuteman in the American Heartland* (Cambridge, MA: Harvard University Press, 2012).

18. Gertner, *Ice at the End of the World*, 62, 94, 120.
19. Most background info comes from Clark, "Camp Century."
20. Bob Mitchell's stories in the Thule Forum describe these failures, including his time spent at Camp Century in 1961 when he reports that they ran out of parts, the steam generator broke, clean clothes and hot showers were hard to come by, they were running out of food supplies, and the cooks did an amazing job with what they could find, namely powdered eggs. A relief plane finally got through in March 1962 with a steam generator and other parts to fix things. Stories by Nate Galbreath and Roy Hansen echo the logistical and weather delays encountered while working up in Greenland. These individual accounts were all available on the Thule Forum, which is no longer being maintained, and the stories are no longer available online. See author notes and saved web pages. Given the paucity of official information about the programs, the lack of showers and limited food supplies cannot be corroborated in official documentation but are entirely in keeping with other descriptions and mishaps across the ice. For mail see Nielsen and Nielsen, *Camp Century*, 167. On whiteouts as like being in a bowl of milk, see Charles W. Lobb's Thule Forum story. On the importance of logistics and potential problems in general, see Wager, *Camp Century*.
21. B. Lyle Hansen, "Ice Tunnel Tuto Area, Greenland 1956," February 1958, SIPRE Technical Report No. 44, ERDC; see also "Under Ice Mining Techniques," SIPRE Technical Report No. 72, January 196, ERDC.
22. Walter Sullivan, "Greenland Ice Cave Given a Test by Scientists in Unplanned Stay," *New York Times*, October 8, 1961, 35; George Barbour, "Conduct Experiments in Survival at 'Cities under the ice,'" *Pittsburgh Courier*, October 26, 1963, 13.
23. Hansen, "Ice Tunnel Tuto Area."
24. T. R. Butkovich and J.K. Landauer, "A Grid Technique for Measuring Ice Tunnel Deformation," SIPRE Special Report No. 24, July 1959; Robert E. Hilty, "Measurements of Ice Tunnel Deformation, Camp Red Rock, Greenland," SIPRE Special Report No. 28, July 1959, B. Lyle Hansen, "Instrumental of Ice-Cap Stations," SIPRE Report 23 (1955), all ERDC.
25. Quoted in Nielsen and Nielsen, *Camp Century*, 132–33.
26. Clark, "Camp Century," 36.
27. Nielsen and Nielsen, *Camp Century*; Clark, "Camp Century," 36.
28. This research was conducted by Liam Colgan, a glaciologist at York University in Canada. For a summary, see Julia Rosen, "Mysterious, Ice-Buried Cold War Military Base May Be Unearthed by Climate Change," *Science*, August 4, 2016.

CHAPTER THIRTEEN

1. Doel, "Constituting the Postwar Earth Sciences," 635–66; Ronald E. Doel, Urban Wråkberg, and Suzanne Zeller, "Science, Environment, and the New Arctic," *Journal of Historical Geography* 44 (April 2014): 2–14.
2. For background information, see Langway Jr., "The History of Early Polar Ice Cores." Quote from Chester C. Langway Jr., "A 400 Meter Deep Ice Core in Greenland: Preliminary Report," *Journal of Glaciology* 3, no. 23 (1958): 217.
3. Chester C. Langway Jr., "Stratigraphic Analysis of a Deep Ice Core from Greenland," CRREL (May 1967), 13, ERDC.
4. Niklas Boers and Martin Rypdal, "Critical Slowing Down Suggests That the Western Greenland Ice Sheet Is Close to a Tipping Point," *PNAS* 118, no. 21 (2021): 1–7.
5. For more comprehensive tales of the ice-core projects and the climate modeling that has come from them, see Gertner, *The Ice at the End of the World*; Richard B. Alley, *The Two-Mile Time Machine: Ice Cores, Abrupt Climate Change, and Our Future* (Princeton, NJ: Princeton University Press, 2000); John McCannon, *A History of the Arctic: Nature, Exploration and Exploitation* (London: Reaktion Books, 2012); Philip Conkling, Richard Alley, Wallace Broecker, and George Denton, *The Fate of Greenland: Lessons from Abrupt Climate Change* (Cambridge, MA: MIT Press, 2013).
6. Alley, *Two-Mile Time Machine*; Daniel Glick, "Ice Cores from Greenland Unlock Ancient Climate Secret," *Scientific America*, April 25, 2011, https://www.scientificamerican.com/article/ice-cores-from-greenland-unlock-ancient-climate-secret/; Gertner, *Ice at the End of the World*, 232.
7. Carl S. Benson interviewed by Karen Brewster, June 2001, Byrd Polar Research Center Archival Program, available at http://kb.osu.edu/items/af811343-7e2d-53fl-a515-cc95899a0b10.
8. Benson interviewed by Brewster, 46.
9. Carl S. Benson, "Stratigraphic Studies in the Snow and Firn of the Greenland Ice Sheet," ii.2. See also Benson interviewed by Brewster, 6; Langway, "History of Early Polar Ice Cores."
10. Benson interviewed by Brewster, 7. See also comments from Mr. John Sater interviewed by Brian Shoemaker, October 2001, Byrd Polar Research Center Archival Program, Byrd Polar Research Center Archival Program; Benson, "Stratigraphic Studies," 9–10; Gertner, *Ice at the End of the World*, 186; Fae L. Korsmo, "Glaciology, the Arctic, and the US Military," in *New Spaces of Exploration: Geographies of Discovery in the Twentieth Century*, edited by Simon Naylor and James R. Ryan, 125–47 (London: I. B. Tauris, 2009).

11. Henri Bader, "Introduction to Ice Petrofabrics," *Journal of Geology* 59, no. 6 (1951): 519–36.
12. Benson interviewed by Brewster. *Firn* and *névé* are often used interchangeably, although *névé* is often more broadly used to describe an area of unconsolidated snow on top of a glacier, and *firn* to describe the snow itself. See Maynard M. Miller, "The Terms 'Névé' and 'Firn,'" *Journal of Glaciology* 2, no. 12 (1952): 150–51. For equipment, Carl S. Benson, "Scientific Work of Party Crystal, 1954," SIPRE Report 24, April 1955, ERDC, 6.
13. Benson, "Stratigraphic Studies."
14. Langway, "History of Early Polar Ice Cores," 6. Langway points out that Bader was building his model on Sorge's 1930s efforts and proof of concept, but to get the real accumulation record, he needed to check across a wide area because of differences in elevation. See Gertner, *Ice at the End of the World*, chap 7 and 8. For Wegener, see Langway, "History of Early Polar Ice Cores." On the history of science on the ice, see the introduction to Doel, Harper, and Heymann, eds., *Exploring Greenland*; Martin-Nielsen, *Eismitte in the Scientific Imagination*; Gertner, *Ice at the End of the World*; Børge Fristrup, "Recent Investigations of the Greenland Ice Cap," *Geografisk Tidsskrift* 58 (January 1959): 1–29.
15. Chris Polashenksi, "Frau at Minus Five and Fabulous on Ze Autobahn," Sunlight Absorption on the Greenland Ice Sheet Experiment (SAGE), April 25, 2014, https://sage2014traverse.wordpress.com/2014/04/. For the clearest description of previous efforts, see Langway, "Stratigraphic Analysis"; Benson, "Stratigraphic Studies in the Snow and Firn of the Greenland Ice Sheet," ii.
16. Ned Rozell, "Tracks across Greenland Ice, 60 Years Apart," Geophysical Institute, University of Alaska, Fairbanks, July 24, 2014, https://www.gi.alaska.edu/alaska-science-forum/tracks-across-greenland-ice-60-years-apart; Carl Benson and Gifford Wong, "Perspectives on Exploring the Greenland Ice Sheet," Arctic Research Consortium of the United States, https://www.arcus.org/witness-the-arctic/2013/2/article/19967; Mike Stewart, "Crispy Cold Cargo!," Sunlight Absorption on the Greenland Ice Sheet Experiment 2013, April 26, 2013, https://sage2013traverse.wordpress.com/page/4/.
17. Langway, "Stratigraphic Analysis of a Deep Ice Core from Greenland."
18. "Polar Ice Coring and IGY 1957–58: An Interview with Dr. Anthony J. Tony Gow," *In Depth: Newsletter of the National Ice Core Laboratory—Science Management Office* 3, no. 1 (2008): 1–5.
19. Hamblin, *Arming Mother Nature*, 90–93.
20. Langway, "History of Early Polar Ice Cores."
21. Paul Mayewski and Frank White, *The Ice Chronicles: The Quest to Understand Global Climate Change* (Lebanon, NH: University of New England

Press, 2002), 1–15. Bess Koffman is quoted in Benjamin J. Hulac, "Ice Lab Gets to the Core of the Climate Crisis," *Roll Call*, February 28, 2023. For background on cores and how scientists read them, see Michon Scott, "What Do Ice Cores Reveal about the Past?," March 24, 2023, National Snow and Ice Data Center, http://nsidc.org/learn/ask-a-scientists/core-climate-history#anchor-drilling-and-storing-ice-cores.

22. Charles R. Bentley and Bruce R. Koci, "Drilling to the Beds of the Greenland and Antarctic Ice Sheets: A Review," *Annals of Glaciology* 47 (2007): 1–9; Gertner, *Ice at the End of the World*, 186; Langway, "History of Early Polar Ice Cores," 8.

23. Langway, "History of Early Polar Ice Cores," 14–16; Langway, "Stratigraphic Analysis of a Deep Ice Core from Greenland," 28. Langway quoted in Brian Williams, "Inventing the Science of Polar Ice Analysis," August 5, 2023, https://www.brianwilliams.us/climate-crash/inventing-the-science-of-polar-ice-analysis.html.

24. Langway, "History of Early Polar Ice Cores," 14.

25. "Willi Dansgaard was truly a pioneer of ice core science. His work on the climatic meaning of oxygen isotopes in precipitation formed the basis for the most fundamental observations that are made on ice cores: those that tell us about past temperature. And he had the vision to see where using these measurements on ice cores could lead." Joseph Cheek, "Willi Dansgaard: Pioneer of Paleoclimate Research," *Science Poles*, February 28, 2011, http://www.sciencepoles.org/article/willi-dansgaard-pioneer-of-paleoclimate-research.

26. Henri Bader, "Scope, Problems, and Potential Value of Deep Core Drilling in Ice Sheets," USA CRREL Special Report 58, December 1962, ERDC https://erdc-library.erdc.dren.mil/jspui/bitstream/11681/11567/1/SR-58.pdf.

27. Herbert T. Ueda and Donald E. Garfield, "Drilling through the Greenland Ice Sheet," USA CRREL Special Report 126, November 1968, https://icedrill.org/sites/default/files/SR-126.pdf.

28. Gertner, *Ice at the End of the World*, 201.

29. Langway, "The History of Early Polar Ice Cores," 27.

30. Langway, "The History of Early Polar Ice Cores," 28; Gertner, *Ice at the End of the World*, 208. The Greenland Ice Core Project took place between 1989 and 1992 and extracted a bedrock core; the North Greenland Ice Core Project followed. Cores have also been drilled in Antarctica. Combined, these led to a more accurate climate record.

31. Andrew Bertain, "OMG: NASA Observes Oceans Melting Greenland at Thule," News, Nellis Air Force Base, November 3, 2020, https://www.nellis.af.mil/News/Article/2404117/omg-nasa-observes-oceans-melting-greenland-at-thule/2.

32. On these networks, see Gertner, *Ice at the End of the World*; Langway, "The History of Early Polar Ice Cores," Nielsen and Nielsen, *Camp Century*; Martin-Nielsen, *Eismitte*.
33. Jessica Stoller-Conrad, "Core Questions: An Introduction to Ice Cores," NASA Vital Signs of the Planet, August 15, 2017, https://climate.nasa.gov/news/2616/core-questions-an-introduction-to-ice-cores/.
34. Stoller-Conrad, "Core Questions."
35. Benjamin J. Hulac, "Ice Lab Gets to the Core of the Climate Crisis," *Roll Call*, February 28, 2023, https://rollcall.com/2023/02/28/ice-lab-gets-to-the-core-of-the-climate-crisis/.
36. Gertner, *Ice at the End of the World*, 283.

PART THREE

1. Arthur C. Clarke, *A Fall of Moondust* (New York: Harcourt, Brace & World, 1961).
2. Clarke, 3–4.

CHAPTER FOURTEEN

1. Lt. Gen Wilson, "The Army Engineers' Role in Space Construction," July 13, 1962, Box XVIII-30, ACEHQ.
2. The literature on the space race is vast, but of particular use to me have been the following: Roger D. Launius, *Reaching for the Moon: A Short History of the Space Race* (New Haven, CT: Yale University Press, 2019); Teasel Muir-Harmony, *Operation Moonglow: A Political History of Project Apollo* (New York: Basic Books, 2020); Walter McDougall, *The Heavens and the Earth: A Political History of the Space Age* (Johns Hopkins Press, 1997); Douglas Brinkley, *American Moonshot: John F. Kennedy and the Great Space Race* (New York: HarperCollins, 2020). I have also learned a great deal from NASA's *Houston, We Have a Podcast*, which interviews people involved in space programs today. It is available at https://www.nasa.gov/podcasts/houston-we-have-a-podcast/. On lunar construction, see Robert Godwin, "The Forgotten Plans to Reach the Moon—before Apollo," *Smithsonian Magazine*, July 19, 2019, https://www.smithsonianmag.com/air-space-magazine/forgotten-plans-reach-moon-apollo-180972695/. See declassified documents on lunar construction programs and an essay by Jeffrey T. Richelson, "Soldiers, Spies, and the Moon: Secret US and Soviet Plans from the 1950s and 1960s," Digital National Security Archive, https://nsarchive2.gwu.edu/NSAEBB/NSAEBB479/; Dwayne Day, "Nuking the Site from Orbit: When the Air Force Wanted a Base on the Moon," *Space Review*,

November 4, 2019, https://www.thespacereview.com/article/3826/1. For Corps activities as part of NASA, Col. Thomas J. Hayes III, "The Corps of Engineers Role in Support of NASA," speech before the Society of American Military Engineers March 21, 1962, Box XVIII-30, ACEHQ. Hayes estimated that the Corps would spend no less than $500 million on space-related projects in four years.

3. On Lowell, see Robert Markley, *Dying Planet: Mars in Science and the Imagination* (Durham, NC: Duke University Press, 2005); Kat Eschner, "The Bizarre Beliefs of Astronomer Percival Lowell," *Smithsonian*, March 13, 2017; Sarah Steward Johnson, *The Sirens of Mars* (New York: Penguin Random House, 2021), 26–27; Robert Crossley, "Percival Lowell and the History of Mars," *Massachusetts Review* 41 (2000): 297–318.

4. Lowell, *Mars and Its Canals* (New York: Macmillan Co, 1906), 16.

5. Lowell, *Mars and Its Canals*, chap. 2.

6. Lowell, *Mars and Its Canals*; K. Maria D. Lang, *Geographies of Mars: Seeing and Knowing the Red Planet* (Chicago: University of Chicago Press, 2010). On colonial discourses of deserts, see Davis, *The Arid Lands*.

7. Markley, *Dying Planet*; Lesley Selander, dir., *Flight to Mars* (Monogram Pictures, 1951); Fred Whipple, "Is There Life on Mars?," *Collier's* (1954). In 1956, von Braun and Ley agreed, describing the fourth planet as "a small planet of which three-quarters is cold desert, with the rest covered with a sort of plant life that our biological knowledge cannot quite encompass." Markley, *Dying Planet*, 171.

8. The Red Planet has been visited by more spacecraft than any other object except the Moon. Along with Europa and possibly Titan, Mars is a favorable planet in the exploration for possible present-day or past life. On the fascination with Mars, see Markey, *Dying Planet*; Johnson, *The Sirens of Mars*.

9. "What Are We Waiting For?," *Collier's*, March 22, 1952, 23.

10. "What Are We Waiting for?" *Collier's*, March 22, 1952, 23.

11. "What to Do with the Moon?" *Popular Science*, April 1952, 164–68.

12. On science fiction and the space race, see Markley, *Dying Planet*; Robert Crossley, *Imagining Mars: A Literary History* (Middletown, CT: Wesleyan University Press, 2011); Howard McCurdy, *Space and the American Imagination* (Washington, DC: Smithsonian Institution, 1997); Megan Prelinger, *Another Science Fiction: Advertising the Space Race 1957–1962* (New York, Blast Books, 2010).

13. "What Are We Waiting For?," *Collier's*, March 22, 1952, 23. On the relative lack of interest in solar system astronomy at midcentury, see Ronald Doel, *Solar System Astronomy in America: Communities, Patronage, and Interdisciplinary Science, 1920–1960* (Cambridge: Cambridge University Press, 1996);

Oliver Morton says that "by the early 20th century almost all astronomers had come to shun the Moon," in *The Moon: A History for the Future* (London: Economist Books, 2019), 12.

14. On space enthusiasts in general, see Markley, *Dying Planet*; McCurdy, *Space and the American Imagination*. On von Braun, see Roger D. Launius, "Interpreting the Moon landings: Project Apollo and the Historians," *History and Technology* 22, no. 3 (2006): 236; Michael Neufeld, *Von Braun: Dreamer of Space, Engineer of War* (New York: Vintage, 2008); Monique Laney, *German Rocketeers in the Heart of Dixie: Making Sense of the Nazi Past during the Civil Rights Era* (New Haven, CT: Yale University Press, 2015).

15. Jared S. Buss, *Willy Ley: Prophet of the Space Age* (Gainesville: University Press of Florida, 2017); Diane Tedeschi, "Willy Ley, Prophet of the Space Age," *Smithsonian Magazine*, December 2017, https://www.smithsonianmag.com/air-space-magazine/Who-Was-Willy-Ley-180967114/.

16. Fred Whipple and Werner von Braun, "The Exploration," *Collier's*, October 25, 1952, 38–40, 44–48; Willy Ley, "Inside the Lunar Base," *Collier's*, October 18, 1952, 58–61.

17. Whipple and von Braun, "Exploration," 38–40, 44–48.

18. Von Braun, "Crossing the Last Frontier," *Collier's*, March 22, 1952.

19. Tom Crouch, "A Saga of a Lunar Landscape," January 22, 2010, Smithsonian National Air and Space Museum, https://airandspace.si.edu/stories/editorial/saga-lunar-landscape. On Bonestell and his influence, see the 2018 documentary *Chesley Bonestell: A Brush with the Future*, directed by Douglass M. Stewart Jr.

20. "What Are We Waiting For?," *Collier's*, March 22, 1952; Crouch, "A Saga of the Lunar Landscape." On Bonestell, see Ron Miller and Frederick C. Durant III, *The Art of Chesley Bonestell* (New York: Paper Tiger, 2001).

21. On the headline, see Day, "Nuking the Site from Orbit"; "What Are We Waiting For?," *Collier's*, March 1952, 23.

22. *Project Horizon Report*, vol. 1, *Summary and Supporting Considerations* (US Army, June 9, 1959), 3; HQ Space Systems Division, AF Systems Command, *Lunar Expedition Plan, LUNEX*, May 1961 section 1.1; https://web.archive.org/web/20060614174558/http://astronautix.com/data/lunex.pdf.

23. On contemporary myths circulating at the time, see "Mythos of the Moon Survive the Ages," *New York Times* September 14, 1959, 17; H. G. Wells, *The First Men in the Moon* (n.p.), https://www.gutenberg.org/ebooks/1013; Don Wilhelms, *To a Rocky Moon: A Geologist's History of Lunar Exploration* (Tucson: University of Arizona Press, 1993), 68. "Resultant lunar surface models ranged from exceedingly fine and porous dust, into which a spacecraft would sink and disappear, to porous but highly cohesive rock." Maj.

Stewart W. Johnson and W. David Carrier III, "Lunar Soil Mechanics," *Military Engineer*, 63, no. 415 (1971): 324–28; "Lunar Area 'Looks Good' for Landing," *Chicago Tribune*, April 23, 1967, 16.
24. E. C. Theiss, M. Mileaf, and F. Egan, *Handbook of Environmental Engineering*, Technical Writing Service (New York: McGraw Hill, 1961).
25. *Handbook of Environmental Engineering* (1961).
26. *Project Horizon Report*, vol. 2, *Technical Considerations & Plans* (US Army, June 9, 1959), 7.
27. *Lunar Construction* (NASA and Office of the Chief of Engineers, April 1963), 9, Corps of Engineers, Space Files, Box XVIII-29, ACEHQ.

CHAPTER FIFTEEN

1. *Lunar Mapping and Construction in Support of Space Programs: Hearing Before the Committee on Science and Astronautics*, US House of Representatives, 86th Congress, 2nd sess., May 10, 1960.
2. See also Thomas J. Hayes Jr., "The Corps of Engineers in the Space Age," Box XVIII-30, ACEHQ; Lt. Gen. Wilson, "The Army Engineers' Role in Space Construction," July 13, 1962, Box XVIII-30, ACEHQ.
3. *Lunar Mapping and Construction in Support of Space Programs*, 2.
4. *Lunar Mapping and Construction in Support of Space Programs*, 14.
5. *Lunar Mapping and Construction in Support of Space Programs*, 11; Wilson, "The Army Engineers' Role in Space Construction"; Sunny Tsiao, *"Read You Loud and Clear!": The story of NASA's Spaceflight Tracking and Data Network* (Washington, DC: NASA History Division, 2008).
6. *Lunar Mapping and Construction in Support of Space Programs*, 12.
7. The Space Science Section was to be part of the Corps' Extraterrestrial Research Agency. Alice Allen, "Memorial to Bruce McCurdy Hall," https://www.geosociety.org/documents/gsa/memorials/v10/Hall-BM.pdf.
8. "Proceedings on the Working Group on Extraterrestrial Resources, Fourth Annual Meeting" (US Air Force Academy, Colorado Springs, CO, 1965), https://ntrs.nasa.gov/citations/19660026216.
9. Allen, "Memorial to Bruce McCurdy Hall."
10. *Lunar Construction*; Maj. William B. Taylor, "Maps for Lunar Exploration," *Military Engineer* 55, no. 364 (1963): 86–87; "Lunar Construction Research," Bruce M Hall, OCE, n.d.; 26. Also identified as key engineering concerns: gravity, magnetism, temperature swings, radiation, and meteors. See also Jack Green, *The Geology of the Lunar Base*, Document Locator Number: 11 00106 XS, JPL History, JPL.
11. Doel, "Evaluating Soviet Lunar Science in Cold War America," *Osiris* 7, no. 1 (1992): 238–64. On US solar system astronomy, see Doel, *Solar System*

Astronomy in America. Shoemaker said, "There was a fair amount of volcanic versus impact debate." Interview of Eugene Shoemaker by Ron Doel, June 17, 1987, Niels Bohr Library & Archives, American Institute of Physics, College Park, MD, www.aip.org/history-programs/niels-bohr-library/oral-histories/5082-3. On contemporaneous theories about formation, see Gerald Schaber, "The US Geological Survey, Branch of Astrogeology—A Chronology of Activities from Conception through the End of Project Apollo (1960–1973)," 26, https://pubs.usgs.gov/of/2005/1190/of2005-1190.pdf.

12. Clyde M. Reedy and Ira A. Junt Jr., "Engineering Problems of a Lunar Exploration," *Military Engineer* 53, no. 352 (1961): 107–10.
13. Gerald Schaber, "The US Geological Survey, Branch Astrogeology—A Chronology from Conception through the End of Project Apollo" (Washington, DC: US Geological Survey, 2005), 17, https://pubs.usgs.gov/of/2005/1190/. Gerard P. Kuiper, one of the few men interested in the Moon and solar system astronomy at midcentury (he has sometimes been called the father of solar system science), instigated the *Photographic Lunar Atlas* in 1958. See interview of Eugene Shoemaker by Ron Doel; Arnold C. Mason and Robert J. Hackman, "Engineer Special Study of the Surface of the Moon" (Washington, DC: US Geological Survey, 1961). On lunar mapping, see also Donald A. Beattie, and Farouk El-Baz, "Apollo Exploration Sites," *Military Engineer* 62, no. 410 (1970): 371–76; *Lunar Mapping and Construction in Support of Space Programs*; Memo: Rosen to Silverstein RE Photography of the Moon, May 9, 1959, Archive, No. 02 00714 XF, JPL. See also William Taylor, "Maps for Lunar Exploration," *Military Engineer* 55, no. 364 (1963): 86–88.
14. On using photos to create better maps, see *Project Horizon Report*, vol. 2.
15. *Lunar Construction*, 131–35.
16. *Lunar Construction*. On similar timeline, see Allyn B. Hazard, "A Plan for Manned Lunar and Planetary Exploration," November 1959, Document Locator Number: 03 00323 XF, JPL.
17. *Lunar Construction*, 9.
18. *Lunar Construction*, 21.
19. *Lunar Construction*, 111.
20. "The Lunar Environment," Technical Release 34-159, November 17, 1960, Document Locator Number: 12 00014 XF, JPL; *Lunar Construction*, 3; Jack Green, "The Geology of the Lunar Base," Document Locator Number: 11 00106 XS, JPL.
21. Reedy and Junt, "Engineering Problems of a Lunar Exploration."
22. *Lunar Construction*, Appendix H-2.
23. Lawrence Taylor, Carle Pieters, and Daniel Britt, "Evaluations of Lunar Regolith Simulants," *Planetary and Space Science*, 126 (July 2016): 1–7. See

also Richard J. Polo, "Lunar Structural Design," *Military Engineer* (1965): 155–58. *Lunar Construction*, 59–63, recommended creating an operations test facility that would be a large hangar-like space covered in simulant for construction training, approximately seventy-five thousand square feet; a lunar environmental research and test facility to simulate lunar conditions as much as possible, and field test sites.

24. For the transcript of Apollo 11 astronauts collecting samples, see "Contingency Sample," Apollo 11 Surface Journal, NASA, https://www.nasa.gov/history/alsj/a11/a11ContingencySample.html. On dust, see L. Sibille and P. Carpenter, "Lunar Regolith Simulant Materials: Recommendations for Standardization, Production, and Usage," September 2006, https://ntrs.nasa.gov/api/citations/20060051776/downloads/20060051776.pdf.

25. Nikki Welch, "Dust: An Out-of-This World Problem," June 8, 2021, https://www.nasa.gov/humans-in-space/dust-an-out-of-this-world-problem/; Daniel Winterhalter, Joel S. Levine, Russel L. Kerschmann, and Timothy K. Brady, *Lunar Dust and Its Impact on Human Exploration: A NASA Engineering and Safety Center (NESC) Workshop* (Hampton, VA: Langley Research Center, September 2020), https://ntrs.nasa.gov/api/citations/20205008219/downloads/20205008219.pdf; Leonard David, "Dusting Off an Old Problem: Impact of Lunar Dust on Human Exploration," February 10, 2020, http://www.leonarddavid.com/dusting-off-an-old-problem-impact-of-lunar-dust-on-human-exploration/. Conrad quoted in "ESA Seeking Dust-Proof Materials for Lunar Return," European Space Agency, October 27, 2020, https://www.esa.int/Enabling_Support/Space_Engineering_Technology/ESA_seeking_dust-proof_materials_for_lunar_return#:~:text=As%20Apollo%2012%20Commander%20Pete,restrictive%20friction%2Dlike%20action%20to.

26. For a digestible overview of these concepts of granularity, see Chad Orzel, "Breakfast Cereal and Sinking Ships: The Billion-Dollar Physics of Granular," *Forbes*, October 9, 2018. See also Carol Clark, "Movers and Shakers: New Evidence for a Unifying Theory of Granular Materials," Emory University, January 6, 2020, https://news.emory.edu/features/2021/01/esc-granular-materials/index.html; R. Allen Wilkerson, Robert Berhringer, James Jenkins, and Michael Louge, "Granular Materials and the Risks They Post for Success on the Moon and Mars," Space Technology and Application Forum, Albuquerque, NM, 2004, https://ntrs.nasa.gov/citations/20050198938. See also Philip Metzger, "Space: The Ultimate Sandbox," April 1, 2015, https://www.philipmetzger.com/granular_matter_in_low_gravity/.

27. "Surveyor's Claw Finds Moon Firm," *New York Times*, April 23, 1967, 5; "Lunar Area 'Looks Good' for Landing," *Chicago Tribune*, April 23, 1967, 16.

28. Crouch, "The Saga of Lunar Landscape." In 2023 it was put up at the Air and Space Museum as part of the *Man and the Moon* exhibit.
29. Rebecca Boyle, *Our Moon: How Earth's Celestial Companion Transformed the Planet, Guided Evolution, and Made Us Who We Are* (New York: Random House, 2024).

CHAPTER SIXTEEN

1. On the Long Island crater field, see "Pictorial Engineering News," *Military Engineer*, 59, no. 388 (1967): 130. On Arizona fields, see "Cinder Lakes Apollo Training Area," US Forest Service, https://www.fs.usda.gov/detail/coconino/?cid=FSEPRD611864. A of the creation of the fields is at "Making Craters: Mare Tranquillitatis in Flagstaff, Arizona," https://astrogeology.usgs.gov/rpif/videos/making-craters.
2. See Mary Hesse on the way that analogies are heuristics that guide scientific studies of the unfamiliar, *Models and Analogies in Science* (Notre Dame, IN: University of Notre Dame Press, 1970). Jordan Bimm's work on Mars Jars is also of interest. See Sarah Scoles, "The Doctor from Nazi Germany and the Search for Life on Mars," *New York Times*, June 24, 2020, https://www.nytimes.com/2020/07/24/science/mars-jars-strughold.html; Matt Farish's work on lab spaces is key, see "Creating Cold War Climates: The Laboratories of American Globalism," in *Environmental Histories of the Cold War*, ed. John Robert McNeill and Corinna R. Unger (Cambridge: Cambridge University Press, 2010), 51–83; Gross, "Layering for a Cold War"; Lisa Messeri, *Placing Outer Space: An Earthly Ethnography of Other Worlds* (Durham, NC: Duke University Press, 2016); Janet Vertesi, *Seeing Like a Rover: How Robots, Teams, and Images Craft Knowledge of Mars* (Chicago: University of Chicago Press, 2015); Maher, *Apollo in the Age of Aquarius*.
3. Scott Kirch, "Cratered Landscapes," *New Geographies* 11 (2019), 104–8; H. J. Moore, "Nevada Test Site Craters Used for Astronaut Training," *Journal of Research of the U.S. Geological Survey*, 5, no. 6 (197): 719–33.
4. Becky Little, "See Photos of How Astronauts Trained for the Apollo Moon Missions," History.com, January 3, 2024, https://www.history.com/news/moon-landing-apollo-11-training-photos; "The Training Regimen of NASA's First Astronauts," YouTube video, Smithsonian Channel, May 31, 2019, https://www.youtube.com/watch?v=3Q2bXBQOdgo.
5. Jordan Bimm, "Pre-NASA Astronauts in the USAF Space Cabin Simulator," in Jeffrey S. Nesbit, ed., *Nature of Enclosure* (New York: Actar Publishers, 2022), 42–54; Matthew Hersch, *Inventing the American Astronaut* (London: Palgrave Press, 2012); Valerie A. Olson, *Into the Extreme: U.S. Environmental*

Systems and Politics beyond Earth (Minneapolis: University of Minnesota Press, 2012); Roger Launius, "Heroes in a Vacuum," *Florida Historical Quarterly* 87, no. 2 (2008): 174–209; Peder Anker, "The Ecological Colonization of Space," *Environmental History* 4, no. 2 (2005): 239–68.

6. On the tension between science and technology and the military, see W. David Compton, *Where No Man Has Gone Before: A History of Apollo Lunar Exploration Missions* (Washington, DC: Government Printing Office, 1989); Donald A. Beattie's *Taking Science to the Moon: Lunar Experiments and the Apollo Program* (Baltimore: Johns Hopkins University Press, 2003); Don E. Wilhelms, *To a Rocky Moon*.

7. More recently, NASA has confirmed the utility of the desert for outer-space planning. "The Arizona desert possesses many characteristics that are analogous to a lunar environment," a report explained. "Including challenging terrain, interesting geology, and minimal communications infrastructure, all of which astronauts will experience near the lunar south pole during Artemis missions," NASA officials said, quoted in Rayne, "NASA to Practice Artemis Moonwalking, Roving Operations in Arizona Desert," NASA, October 3, 2022, https://www.nasa.gov/humans-in-space/nasa-to-practice-artemis-moonwalking-roving-operations-in-arizona-desert/.

8. Gerald Schaber, "The U.S. Geological Survey, Branch of Astrogeology—A Chronology of Activities from Conception through the End of Project Apollo (1960–1973)," https://pubs.usgs.gov/of/2005/1190/of2005-1190_figs_040-059.pdf.

9. On Shoemaker, see David Levy, *Shoemaker by Levy: The Man Who Made an Impact* (Princeton, NJ: Princeton University Press, 2002); Doel, *Solar System Astronomy in America*, 176–80.

10. Levy, *Shoemaker by Levy*; Doel, *Solar System Astronomy in America*; Interview of Eugene Shoemaker by Ron Doel, June 17, 1987.

11. Paul Spudis, *The Once and Future Moon* (Washington, DC: Smithsonian Institution, 1998), 20–21; Paul D. Lowman Jr., "Scientific Missions for a Lunar Base," NASA, Goddard Space Flight Center, 1965, 128, https://ntrs.nasa.gov/api/citations/19660026216/downloads/19660026216.pdf; *Proceedings of the Working Group on Extraterrestrial Resources, Fourth Annual Meeting* (Colorado Springs, CO: US Air Force Academy, 1965), https://ntrs.nasa.gov/citations/19660026216. On military interests that would further Shoemaker's mapping abilities, see Doel, *Solar System Astronomy*, 180. Doel also highlights the fact that NASA had limited interested in working with geologists.

12. Levy, *Shoemaker by Levy*, 85.

13. Long overlooked in histories, these pioneering women have recently been given more attention. See Swapna Krishna, "The Mercury 13: The Women Who Could Have Been NASA's First Female Astronauts,"

July 24, 2020, Space.com, https://www.space.com/mercury-13.html; Margaret Weitekamp, *Right Stuff, Wrong Sex: America's First Women in Space Program* (Baltimore, MD: Johns Hopkins University Press, 2005); Launius, *Reaching for the Moon*, 59. The Army was clear that space gear was to be built for the "standard man": 72 inches tall (67.7-inch eye height), 175 pounds with 74.5-inch arm span. *Project Horizon Report*, 2:22, table 11-3.
14. Green, *The Geology of the Lunar Base* (1961).
15. "Before Moon Landing," *Salt Lake (UT) Tribune*, July 8, 2019.
16. "The Toxic Side of the Moon," European Space Agency, 2018, https://www.esa.int/Science_Exploration/Human_and_Robotic_Exploration/The_toxic_side_of_the_Moon. See also Muir-Harmony, *A New Apollo*.
17. From the "Apollo Surface Journal: Appendix E. Geology Field Exercises: Early Training," https://www.hq.nasa.gov/alsj/ap-geotrips.pdf.
18. L. Sibile and P. Carpenter, "Lunar Regolith Simulant Materials: Recommendations for Standardization, Production, and Usage," NASA STI Program Office, September 2006 https://ntrs.nasa.gov/api/citations/20060051776/downloads/20060051776.pdf.
19. Caela Barry, "Genesis Rock," NASA, September 22, 2017, https://moon.nasa.gov/resources/14/genesis-rock.
20. On resource extraction, see Jack Green, "The Geology of the Lunar Base."
21. Monthly Report for Director and Secretary, E. M. Shoemaker to V. R. Wilmarth, Branch of Astrogeology, Geological Survey, October 31, 1963, https://catalog.archives.gov/OpaAPI/media/1919479/content/nrhla/1963_10_monthly_report.pdf; Shoemaker, "Proposal for a Lunar Geological Observatory," in Second Annual Meeting of the Working Group on Extraterrestrial Resources (Alamogordo, NM, 1964), https://ntrs.nasa.gov/citations/19670011083.
22. Judith Haley Alton, "Catalog of Apollo Lunar Surface Geological Samling Tools and Containers," NASA, March 1989, https://history.nasa.gov/alsj/tools/Welcome.html
23. Alton, "Catalog of Apollo Lunar Surface Geological Samling Tools and Containers"; Mike Tolson, "Misplaced from Space," *Houston Chronicle*, May 7, 2010; Jamie Svirzdin, "I Fell under the Spell of NASA's Most Notorious Thief," *The Atlantic*, February 15, 2019, https://www.theatlantic.com/science/archive/2019/02/moon-rock-thief-thad-roberts/582757/; "Lunar Rocks and Soils from Apollo Missions," Curation/Lunar, NASA, https://curator.jsc.nasa.gov/lunar/; "Soviet-Era 'Moon Rocks' Sell for $855,000 at New York Auction," Reuters, November 29, 2018, https://www.reuters.com/article/idUSKCN1NY2U8/.
24. "50 Years after NASA's Apollo Mission, Moon Rocks Still Have Secrets to Reveal," Oak Ridge National Laboratory, March 31, 2023, https://www.ornl

.gov/news/50-years-after-nasas-apollo-mission-moon-rocks-still-have-secrets-reveal; "Lunar Rock Stars: NASA Selects Geology Team for the First Crewed Artemis Moon Landing," NASA, August 30, 2023, https://scitechdaily.com/lunar-rock-stars-nasa-selects-geology-team-for-the-first-crewed-artemis-moon-landing/.

25. Lawrence A Taylor, Carle M. Pieters, and Daniel Britt, "Evaluations of Lunar Regolith Simulants," *Planetary and Space Science* 126 (July 2015): 1–7.
26. "NASA to Practice Artemis Moonwalking, Roving Operations in Arizona," NASA, https://www.nasa.gov/feature/nasa-to-practice-artemis-moonwalking-roving-operations-in-arizona-desert.
27. James R. Zimbelmann and Haim Tsoar, "Learning about Planets through Studying Wind-Related Processes on Earth," *Journal of Geophysical Research Planets* 123, no. 5 (2018): 1003–6, https://agupubs.onlinelibrary.wiley.com/doi/full/10.1029/2018JE005548.
28. Zimbelman and Tsoar, "Learning about Planets through Studying Wind-Related Processes on Earth."
29. Zimbelman and Tsoar.
30. A few years later he would add his name to the foreword of a NASA-sponsored book about how the Western Desert of Egypt could provide interesting analogs to Martian landscapes. Farouk El-Baz and Ted Maxwell, eds., *Desert Landforms of Southwestern Egypt: A Bases of Comparison with Mars* (Washington, DC: NASA, 1982), https://ntrs.nasa.gov/api/citations/19830008725/downloads/19830008725.pdf; Ralph A. Bagnold, *Sand, Wind, and War: Memoirs of a Desert Explorer* (Tucson: University of Arizona Press, 1991), 190.
31. See *Sand, Wind, and War*. No Martian rocks have been returned to Earth. NASA hopes to bring samples back sometime in the 2030s with remote missions that can scoop materials on the surface and somehow get them back (the "how" remains a mystery).

CONCLUSION

1. "2023 Was the World's Warmest Year on Record, by Far," NOAA, January 12, 2024, https://www.noaa.gov/news/2023-was-worlds-warmest-year-on-record-by-far.
2. "Climate Change," Joshua Tree National Park, https://www.nps.gov/jotr/learn/nature/climate-change.htm.
3. Kate D. Ramsayer, "Emissions Could Add 15 Inches to 2100 Sea Level Rise, NASA-Led Study Finds," NASA, September 17, 2020, https://www.nasa.gov/centers-and-facilities/goddard/emissions-could-add-15-inches-to-2100-sea-level-rise-nasa-led-study-finds/#:~:text=The%20IPCC%20report%20projected%20that,(3%20to%2028%20cm); Chris Mooney, "Greenland

Ice Sheet Set to Raise Sea Levels by Nearly a Foot, Study Finds," *Washington Post*, August 29, 2022, https://www.washingtonpost.com/climate-environment/2022/08/29/greenland-ice-sheet-sea-level/.
4. "The Strange Things Humans Have Left on the Moon," Royal Museums Greenwich, https://www.rmg.co.uk/stories/topics/strange-things-humans-have-left-on-moon.
5. The NOAA also estimates that significant increase in rocket activity will have a negative effect on the ozone layer, which protects Earth from harmful ultraviolet rays. "Projected Increase in Space Travel May Damage Ozone Layer," NOAA, June 21, 2022, https://research.noaa.gov/2022/06/21/projected-increase-in-space-travel-may-damage-ozone-layer/.
6. Barry Lopez, *Arctic Dreams: Imagination and Desire in a Northern Landscape* (New York: Charles Scribner's Sons, 1986); Edward Abbey, *Desert Solitaire: A Season in the Wilderness* (New York: McGraw Hill, 1968), 210.
7. Quoted in Michael T. Klare, *All Hell Breaking Loose: The Pentagon's Perspective on Climate Change* (New York: Metropolitan Books, 2019), 3.
8. David Vine, "Where in the World Is the US military?" *Politico*, July–August 2015, https://www.politico.com/magazine/story/2015/06/us-military-bases-around-the-world-119321. On military response to catastrophe, see Julia Irwin, *Catastrophic Diplomacy: US Foreign Disaster Assistance in the American Century* (Chapel Hill: University of North Carolina Press, 2024).
9. Shatner quoted in Tony Sokol, "Shatner in Space: What William Shatner Learned during His Trip to the Final Frontier," *Den of Geek*, December 13, 2021, https://www.denofgeek.com/culture/william-shatner-space-final-frontier/. For the overview effect, Marina Koren, "Seeing Earth from Space Will Change You," *The Atlantic*, January–February 2023, https://www.theatlantic.com/magazine/archive/2023/01/astronauts-visiting-space-overview-effect-spacex-blue-origin/672226/.
10. For science projects and teams in Greenland, see NSF/Battelle, "Arctic Research Support & Logistics Services Gateway," https://battellearcticgateway.org/program-locations/greenland/pituffik-space-base.
11. "New Name for Thule Air Base in Greenland," *High North News*, April 12, 2003, https://www.highnorthnews.com/en/new-name-thule-air-base-greenland. For speeches and ceremony, see the video "Pituffik Base Renaming Ceremony CSO Speech," Defense Visual Information Distribution Service, April 6, 2023, https://www.dvidshub.net/video/879112/pituffik-base-renaming-ceremony-cso-speech.

SELECTED BIBLIOGRAPHY

Abbey, Edward. *Desert Solitaire: A Season in the Wilderness*. New York: McGraw-Hill, 1968.
Adas, Michael. *Dominance by Design: Technological Imperatives and America's Civilizing Mission*. Cambridge, MA: Harvard University Press, 2009.
Akins, Damon B., and William J. Bauer Jr. *We Are the Land: A History of Native California*. Berkeley: University of California Press, 2022.
Alley, Richard B. *The Two-Mile Time Machine: Ice Cores, Abrupt Climate Change, and Our Future*. Princeton, NJ: Princeton University Press, 2000.
Anker, Peder. "The Ecological Colonization of Space." *Environmental History* 10, no. 2 (April 2005): 239–68.
Arnold, David. "Europe, Technology, and Colonialism in the 20th Century." *History and Technology* 21, no. 1 (2005): 85–106.
Atkinson, Rick. *An Army at Dawn: The War in North Africa, 1942–1943*. New York: Henry Holt & Co., 2002.
Austin, Mary Hunter. *The Land of Little Rain*. Boston: Houghton Mifflin, 1903.
———. *Lost Borders, the People of the Desert*. New York; Harper & Bros, 1909.
Bagnold, Ralph A. *The Physics of Blown Sand and Desert Dunes*. London: Methuen & Co., 1941.
———. *Sand, Wind, and War: Memoirs of a Desert Explorer*. Tucson: University of Arizona Press, 1991.
Balchen, Bernt. *Come North with Me: An Autobiography*. New York: E. P. Dutton, 1958.
Barton, Jennifer. "Branding the Earth: Selling Earth System Science in the United States, 1983–1988." *Social Studies of Science* 53, no. 1 (2023): 49–80.
Beiser, Vince. *The World in a Grain: The Story of Sand and How It Transformed Civilization*. New York: Riverhead Books, 2018.

Benson, Keith R., and Helen M. Rozwadowski, eds. *Extremes: Oceanography's Adventures at the Poles*. Sagamore Beach, MA: Science History Publications, 2007.

Bimm, Jordan. "Subjects of Enclosure: Pre-NASA Astronauts in the USAF Space Cabin Simulator." In *Nature of Enclosure*, edited by Jeffrey S. Nesbit, 40–55. New York: Actar Publishers, 2022.

Black, Megan. *The Global Interior: Mineral Frontiers and American Power*. Cambridge, MA: Harvard University Press, 2018.

Blaker, James R. *United States Overseas Basing: An Anatomy of a Dilemma*. New York: Praeger Publishers, 1990.

Bocking, Stephen, and Daniel Heidt, eds. *Cold Science Environmental Knowledge in the North American Arctic during the Cold War*. New York: Routledge, 2019.

Boggs, S. Whittemore. "An Atlas of Ignorance: A Needed Stimulus to Honest Thinking and Hard Work." *Proceedings of the American Philosophical Society* 93, no. 3 (1949): 253–58.

Blower, Brooke. "From Isolationism to Neutrality: A New Framework for Understanding American Political Culture, 1919–1941." *Diplomatic History* 38, no. 2 (April 2014): 345–76.

Brady, Lisa. "War from the Ground Up: Integrating Military and Environmental Histories." In *A Field on Fire: The Future of Environmental History*, edited by Mark Hersey and Ted Steinberg, 250–62. Tuscaloosa: University of Alabama Press, 2019.

Brøsted, Jens, and Mads Faegteborg. "Expulsion of the Great People When US Air Force Came to Thule: An Analysis of Colonial Myths and Actual Incidents." In *Native Power: The Quest for Autonomy and Nationhood of Indigenous Peoples*, edited by Jens Brøsted, Jens Dahl, Andrew Gray, Hans Christian Gulløv, Georg Henriksen, Jørgen Brøchner Jøgensen, and Inge Kleivan, 213–38. Oslo: Universitetsforlaget AS, 1985.

Brown, Kate. *Plutopia: Nuclear Families, Atomic Cities, and the Great Soviet and American Plutonium Disasters*. New York: Oxford University Press, 2013.

Buss, Jared S. *Willy Ley: Prophet of the Space Age*. Gainesville: University Press of Florida, 2017.

Butcher, Harry C. *My Three Years With Eisenhower: The Personal Diary of Captain Harry C. Butcher, USNR, Naval Aide to General Eisenhower, 1942 to 1945*. New York: Smith & Schuster, 1946.

Cabot, Mabel H. *Vanished Kingdoms: A Woman Explorer in Tibet, China, and Mongolia, 1921–1925*. New York: Aperture, 2005.

Chakrabarty, Dipesh. *The Climate of History in a Planetary Age*. Chicago: University of Chicago Press, 2021.

Chaplin, Joyce. "Planetary Power? The United States and the History of Around-the-World Travel." *Journal of American Studies* 47, no. 1 (2013): 1–21.

———. *Round about the Earth: Circumnavigation from Magellan to Orbit*. New York: Simon & Schuster, 2012.

Clarke, Arthur C. *A Fall of Moondust*. New York: Harcourt, 1961.

Clayton, James L., ed. *The Economic Impact of the Cold War: Sources and Readings*. New York: Houghton Mifflin Harcourt, 1970.

Compton, William David. *Where No Man Has Gone Before: A History of Apollo Lunar Exploration Missions*. Derby, PA: DIANE Publishing, 1996.

Conkling, Philip, Richard Alley, Wallace Broecker, and George Denton. *The Fate of Greenland: Lessons from Abrupt Climate Change*. Cambridge, MA: MIT Press, 2013.

Corn, Joseph J., ed. *Imagining Tomorrow: History, Technology, and the American Future*. Cambridge, MA: MIT Press, 1988.

Cosgrove, Denis. *Apollo's Eye: A Cartographic Genealogy of the Earth in the Western Imagination*. Baltimore: Johns Hopkins University Press, 2001.

———. "Contested Global Visions: One-World, Whole-Earth, and the Apollo Space Photographs." *Annals of the Association of American Geographers* 84, no. 2 (1994): 270–94.

Crawford, Neta C. *The Pentagon, Climate Change, and War: Charting the Rise and Fall of U.S. Military Emissions*. Cambridge, MA: MIT Press, 2022.

Cronin, Marionne, "Polar Horizons: Images of the Arctic in Accounts of Amundsen's Polar Expeditions." *Scientia Canadensis* 33, no. 2 (2010): 99–120.

Cronon, William. "The Trouble with Wilderness; or, Getting Back to the Wrong Nature." In *Uncommon Ground: Rethinking the Human Place in Nature*, edited by William Cronon, 69–90. New York: W. W. Norton, 1995.

Cullather, Nick. *The Hungry World: America's Cold War Battle against Poverty in Asia*. Cambridge, MA: Harvard University Press, 2010.

Davis, Diana K. *The Arid Lands: History, Power, Knowledge*. Cambridge, MA: MIT Press, 2016.

Davis, Diana K., and Edmund Burke, eds. *Environmental Imaginaries of the Middle East and North Africa*. Athens: Ohio University Press, 2011.

Demuth, Bathsheba. *Floating Coast: An Environmental History of the Bering Strait*. New York: Norton, 2020.

Dennis, Michael Aaron. "Reconstructing Sociotechnical Order: Vannevar Bush and US Science Policy." In *States of Knowledge: The Co-Production of Science and the Social Order*, edited by Sheila Jasanoff, 225–53. New York: Routledge, 2004.

Doel, Ronald E. "Constituting the Postwar Earth Sciences: The Military's Influence on the Environmental Sciences in the USA after 1945." *Social Studies of Science* 33, no. 5 (2003): 635–66.

———. "Evaluating Soviet Lunar Science in Cold War America." *Osiris* 7 (1992): 238–64.

———. "Scientists as Policymakers, Advisors and Intelligence Agents: Linking Contemporary Diplomatic History with the History of Contemporary Science." In *The Historiography of Contemporary Science and Technology*, edited by Thomas Söderqvist, 215–44. New York: Routledge, 1997.

———. *Solar System Astronomy in America: Communities, Patronage, and Interdisciplinary Science, 1920–1960*. New York: Cambridge University Press, 1996.

Doel, Ronald E., Kristine C. Harper, and Matthias Heymann, eds. *Exploring Greenland: Cold War Science and Technology on Ice*. New York: Palgrave Macmillan, 2016.

Doel, Ronald E., Tanya J. Levin, and Mason K. Marker. "Extending Modern Cartography to the Ocean Depths: Military Patronage, Cold War Priorities, and the Heezen-Tharp Mapping Project, 1952–1959." *Journal of Historical Geography* 32, no. 3 (2006): 605–26.

Doel, Ronald E., and Naomi Oreskes. "The Physics and Chemistry of the Earth." In *The Cambridge History of Science: Part V—Mathematics, Astronomy, and Cosmology since the Eighteenth Century*, edited by Mary Jo Nye, 538–58. Cambridge: Cambridge University Press, 2002.

Dorsey, Kurk. "Dealing with the Dinosaur (and Its Swamp): Putting the Environment in Diplomatic History." *Diplomatic History* 29, no. 4 (2005): 573–87.

Dorsey, Kurk, and Mark Lytle. "Introduction." *Diplomatic History* 32, no. 4 (2008): 517–18.

Edwards, Paul. *A Vast Machine: Computer Models, Climate Data, and the Politics of Global Warming*. Cambridge, MA: MIT Press, 2013.

Farish, Matthew. *The Contours of America's Cold War*. Minneapolis: University of Minnesota Press, 2010.

———. "Creating Cold War Climates: The Laboratories of American Globalism." In *Environmental Histories of the Cold War*, edited by J. R. McNeill and Corinna R. Unger, 51–83. New York: Cambridge University Press, 2010.

———. "Frontier Engineering: From the Globe to the Body in the Cold War Arctic." *Canadian Geographies* 50, no. 2 (June 2006): 177–96.

———. "The Lab and the Land: Overcoming the Arctic in Cold War Alaska." *Isis* 104, no. 1 (March 2013): 1–29.

Ferguson, Eugene S. "The Mind's Eye: Nonverbal Thought in Technology." *Science* 197, no. 4306 (1977): 827–36.

Fernlund, Kevin J., ed. *The Cold War American West: 1945–1989*. Albuquerque: University of New Mexico Press, 1998.

Forsyth, Isla. "Desert Journeys: From Exploration to Covert Operations." *Geographical Journal* 182, no. 3 (September 2016): 226–35.

Fowler, Catherine S. "Reconstructing Southern Paiute-Chemehuevi Trails in the Mojave Desert of Southern Nevada and California: Ethnographic Perspectives from the 1930s." In *Landscapes of Movement: Trails, Paths, and Roads in Anthropological Perspective*, edited by James E. Snead, Clark L. Erickson, and J. Andrew Darling, 84–105. Philadelphia: University of Pennsylvania Press, 2009.

Fox, Sarah Alisabeth. *Downwind: A People's History of the Nuclear West*. Lincoln: University of Nebraska Press, 2014.

Freed, Libbie. "Networks of (Colonial) Power: Roads in French Central Africa after World War I." *History and Technology* 26, no. 3 (2010): 203–23.

Fristrup, Børge. *The Greenland Ice Cap*. Seattle: University of Washington Press, 1967.

Gertner, Jon. *The Ice at the End of the World: An Epic Journey into Greenland's Buried Past and Our Perilous Future*. London: Icon Books, 2019.

Gillem, Mark L. *America Town: Building the Outposts of Empire*. Minneapolis: University of Minnesota Press, 2007.

Goodman, Jordan, Anthony McElligott, and Lara Marks, eds., *Useful Bodies: Humans in the Services of Medical Science in the Twentieth Century*. Baltimore: Johns Hopkins University Press, 2008.

Goodwin, Doris Kearns. *No Ordinary Time: Franklin & Eleanor Roosevelt: The Home Front in World War II*. New York: Simon & Schuster, 1994.

Gregory, Derek. "The Natures of War." *Antipode* 48, no. 1 (2015): 3–56.

Grønnow, Bjarne. "Living at a High Arctic Polynya: Inughuit Settlement and Subsistence around the North Water during the Thule Station Period, 1910–53." *Arctic* 69, no. 1 (2016): 1–15.

Gross, Rachel S. "Layering for a Cold War: The M-1943 Combat System, Military Testing, and Clothing as Technology." *Technology and Culture* 60, no. 2 (2019): 378–408.

Hamblin, Jacob Darwin. *Arming Mother Nature: The Birth of Catastrophic Environmentalism*. New York: Oxford University Press, 2013.

Harper, Kristine C. *Make It Rain: State Control of the Atmosphere in Twentieth-Century America*. Chicago: University of Chicago Press, 2017.

Heefner, Gretchen. *The Missile Next Door: The Minuteman in the American Heartland*. Cambridge, MA: Harvard University Press, 2012.

———. "Overseas Bases and the Expansion of U.S. Military Presence." In *The Cambridge History of America and the World*, vol. 4, *1945 to the Present*, edited by David C. Engerman, Max Paul Friedman, and Melani McAlister, 55–79. Cambridge: Cambridge University Press, 2022.

———. "'A Slice of Their Sovereignty': Negotiating the U.S. Empire of Bases, Wheelus Field, Libya, 1950–1954." *Diplomatic History* 41, no. 1 (2017): 50–77.

———. "'A Fighter Pilot's Heaven': Finding Cold War Utility in the North African Desert." *Environmental History* 22, no. 1 (2017): 50–76.

———. "Building the Bases of Empire: The U.S. Army Corps of Engineers and Military Construction during the Early Cold War. In *The Military and the*

Market, ed. Jennifer Mittelstadt and Mark Wilson. Philadelphia: University of Pennsylvania Press, 2022.

Hegge, Vanessa. "Blood, Race, and Indigenous Peoples in Twentieth Century Extreme Physiology." *History and Philosophy of Life Sciences* 41, no. 2 (2019).

Hersch, Matthew H. *Inventing the American Astronaut*. New York: Palgrave Macmillan, 2012.

Herzberg, Julia, Christian Kehrt, and Franziska Torma, eds., *Ice and Snow in the Cold War: Histories of Extreme Climatic Environments*. New York: Berghahn Books, 2018.

Hesse, Mary B. *Models and Analogies in Science*. Notre Dame, IN: University of Notre Dame Press, 1963.

Higuch, Toshihiro. *Political Fallout: Nuclear Weapons Testing and the Making of a Global Environmental Crisis*. Stanford, CA: Stanford University Press, 2020.

Hirshberg, Lauren. "Navigating Sovereignty under a Cold War Military Industrial Colonial Complex: US Military Empire and Marshallese Decolonization," *History and Technology* 31, no. 3 (2015): 259–74.

———. *Suburban Empire: Cold War Militarization in the US Pacific*. Berkeley: University of California Press, 2022.

Höhn, Maria, and Seungsook Moon, eds., *Over There: Living with the U.S. Military Empire from World War Two to the Present*. Durham, NC: Duke University Press, 2010.

Holle, Marie-Louise. "The Forced Relocation of Indigenous Peoples in Greenland—Repercussions in Tort Law and Beyond." Working Paper No. 1940, Copenhagen Business School, November 2019.

Holmes, Amy Austin. *Social Unrest and American Military Bases in Turkey and Germany since 1945*. Cambridge: Cambridge University Press, 2014.

Howkins, Adrian. *The Polar Regions: An Environmental History*. Cambridge, UK: Polity Press, 2016.

Huf, Ben, Glenda Sluga, and Sabine Selchow. "Business and the Planetary History of International Environmental Governance in the 1970s." *Contemporary European History* 31, no. 4 (2022): 553–69.

Igler, David. "On Vital Areas, Categories, and New Opportunities." *Journal of American History* 100, no. 1 (June 2013): 120–23.

Immerwahr, Daniel. *How to Hide an Empire: A History of the Greater United States*. New York: Farrar, Straus & Giroux, 2019.

Irwin, Julia. *Catastrophic Diplomacy: US Foreign Disaster Assistance in the American Century/* Chapel Hill: University of North Carolina Press, 2024.

Johnson, Andi. "'They Sweat for Science': The Harvard Fatigue Laboratory and Self-Experimentation in American Exercise Physiology." *Journal of the History of Biology* 48, no. 3 (August 2015): 425–54.

Klare, Michael T. *All Hell Breaking Loose: The Pentagon's Perspective on Climate Change*. New York: Metropolitan Books, 2019.

Koch, Natalie. *Arid Empire: Entangled Fates of Arizona and Arabia*. New York: Verso Books, 2023).

Krige, John, ed. *How Knowledge Moves: Writing the Transnational History of Science and Technology*. Chicago: University of Chicago Press, 2019.

Kuletz, Valerie L. *The Tainted Desert: Environmental and Social Ruin in the American West*. New York: Routledge, 1998.

Landsberg, Paul. "Deploy Globally, Train Locally: the U.S. Army and the Global Environment." PhD diss., University of Kansas, 2022.

Lane, K. Maria D. *Geographies of Mars: Seeing and Knowing the Red Planet*. Chicago: University of Chicago Press, 2010.

Laney, Monique. *German Rocketeers in the Heart of Dixie: Making Sense of the Nazi Past during the Civil Rights Era*. New Haven, CT: Yale University Press, 2015.

Latour, Bruno. *Science in Action: How to Follow Scientists and Engineers through Society*. Cambridge, MA: Harvard University Press, 1988.

Launius, Roger D. "Interpreting the Moon Landings: Project Apollo and the Historians." *History and Technology* (2006): 225–55.

———. *Reaching for the Moon: A Short History of the Space Race*. New Haven, CT: Yale University Press, 2019.

———. "Writing the History of Space's Extreme Environment." *Environmental History* 15, no. 3 (2010): 526–32.

Launius, Roger D., James Rodger Fleming, and David H. DeVorkin, eds. *Globalizing Polar Science: Reconsidering the International Polar and Geophysical Years*. New York: Palgrave Macmillan, 2010.

Leffler, Melvin P. *A Preponderance of Power: National Security, the Truman Administration, and the Cold War*. Stanford, CA: Stanford University Press, 1993.

Leslie, Stuart W. *The Cold War and American Science: The Military-Industrial-Academic Complex at MIT and Stanford*. New York: Columbia University Press, 1994.

Ley, Willy. *Engineers' Dreams: Great projects that could come true*. New York: Viking, 1954.

Limerick, Patricia Nelson. *Desert Passages: Encounters with the American Deserts*. Albuquerque: University of New Mexico Press, 1985.

Lopez, Barry. *Arctic Dreams: Imagination and Desire in a Northern Landscape*. New York: Charles Scribner's Sons, 1986.

Lowell, Percival. *Mars and Its Canals*. New York: Macmillan, 1906.

Lutz, Catherine, ed. *The Bases of Empire: The Global Struggle Against U.S. Military Posts*. New York: New York University Press, 2009.

Mack, Pamela E. *Viewing the Earth: The Social Construction of the Landsat Satellite System*. Cambridge, MA: MIT Press, 1990.
MacLeod, Roy. Introduction to "Nature and Empire: Science and the Colonial Enterprise." Special issue, *Osiris* 15, no. 1 (2000): 1–13.
Maher, Neil M. *Apollo in the Age of Aquarius*. Cambridge, MA: Harvard University Press, 2017.
———. "Neil Maher on Shooting the Moon." *Environmental History* 9, no. 3 (2004): 526–31.
Malaurie, Jean. *The Last Kings of Thule*. Translated by Adrienne Foulke. Boston: E. P. Dutton, 1982.
Markley, Robert. *Dying Planet: Mars in Science and the Imagination*. Durham, NC: Duke University Press, 2005.
Martini, Edwin A., ed. *Proving Grounds: Militarized Landscapes, Weapons Testing, and the Environmental Impact of U.S. Bases*. Seattle: University of Washington Press, 2017.
Martin-Nielsen, Janet. *Eismitte in the Scientific Imagination: Knowledge and Politics at the Center of Greenland*. New York: Palgrave Macmillan, 2013.
McCaffrey, Colleen T. *Military Power and Popular Protest: The U.S. Navy in Vieques, Puerto Rico*. New Brunswick, NJ: Rutgers University Press, 2012.
McCannon, John. *A History of the Arctic: Nature, Exploration and Exploitation*. London: Reaktion Books, 2012.
McCoy, Alfred W. *In the Shadows of the American Century: The Rise and Decline of US Global Power*. Chicago: Haymarket Books, 2017.
McCray, W. Patrick. *The Visioneers: How a Group of Elite Scientists Pursued Space Colonies, Nanotechnologies, and a Limitless Future*. Princeton, NJ: Princeton University Press, 2012.
McCurdy, Howard E. *Space and the American Imagination*. Baltimore: Johns Hopkins University Press, 2011.
McNeill, J. R., and Peter Engelke, *The Great Acceleration: An Environmental History of the Anthropocene since 1945*. Cambridge, MA: Harvard University Press, 2016.
McNeill, J. R., and Corinna R. Unger, eds. *Environmental Histories of the Cold War*. New York: Cambridge University Press, 2010.
Mendell, W. W., ed. *Lunar Bases and Space Activities of the 21st Century*. Houston, TX: Lunar and Planetary Institute, 1985.
Messeri, Lisa. *Placing Outer Space: An Earthly Ethnography of Other Worlds*. Durham, NC: Duke University Press.
Middleton, Nick. *Deserts: A Very Short Introduction*. New York: Oxford University Press, 2007.
Miller, Ron, and Frederick C. Durant III. *The Art of Chesley Bonestell*. New York: Paper Tiger, 2001.

Mitchell, Timothy. *Rule of Experts: Egypt, Techno-Politics, Modernity*. Berkeley: University of California Press, 2002.

Mittelstadt, Jennifer. *The Rise of the Military Welfare State*. Cambridge, MA: Harvard University Press, 2015.

Mittelstadt, Jennifer, and Mark R. Wilson, eds. *The Military and the Market*. Philadelphia: University of Pennsylvania Press, 2022.

Montoya, Maria. "Landscapes of the Cold War West." In *The Cold War American West: 1945–1989*, edited by Kevin J. Fernlund, 9–28. Albuquerque: University of New Mexico Press, 1998.

Montgomery, Scott L. *The Moon and the Western Imagination*. Tucson: University of Arizona Press, 2001.

Morrissey, Katherine G., and Marcus A. Burtner. "Global Imaginary of Arid Lands: Early Twentieth-Century United States Botanists in Africa." *Global Environment* 12, no. 1 (March 2019): 102–33.

Morton, Oliver. *The Moon: A History for the Future*. London: Economist Books, 2019.

Murphy, Robert. *Diplomat among Warriors*. New York: Doubleday & Co., 1964.

Nabhan, Gary Paul. *Arab/American: Landscape, Culture, and Cuisine in Two Great Deserts*. Tucson: University of Arizona Press, 2008.

———. *Gathering the Desert*. Tucson: University of Arizona Press, 1986.

Nash, Gerald. *The Federal Landscape: An Economic History of the 20th Century West*. Tucson: University of Arizona Press, 1999.

Nash, Linda. "The Agency of Nature of the Nature of Agency?" *Environmental History* 10, no. 1 (2005): 67–69.

———. "From Safety to Risk: The Cold War Contexts of American Environmental Policy." *Journal of Policy History* 29, no. 1 (2017): 1–33.

———. "Furthering the Environmental Turn." *Journal of American History* 100, no. 1 (2013): 131–35.

———. *Inescapable Ecologies: A History of Environment, Disease, and Knowledge*. Berkeley: University of California Press, 2007.

———. "Traveling Technology? American Water Engineers in the Columbia Basin and the Helmand Valley." In *Where Minds and Matters Meet: Technology in California and the West*, edited by Volker Janssen, 123–58. Berkeley: Huntington Library and University of California Press, 2012.

Nasht, Simon. *The Last Explorer: Hubert Wilkins Australia's Unknown Hero*. Sydney: Hodder, 2005.

Needell, Allan A. *Science, Cold War, and the American State: Lloyd V. Berkner and the Balance of Professional Ideals*. Amsterdam: Harwood, 2000.

Nesbitt, Paul H. "A Brief History of the Arctic, Desert, and Tropic Information Center and Its Arctic Research Activities." In *United States Polar Exploration*, edited by Herman R. Friis and Shelby G. Bale, 134–45. Athens: Ohio University Press, 1970.

Neufeld, Michael J. *Von Braun: Dreamer of Space, Engineer of War*. New York: Vintage Books, 2008.

Nielsen, Kristian H., and Henry Nielsen. *Camp Century: The Untold Story of America's Secret Arctic Military Base Under the Greenland Ice*. New York: Columbia University Press, 2021.

Noble, David F. *America by Design: Science, Technology and the Rise of Corporate Capitalism*. New York: Oxford University Press, 1979.

Nye, David E. *America as Second Creation: Technology and Narratives of New Beginnings*. Cambridge, MA: MIT Press, 2004.

Olson, Valerie. *Into the Extreme: U.S. Environmental Systems and Politics beyond Earth*. Minneapolis: University of Minnesota Press, 2018.

Oreskes, Naomi. *Science on a Mission: How Military Funding Shaped What We Do and Don't Know about the Ocean*. Chicago: University of Chicago Press, 2021.

Palka, Eugene J. "Military Geography in the US: History, Scope, and Recent Developments." In *Modern Military Geography*, edited by Francis A. Galgano and Eugene J. Palka, 5–20. New York: Routledge, 2011.

Petersen, Nikolaj. "The Iceman That Never Came: 'Project Iceworm,' the Search for a NATO Deterrent, and Denmark, 1960–1962." *Scandinavian Journal of History* 33, no. 1 (2008): 75–98.

———. "SAC at Thule: Greenland in the U.S. Polar Strategy." *Journal of Cold War Studies* 13, no. 2 (2011): 90–115.

Pikialasorsuaq Commission. *People of the Ice Bridge: The Future of the Pikialasorsuaq*. Ottawa: Inuit Circumpolar Council Canada, 2017.

Platoff, Anne M. *Eyes on the Red Planet: Human Mars Mission Planning, 1952–1970*. Houston, TX: NASA Johnson Space Center, 2001.

Poole, Robert. *Earthrise: How Man First Saw the Earth*. New Haven, CT: Yale University Press, 2008.

Prelinger, Megan. *Another Science Fiction: Advertising the Space Race 1957–1962*. New York: Blast Books, 2010.

Pyle, Ernie. *Here Is Your War*. New York: Henry Holt & Co., 1943.

Pyne, Steve. "Extreme Environments." *Environmental History* 15, no. 3 (2010): 509–13.

Rand, Lisa Ruth. "Falling Cosmos: Nuclear Reentry and the Environmental History of Earth Orbit." *Environmental History* 24, no. 1 (2019): 78–103.

Rankin, William. *After the Map: Cartography, Navigation, and the Transformation of Territory in the Twentieth Century*. Chicago: University of Chicago Press, 2016.

Ribeiro, Sofia, Audrey Limoges, Guillaume Massé, Kasper L. Johansen, William Colgan, Kaarina Weckström, Rebecca Jackson, et al. "Vulnerability of the North Water Ecosystem to Climate Change." *Nature Communications* 12, no. 4475 (2021): 1–12.

Robertson, Thomas. "Cold War Landscapes: Toward an Environmental History of U.S. Development Programs in the 1950s and 1960s." *Cold War History* 16, no. 4 (2016): 417–41.

Robinson, Michael F. *The Coldest Crucible: Arctic Exploration and American Culture*. Chicago: University of Chicago Press, 2006.

Sandars, Christopher T. *America's Overseas Garrisons: The Leasehold Empire*. Oxford: Oxford University Press, 2000.

Scott, David Meerman, and Richard Jurek. *Marketing the Moon: The Selling of the Apollo Lunar Program*. Cambridge, MA: MIT Press, 2014.

Selcer, Perrin. *The Postwar Origins of the Global Environment: How the United Nations Built Spaceship Earth*. New York: Columbia University Press, 2018.

Shreve, Forrest. "Ecological Aspects of the Deserts of California." *Ecology* 6, no. 2 (April 1925): 93–103.

———. "The Edge of the Desert." *Yearbook of the Association of Pacific Coast Geographers* 6 (1940): 6–11.

Shreve, Forrest, and Ira L. Wiggins. *Vegetation and Flora of the Sonoran Desert*. Stanford, CA: Stanford University Press, 1964.

Smith, Susan L. *Toxic Exposures: Mustard Gas and the Health Consequences of World War II in the United States*. New Brunswick, NJ: Rutgers University Press, 2019.

Solnit, Rebecca. *Savage Dreams: A Journey into the Hidden Wars of the American West*. Berkeley: University of California Press, 1994.

Spiermann, Ole. "Hingitaq 53, Qajutaq Petersen, and Others v. Prime Minister's Office (Qaanaaq Municipality and Greenland Home Rule Government Intervening in Support of the Appellants)." *American Journal of International Law* 98, no 3. (2004): 572–78.

Spudis, Paul D. *The Once and Future Moon*. Washington, DC: Smithsonian Institution, 1998.

Steffen, Will, Katherine Richardson, Johan Rockström, Hans Joachim Schellnhuber, Opha Pauline Dube, Sébastian Dutreuil, Timothy M. Lenton, and Jane Lubchenco. "The Emergence and Evolution of Earth System Science." *Nature Reviews Earth & Environment* 1 (January 2020): 54–63.

Steiger, Eric John. "Engineering the Desert: American Expansion and Global Expertise in the Colorado Desert, 1847–1920." PhD diss., University of California, Irvine, 2013.

Stewart, Douglass M., Jr., dir. *Chesley Bonestell: A Brush with the Future*. DMS Production Services, 2020. https://www.chesleybonestell.com/home.html.

Stroud, Ellen. "Does Nature Always Matter? Following Dirt through History." *History and Theory* 42, no. 4 (2003): 75–81.

Sutter, Paul S. "The World with Us: The State of American Environmental History." *Journal of American History* 100, no. 1 (2013): 94–119.

Tobin, James. *Ernie Pyle's War: America's Eyewitness to World War II*. New York: Free Press, 2006.

Tucker, Richard P., and Edmund Russell, eds., *Natural Enemy, Natural Ally: Toward an Environmental History of War*. Corvallis: University of Oregon Press, 2004.

Turchetti, Simone, and Peder Roberts, eds. *The Surveillance Imperative: Geosciences during the Cold War and Beyond*. New York: Palgrave Macmillan, 2014.

Van Vleck, Jenifer. *Empire of the Air: Aviation and the American Ascendancy*. Cambridge, MA: Harvard University Press, 2013.

Vertesi, Janet. *Seeing Like a Rover: How Robots, Teams, and Images Crat Knowledge of Mars*. Chicago: University of Chicago Press, 2015.

Villard, Henry Serrano. *Libya: The New Arab Kingdom of North Africa*. Ithaca, NY: Cornell University Press, 1956.

Vincenti, Walter G. *What Engineers Know and How They Know It: Analytical Studies from Aeronautical History*. Baltimore: Johns Hopkins University Press, 1993.

Vine, David. *Base Nation: How U.S. Military Bases Abroad Harm America and the World*. New York: Macmillan, 2015.

Wager, Walter. *Camp Century: City under the Ice*. Boston: Chilton Books, 1961.

Walker, Lawrence R., and Frederick H. Landau. *A Natural History of the Mojave Desert*. Tucson: University of Arizona Press, 2018.

Warde, Paul, Libby Robin, and Sverker Sörlin. *The Environment: A History of the Idea*. Baltimore: Johns Hopkins University Press, 2021.

Warkentin, Benno P. *Footprints in the Soil: People and Ideas in Soil History*. Amsterdam: Elsevier, 2006.

Weart, Spencer R. *The Discovery of Global Warming, Revised and Expanded Edition*. Cambridge, MA: Harvard University Press, 2008.

Webb, George F. *Science in the American Southwest: A Topical History*. Tucson: University of Arizona Press, 2002.

Weitekamp, Margaret. *Right Stuff, Wrong Sex: America's First Women in Space Program*. Baltimore: Johns Hopkins University Press, 2005.

Welland, Michael. *The Desert: Lands of Lost Borders*. London: Reaktion Books, 2015.

———. *Sand: The Never-Ending Story*. Berkeley: University of California Press, 2009.

Welsch, Michael. "The Legacy of Containment." In *The Cold War American West: 1945–1989*, edited by Kevin J. Fernlund, 87–100. Albuquerque: University of New Mexico Press, 1998.

Whitaker, Ewen A. *Mapping and Naming the Moon: A History of Lunar Cartography and Nomenclature*. New York: Cambridge University Press, 2003.
White, Richard. *"It's Your Misfortune and None of My Own": A New History of the American West*. Norman: University of Oklahoma Press, 1993.
———. *The Organic Machine: The Remaking of the Columbia River*. New York: Hill & Wang, 1996.
Wilhelms, Don E. *To a Rocky Moon: A Geologist's History of Lunar Exploration*. Tucson: University of Arizona Press, 1993.
Woodward, Rachel. "From Military Geography to Militarism's Geographies: Disciplinary Engagements with the Geographies of Militarism and Military Activities." *Progress in Human Geography* 29, no. 6 (December 2005): 718–40.
Wulff, Steen. "The Legal Bases for the Inughuit Claim to their Homeland." *International Journal on Minority and Group Rights* 12, no. 1 (2005): 63–91.
Young, Anthony. *The Apollo Lunar Samples: Collection Analysis and Results*. New York: Springer, 2017.

INDEX

Page numbers in italics refer to figures.

abandonment: of air bases, 190, 211; of Camp Century, 226; of contaminated sites, 11; of ice tunnels, 17; of radar Sites I and II, 214. *See also* trash and waste
Abbey, Edward, 300
aerial surveillance, 125, 135, 136
airfields. *See* runways and taxiways
Air Force Aeronautical Chart and Information Center, 270
air force bases, US. *See individual base names*
Air Force Engineer, The, 52
Alaska, defense installations in, 143, 172; and indigenous peoples, 32, 80; as more accessible, 173; strategic importance of, 174; in World War II, 33
Algeria, 21, 22, 40, 52, 54, 126
Alley, Richard, 232
Andrews, Roy Chapman, 29, 31–33, 35, 44
animals, 11, 68, 151–52, 183, 187, 190
Antarctic exploration, 94, 153, 192, 215, 238, 239, 241, 243, 245
Apollo missions, 9, 17, 262, 276, 277–78, 280–82, *281*, 285, 288–89, 291–92

Apollo-Soyuz Test Project, 135
Arctic, 125, 135–36, 197, 208, 302; changes to, 11, 192; clothing for, 17, 81, 83, 96, 126; compared to deserts, 9, 202–4, *203*, 213; definitions and terminology for, 169–70, 197–98, 215; expertise in, 32, 34, 80, 95, 149, 180, 199, 208, 232, 262; mapping of, 166, 174, *181*; and outer space, 258–59, 262, 270, 272, 293; seasons in, 141, 187; Soviet expertise in, 4, 255; as strategic problem, 4, 34, *148*, 149–50, 155, 163–65, *164*, 194, 212; weather stations, 155–59
Arctic, Desert, and Tropic Information Center (ADTIC), 33–34, 60–61, 91, 101, 169, 170
Arctic Circle, 6, 146, 149, 150, 151, 210
Arctic Dreams (Lopez), 10
Arctic fox, 190; importance to Inughuit, 183; pelts of, 153
Arctic Research and Information Center. *See* Arctic, Desert, and Tropic Information Center (ADTIC)
Arizona, 58–59, 68, 101, 118, 119, 126, 133, 251, 280, 283, 285, 288, 293

Armstrong, Neil, 8–9, 16, 266, 275, 280, 288
Arnold, Henry "Hap," 149, 154
artwork, space. *See* Bonestell, Chesley
astrogeology, *284*, 286, 289
astronauts, 9, 10, 135, 136, 261, 266, 273, 275–76, 278, *281*, 282–92, 293
Atacama Desert, 284, 293
Austin, Mary Hunter, 65, 67, 71, 294
Aviation Engineer Notes, 52
aviation engineers, 21, 45, 47, 50, 51, 52–53, 167, 288, 297

Bader, Henri, 139, 141–42, 143, 188–89, 194–95, 197–200, 205, 214–15, 222, 229, 232, 233, 238–41. *See also* Benson, Carl
Bagnold, Ralph A., 166; expertise in deserts and sand behavior, 64, 114, 116–17, 293–95; on extraterrestrial matters, 295–97; scientific pursuits of, 114–15; on water rations, 65; wind tunnel experiments of, 115
Balchen, Bernt, 143, 153–54
Basic Field Manual Desert Operations, 64
Belding, Hardwood "Woody," 84–85
Benson, Carl, 139, 141–42, 188, 212, 222, 232–37. *See also* Bader, Henri
Boggs, Samuel W., 165–69, 175, 176
Bonestell, Chesley, 246, 258–59, 266, 278, 285
botanical research, 58–60, 66–68, 69, 116, 131, 170. *See also* Hoyt, Minerva Hamilton; Shreve, Forrest
Buckner, Simon B., Jr., 32, 33
bulldozer, 22, 51, 56, 180, 181, 267, 280, 297
Bush, Vannevar, 91–92, 95–96, 97–98

California, 13, 28, 29, 36, 44, 56, 57, 59, 63, 66, 68, 70, 119, 126, 182, 249, 258, 269, 283, 284, 294
Camp Century, 217–20, *221*, 223, 240, 241; compared to lunar bases, 266, 267, 272–73; daily life at, 1, *210*, 224, 226; ice cores extracted at, 240–42. *See also* ice cores; nuclear weapons
Camp Fistclench, 217, 218, 220, 229, 238
Camp Redrocks, 208
Camp Tuto, 143, *150*, 206, 207–8, 211, 223, 224, 238, 239, 263
cartohypnosis, 166
Casagrande, Arthur, 48–49, 273, 288; soil system of, 199
Clarke, Arthur C., 245–46, 247, 255, 261
climate chambers, 35, 72, 74, 75, 84, 262, *268*, 282–83
climate change, 67, 174, 227, 231, 282, 299–302, 304–5
clothing, 45, 95–96, 102, 126, 159, 199, 241; boots and shoes, 14, 56, 63, 75, 82, 83, 152, 261, 275, 300; camouflage patterns for, 123, 127, 129, 273; metallic suit, 82–83; quantification of ("clo"), 74, 79, 84, 97, 287; for "standard man," 367n13; underwear, 43, 45, 87. *See also* thermal manikins
Cold Regions Research and Engineering Laboratory. *See* Snow, Ice, and Permafrost Research Establishment (SIPRE)
Cold War, *148*, 213, 283, 285, 301; absurdist logic of, 189, 221; buildup of bases, 145; effects on military bases, 105, 110; as fought on margins, 102; and globalism, 137, 168; influence on science and technology, 98; strategic thinking about, 15, 90, 93, 147, 255; tensions, 101–2, 109, 148–49, 194, 260; uncertainties during, 14
Collier's, 251, 253, 255, 256–60
Committee on Geographical Exploration. *See* Research and Development Board (RDB)
concrete, 6, 13, 15, 104, 110, 119, 120, 171, 173, 175, 187, 215, 224, 290
Conquest of Space, The (Ley), 257, 259
conservationists, 66, 69, 135, 300–301

Corps of Engineers, US Army, 105, 160, 161, 170, 173, 174, 175, 206; and environmental intelligence, 129, 173, 195; and lunar construction, 227; and reliability maps, 174–75, 328n24; reorganization of, 120; Yuma Test Station, 101, 118, 126, 129, 130, 131–34. *See also* Snow, Ice, and Permafrost Research Establishment (SIPRE)

craters, 280, *281*, 282, 285, 287, 288, 289

creosote, 59, 60, 61–62

crevasse: in ice, 139, 141, 142, 223; in space, 257, 270

Cronkite, Walter, 220, 224, 227

Crow-Steers-Shepherd, 109. *See also* Libya

Crystal Party, 208

Dansgaard, Willi, 240, 358n25

Davison, Donald A., 21, 40, 47, 49–50, 51–52, 53, 56, 167, 297

Death Valley, 13, 115–16, 118

dehydration, 62, 65, 77, 78, 128, 202, 204, 273

Demorest, Max, 142

Desert Laboratory (Carnegie Institution), 58

deserts: and climate change, 11, 299; compared to Arctic, 9, 202–4, *203*, 213; compared to outer space, 251–52, 253, 258, 259, 262, 270, *281*, 283–84, 294–97, 300; efforts to quantify terrain, 131, 132–33, *133*, 135; as expendable, 10, 29, 36–37, 68, 70, 105, 106–7, 108, 118; expertise in, 32, 70–71, 100, 291; as good for astronomy, 283; as regions, 119–22, *121*, 125–26, 131, 136, 208, 271, 302; as space-age destinations, 283–84, 293, *294*; as strategic problem, 9, 14, 19–23, 31, 44, 56, 81, 90, 100, 105, 107, 143, 173, 272

Desert Solitaire (Abbey), 300

desert studies and research activities, 2, 3, 14, 44, 58, 62–64, 68–70, 72, 75, 79, 100, 101, 118, 130–31, 133, 199. *See also* Bagnold, Ralph A.

Desert Training Center (DTC), 29, 37–38, 56, 57, 59, 62, 77, 82, 91, 95, 101, 126, 130, 304

Destination Moon (film), 259

Dill, David Bruce, 72, 73, 74–75, 76, 79, 81, 88, 195

Doel, Ronald, 14, 92

Dogsled, 208

drifts: of sand, 7, 102, 118; of snow, 201, 214. *See also* dunes

dunes: Bagnold and, 113, 117; difficulties related to, 15–17, 56, 100, 117–19; formation of, 116; on Mars, 117, 294–96, *294*; paving over, 56, 119; sand, 114–16, 117–18, 125, 128; science of, 114–16, 125, 128, 271, 291; snow, 217, 235, 271

Eagle Scouts, at Camp Century, 218–20, 225–26

Earth Observations and Photography Experiment. *See* aerial surveillance

Eglin Field climate chamber. *See* McKinley Climatic Laboratory

Egypt, 29, 31, 39, 44, 101, 113, 114, 118, 126, *133*, 136

Eisenhower, Dwight D., 22, 42–43, 45, 46, 52

Eismitte, 212, 234, 232

El-Baz, Farouk, 135, 136. *See also* aerial surveillance

El Uotia bombing range (Libya), 123, 271

empire. *See* military empire, US

environmental expertise, 4–5, 44, 137, 229

environmental intelligence, 9, 15, 16, 22, 27, 35, 47, 53, 55–56, 80, 94, 103, 108, 125, 129, 132, 133, 148, 238, 262, 301; and analog maps, 135; costs of poor or incomplete, 22, 34, 40, 135, 166, 173;

environmental intelligence (*cont.*) driven by strategic needs, 137–38; and Greenland, 148; importance of, 4, 16, 28, 46, 53, 81, 130, 297, 302; shortcomings of, 112. *See also* mapping; maps

Evans, Thomas, 220

expansion: of bases, 107, 143, 180; into space, as next frontier, 254, 259, 279; of US military power, 4, 15, 100, 109, 112, 151, 189, 279; westward, 67, 259

experimentation on human subjects. *See* Harvard Fatigue Laboratory

explorers, 77, 153, 222, 251; in Greenland, 152, 155, 204; importance of to wartime work, 153–54; and modern military, 81, 82, 83, 94, 96, 97

Extraterrestrial Engineering Center, 267, *268*, 274

Extraterrestrial Research Agency, 269

Fall of Moondust, A (Clarke), 245–46, 255, 261, 262

feedback mechanisms, 174, 231, 244

firn, 169, 198, 215, 234, 238

food and nutrition, 14, 23, 77–78, 87–88, 142, 179, 182, 183, 224, 225, 333n8

Forbes, William, 76, 77, 87, 88, 90

Frémont, John C., 67

frostbite, 159, 204

Frost Effects Laboratory. *See* Corps of Engineers, US Army

Geological Survey, US, 232, 280; Astrogeological Branch, 284, 286; lunar mapping project, 270

glaciology, 14, 229, 285

globalism, 3, 5, 7, 11–12, 15, 27, 34, 45, 49, 54–55, 78, 85, 91, 100, 102, 108, 123–27, 137, 147–48, 168, 174; and cold environments, 200, 230, 231; and expertise, 229; and mapping, 172, 265; and network of facilities, 266; and US military needs, 202

global warming, 174, 237, 244, 299, 301; US military as contributor to, 11, 301

Gobi Desert, 31, 44; simulated in United States, 133

Grand Canyon, 287

granular materials, 116, 172, 201, 202, 270, 276, 278, 296; importance of, 8; physics of, 7, 276; on runways, 48; sand and moondust, 270; sand and snow, 201. *See also* sand

Greenland: and environmental intelligence, 148; explorers in, 152, 155, 204; nuclear reactors in, 1, 8, 196, 218–20, *219*, 223, 226–27, 240; nuclear weapons in, 11, 147, 190–91; private companies in, 159–60; weather in, 158, 159–60, 179, 223, 241, 273. *See also* Thule Air Base (Greenland)

Gregersen, Søren, 225

Hall, Bruce M., 268–69, 271, 273, 290–91

Hamblin, Jacob Darwin, 99

Harvard Fatigue Laboratory, 73, 74–76, 84, *85*, 87, 88, 91; researchers and leadership, 76, 77–78, *85*, 87, 93

hazardous waste, 11, 190–91, 226–27. *See also* trash and waste

Heald, Weldon F., 63–64, 77, 81

heat, 128, 214; extreme, 36, 62, 63; general concerns about, 28; response of materials to, 6, 63, 119; response of soldiers to, 64, 75, 76, 77. *See also* dehydration

Hoyt, Minerva Hamilton, 65–66, 67, 68–69, 70, 295, 300–301, 332n30. *See also* botanical research

human subject testing, 92; ethics of, 80; quantification of, 64, 77–78, 79. *See also* Harvard Fatigue Laboratory

ice cores, 229, *230*, 231–32, *237*, 238–44

Iceland, 50, 143, 283

ice tunnels, 16, 17, 189, 209, 211, 212–13, 224–25

Ice Worm, 220–21
igloo, 202, 215, *216*
Infantry Division, Ninety-Third, 36, 79
infrastructure: astronomical, 284; global military, 5, 12, 15, 28, 56, 123, 147, 156, 173, 204, 222, 242, 244, 265–66, 301, 303, 304; for outer space, 259, 268, 283
Intergovernmental Panel on Climate Change, 300
International Deserts Conservation League, 69
International Geophysical Year (IGY), 238–39
Inughuit, 146, 151, 177; Council of Hunters, 190; destruction of way of life, 178, 180, 187–90; distinction from Inuit, 344n3; as guides for Western explorers, 152–53; interactions with US military, 177–91; practices and methods, 152, 167, 177, 182, 183, 184; relationship with snow and ice, 142; in Thule area, 142, 146–47, 152–53, 162, 177–84. *See also* Inuit
Inuit, 146, 152; clothing as model for explorers, 95–97; descriptions of, 97; distinction from Inughuit, 344n3. *See also* Inughuit
Inuit Pikialasorsuaq Commission, 152
Itschner, Emerson C., 264–67, 271, 274

Johnson Space Center, 292, 293
Joint Research and Development Board (JRDB). *See* Research and Development Board (RDB)
Joshua tree, 66–68, 70, 299
Joshua Tree National Monument, *30*, 57, 66, 299

Keys, Ancel, 77, 92, 93, 333n8, 334n13
Kon-Tiki, (film), 184
Korea, 101, 181; war in as catalyst for research, 102
Krogh, Torben, 185–86, 187

labor, 52, 56, 76, 119, 128, 161
Land of Little Rain (Austin), 57
Langway, Chester C., Jr., 229, 233, 238–40, 242, 243
LeGrande, Allegra, 243
Levy, Dave, 285–86
Ley, Willy, 3, 257, 259
Libya, 104, 105–6, 107, 108–12, 119, 120, 122, 127; El Uotia, 123, 271. *See also* runways and taxiways
"Little Libya." *See* Desert Training Center (DTC)
Lopez, Barry, 10, 300
Lowell, Percival, 251–52, 283–84
Lunar Construction, *250*, 269, 271–72, 273, 274, 278, 290
Lunar Landscape, A (Bonestell), *246*, 258–59, 278
lunar resources, 289–93
Lunex, 260

maars, 287
Malaurie, Jean, 142, 147, 151, 152, 178–79, 181–88, 211
manikins, thermal, 84–85
mapping: aerial, 135, 171, 302; analog, 123, *124*, 132, *133*, 134, 135, 136–37, 155, *164*, 262; lunar, 270–71, 288; necessity of, 54, 131, 165–67, 172; of plant populations, 58, *59*, 69; poor commitment to by United States, 95; wartime, 121
maps, 3, *22*, 25, 26–27, 51, 53, 133, *148*, 196, 208, 264, 285, 303; Arctic, *150*, 169, 172, 185, 196, 233, 235, 302; desert, 57, 58, 69, *121*, 131; of global environments, 123, *124*, 135, 208, 265; of Libya, 111; lunar, 256, 258, 264, 265, 270, 271, 278, 286, 287, 289, 291; reliability, 174–76; shortcomings of, 51, 114, 167, 168, 185; soil, 91, 167
Mariner missions, *294*, 295
Mars, 13, 117, 251–53, 254, 293–97; bridge to, 1, *2*, 3, 259

Marshall Plan, 89–90
Martian, The (film), 273
McKinley Climatic Laboratory, 84
Mercury 13 (program), 286
Meteor Crater, 285, 288
militarism, 282; radiating effects of, 15; violence of, 6, 112, 131, 304
military empire, US, 5, 9, 102–3, 186, 195, 283
Military Engineer, 2, 3, 47, 52, 53, 56, 90, 121, 129, 201–2, *203*
military engineers: attempts to regularize environments, 8; challenges confronted by, 7–8, 13, 56, 118, 161, 172, 207–9; experience and expertise of, 7, 9, 12, 14–15, 88–89, 108, 120, 161; mistakes by, 3
Military Evaluation of Geographic Areas program, 132, 134, 166, 271. *See also* mapping; maps
military logistics, 3, 5, 17, 55, 158, 223, 227, 265; effects on research programs, 15; global structure of, 206, 272, 283; importance for civil life, 89, 229; importance for military life and activities, 82, 126, 148, 199, 204, 238; involving outer space, 259, 272; and routes, 108, 126–27, *157*, 180, 259, 272, 290; shortcomings and limits of, 41, 42, 134, 158–59, 182, 209, 297
Mojave Desert, 11, 294; climate change and, 11; compared with Sahara, 38, 45; damage to, 10–11, 68, 70–71; descriptions of, 11, 29, 36–37; hazards of, 65; Mars conference held in, 294; Native Americans in, 56–57, 61–62; nuances of, 39; plants in, 60–62, 68–69; as training area for military, 29
Montbach, Erica, 275
Moon: approximating conditions on, 9; designing bases on, 8, 249, *250*, 266–69; geology of, 260–63, 273–75, 287–89; lunar simulants, 274, 284; and military needs, 255, 259–60; Ranger program, 277; Surveyor program, 277–78, *277*, 287; US ambitions regarding, 8, 85. *See also* Project Horizon
mooncrete, 15
moondust, 8, 246, *268*, 269, 273, 277–78, 290, 291; dangers of, 275–76; fake, 274–75; odor of, 288. *See also* regolith
Moon rocks, 273, 278, 289, 290, 292, 293
Morocco, 101, 105, 107, 108, 112, 120
mud: descriptions of, 19; in North Africa, 40, *41*, 45; as operational problem, 21, 42, 52; postwar plans for, 99
Murphy, Robert, 42–43

NASA, 261; and Army Corps of Engineers, 263, 265, 266; and Earth-based training locations, 280, *281*, 282, *284*, 293, *294*; interest in snow and ice conditions, 251–54; Lunar Exploration Advisory Group, 293. *See also* Operation IceBridge
national security, 15, 81, 92, 100, 301
National Security Act, 92
Native Americans, 32, 38, 56–57, 62, 65, 181, 184; in Alaska, 32, 80; Bureau of Native Arts and Crafts, 32; expertise of, 56, 61–62; ignored in strategic planning, 32, 56, 62; interactions with, 153, 185–87; relocating, 189–90; in Southwest, 56–57; in Western stories, 37–38
Nevada Test Site, 282, 285
New York Times, 27, 114, 155, 252
Ninety-Third Infantry Division, 36, 79
North Africa, 3, 5, 19, 20–21, 27, 29, 31, 37, 39, 43, 160; lessons learned from, 52–53, 61, 108–9, 118, 119; poor environmental intelligence about, 22–23, 40–42, 45, 50, 51, 104
nuclear fallout, 156
nuclear reactors: at Camp Century, 1, 218, *219*, 220, 223, 226, 240; in Greenland, 1, 8, 196, 218–20, *219*, 223, 226–27, 240; on Moon, 8, 227, 255, 267, 272

nuclear weapons, 174, 220–21, 265; craters from testing, 282; as environmental agents, 11; lost in Greenland, 11, 147, 190–91; simulated practice with, 106, 131. *See also* Ice Worm

Odaaq, 177–78, 180, 183, 185, 349n1
Office of Scientific Research and Development, US, 91, 92
Olmsted, Frederick Law, Jr., 70
Operation IceBridge, 242, 303
Operation Paperclip, 251, 256
Operation Torch, 19, *20*, 22, 31, 39, 40, 41–42, 43
Oreskes, Naomi, 15

Panel on Arctic Environments. *See under* Research and Development Board (RDB)
Panel on Cartography and Geodesy. *See under* Research and Development Board (RDB)
Panel on Desert and Tropical Environments. *See under* Research and Development Board (RDB)
Panel on Expeditions. *See under* Research and Development Board (RDB)
Panel on Exploratory Physiology. *See under* Research and Development Board (RDB)
Panel on Geographic Environments. *See under* Research and Development Board (RDB)
Panel on Geographic Research Techniques. *See under* Research and Development Board (RDB)
Panel on the Atmosphere. *See under* Research and Development Board (RDB)
Patton, George, Jr., 27, 28–29, 31–32, 33, 34–36, 38, 39, 44, 62; on acclimatizing troops, 64, 70
Peary, Robert, 152, 153, 222
Pentagon, 14, 94, 98–100, 105, 123, 145, 163, 172–73, 194

perma-crete, 15, 173, 224, 290
permafrost: definitions and descriptions of, 170–72, 173–74; engineering challenges of, 6, 9, 160; military interest in and research on, 4, 173. *See also* Snow, Ice, and Permafrost Research Establishment (SIPRE)
Philippines, 33, 54
photography. *See* aerial surveillance
Pick, Lewis, 148, 161
Pikialasorsuaq, 151
Pituffik. *See* Thule Air Base (Greenland)
planetary knowledge and power, 9, 17, 168; limits to, 17, 291; local projects leading to, 108; physical nature of, 316n8; role of IGY, 238; of United States, 4, 12, 265, 302
plants. *See* botanical research; Joshua tree
Pocket Guide to North Africa, A, 19, 20, 37
pollution. *See* trash and waste
polynya, North Water, 151–52
Priebe, Harold "Oakie," 181, 182
private companies: in Greenland, 159–60; in Libya, 109; and military research, 83, 88–89, 91–93, 98–103, 109, 173, 200, 214, 217
Project Horizon, 14, 262, *266*
Pyle, Ernie, 19, 21, 39, 45
Pyne, Steve, 10

Qaanaaq, 189
Quartermaster Corps, US Army, 14, 33, 34, 44, 63, 77, 83, 91, 96, 97–98

radar installations, 107, *108*, 139, 141, 173, *178*, 189, 200, 205, 214–15, 216, 242
radiation experiments, 80
Ranger program, 277
Rasmussen, Knud, 152–53, 222
regolith, 8, 267, 270, 273–74, 278; fake, 274–75; first samples, 275; simulants, 275, 363n23; use on Mars, 276

Research and Development Board (RDB), 92, 93, 94–101, 149, 163, 164, 167, 168, 169, 170, 172, 194, 198, 204, 256, 291; Committee on Geographical Exploration, 92, 93; Panel on Arctic Environments, 163–64, 165, 166, 169–70; Panel on Cartography and Geodesy, 99; Panel on Desert and Tropical Environments, 97–100; Panel on Expeditions, 93; Panel on Exploratory Physiology, 94; Panel on Geographic Environments, 94; Panel on Geographic Research Techniques, 99; Panel on the Atmosphere, 98–99
resistance to military bases. *See* Inughuit: interactions with US military; Wheelus Air Force Base (Libya): resistance to
Reybold, Eugene, 88–89, 99, 103
Ride, Sally, 286
Roosevelt, Franklin D., 22, 25, *26*, 31, 40–41, 46, 52, 69
Roosevelt, Theodore, III, 21, 45
Roster of Arctic Specialists, 169
runways and taxiways, 5; in Arctic conditions, 91, 104, 156, 159, 164–65, 188–89; in desert conditions, 45, 51–52, 104–6, 108, 110, 117–18, 119, 127; design and creation of, 4, 8, 9, 28, 40, 46, 47, 48–50, 53, 91, 201

Sahara: compared to extraterrestrial bodies, 245, 252; conditions in, 4, 29, 107; construction in, 7, 108; fighting in, 29; homogeneity of, 22–23, 51, 61; military activities in, 107; strategic importance of, 60; Wheelus Air Fore Base at edge of, 2
sand: characteristics and behavior of, 115–17, 118; as cognate to Mars soil, 294–97; compared to moondust, 17, 270, 275; compared to snow, 201, 208; imported for concrete construction, 120; and scarcity, 104

sandstorms, 31, 85, 116–17, 118
satellites, 53, 135, 155, 244, 249, 260, 264–65
Saturday Evening Post, 22, 29, 37
Saudi Arabia, 101, 105, 107, 109, 119, 120, 128
Scott, David, 289
selenodesy, 271
Shatner, William, 303
Shoemaker, Eugene, 284–88, 290–91
Shreve, Edith, 58
Shreve, Forrest, 58–59, 69, 116, 118, 131, 135
Siberia, 4, 33
simulations, 7, 73, 74, 84, 130, 213; and ice cores, 243; Johnson Space Center chamber, 293; limitations of, 297; lunar, *268*, 274, 280–82; of Mojave Desert, 68; and nuclear craters, 282; of outer space, 282–83; Yuma Test Station, 130
Siple, Paul, 74, 86, 94, 96, 97, 149, 153, 262
SIPRE. *See* Snow, Ice, and Permafrost Research Establishment (SIPRE)
SnoComp, 208
snow: characteristics and behavior of, 139, 141, 143, 160, 193, 194, 196, 198–200, 209, 214, 228; compared to moondust, 267, 296; compared to sand, 201, 208; as construction and engineering problem, 15, 17, 28, 99, 138, 139–43, 160, 170, 187–88, 189, 193–98, *197*, 200–205; disappearing in Arctic, 11; expertise in, 194, 195–96, 211; on Mars, 13; moving across, 200, 201; snowcrete, 15, 215; testing gear in, 34. *See also* Bader, Henri; ice cores
Snow, Ice, and Permafrost Research Establishment (SIPRE): classification systems, 198; formation of, 150; need for, 195, 215; work undertaken by, 196, 199, 206, 217, 221, 224, 229, 233, 234. *See also* Bader, Henri
snowcrete, 15, 215
Solomon, Morton, 145–47, 150, 159–61, 171, 178–79, 183, 185, 188

Sonora, 58, *59*, 112, 131, 132, 133, 134, 252, 283
Sorge, Ernst, 212, 213, 234
Soviets: and space race, 255, 259–60, 263; US cooperation with, 135; US deterrence of through bases, 105, 145, 147, 148; expertise in Arctic, 4, 33, 149, 163–64, 171, 194
space boosters, 251, *254*, 255–56, 258, 271
space race, 135, 255, 259–60, 263, 286
Sputnik, 260
Stars and Stripes, 41, 45
Stefánsson, Vilhjálmur, 34, 83, 95–97, 208
Surveyor program, 277–78, *277*, 287

Tereshkova, Valentina, 286
terrain assessments, 56, 174, 270–71
thermal manikins, 84, 132, 287
Thule Air Base (Greenland), 139, 141–43, 145–48, 150, 157–59, 161–62, 163, 179, 181, 211, *216*. *See also* Camp Tuto
Thule region, 13, 150–51, 153–54, 161, 177, 180, 182, 189
trash and waste, 165, 184, 188, 190, 209–11, 300, 302, 317n11. *See also* hazardous waste
Troxler, Paul D., 108–11, 119, 256, 297
Truman Doctrine, 89–90
Tunisia, 19, 21, 22, 41, 45, 46, 52

uranium deposits, 285
Uummannaq, 150, 153, 177, 180, 183, 186, 189. *See also* Inughuit

V-2, 256, 285
Victor, Paul-Émile, 207, 212, 213
Villard, Henry, 110, 112
Villumsen, Rasmus, 234
von Braun, Wernher, 255–58, 260, 262

War of the Worlds (film), 261
Waterways Experiment Station. *See* Corps of Engineers, US Army

weather: changing, 91, 94, 98–99; extreme, 299; forecasting, 98–99, 154–56, 168, 205, 231; and gear, 32–33; in Greenland, 158, 159–60, 179, 223, 241, 273; and ice sheets, 232; importance to military needs, 52, 154–55, 160, 231, 283; in North Africa, 19, 22, 42, 46, 52; protection against in space, 275, 295; stations, 99, 145–47, *150*, 154–62, *157*, 172, 182, 234, 272; weather-proofing, 74, 204
Wegener, Alfred, 234
Wheelus Air Force Base (Libya), 2, 104–6, 107, 110–11, 117, 119, 127; daily life at, 2, 111; resistance to, 112–13, 211, 304
Whipple, Fred, 253, 255, 257
White, Richard, 105
Wilkins, Hubert: as assessor of extreme heat, 63–64, 77; interest in color swatches and camouflage, 123, 125–26, *127*, 128, 129–30, 132, 302; interest in Arctic, 153, *164*, 189, 202, 208; and maps, *124*, 164, 262, 283, 303; and US military, 35, 77, 81, 93, 100; as war photographer, 34
women: excluded from studies, 74; not consulted for environmental expertise, 44
Working Group on Extraterrestrial Resources, 290, 291. *See also* Hall, Bruce M.; lunar resources
World War II, 15, 19, 33, 34, 47, 74, 102, 153, 160, 195, 207, 213, 251, 270, 273
Wulsin, Frederick, 44–45, 78, 79, 82–83
Wulsin, Janet, 44

yucca moth, 67
Yuma Test Station. *See under* Corps of Engineers, US Army

Zerzura Dinner Club, 114, 253. *See also* Bagnold, Ralph A.